Nanostructure Science and Technology

Volume 170

AF815406

Series editor

David J. Lockwood, Ottawa, Canada

For further volumes:
http://www.springer.com/series/6331

Alessandro Lavacchi · Hamish Miller
Francesco Vizza

Nanotechnology in Electrocatalysis for Energy

Springer

Alessandro Lavacchi
Hamish Miller
Francesco Vizza
ICCOM-CNR
Sesto Fiorentino, FI
Italy

ISSN 1571-5744 ISSN 2197-7976 (electronic)
ISBN 978-1-4899-8058-8 ISBN 978-1-4899-8059-5 (eBook)
DOI 10.1007/978-1-4899-8059-5
Springer New York Heidelberg Dordrecht London

Library of Congress Control Number: 2013955057

© Springer Science+Business Media New York 2013
This work is subject to copyright. All rights are reserved by the Publisher, whether the whole or part of the material is concerned, specifically the rights of translation, reprinting, reuse of illustrations, recitation, broadcasting, reproduction on microfilms or in any other physical way, and transmission or information storage and retrieval, electronic adaptation, computer software, or by similar or dissimilar methodology now known or hereafter developed. Exempted from this legal reservation are brief excerpts in connection with reviews or scholarly analysis or material supplied specifically for the purpose of being entered and executed on a computer system, for exclusive use by the purchaser of the work. Duplication of this publication or parts thereof is permitted only under the provisions of the Copyright Law of the Publisher's location, in its current version, and permission for use must always be obtained from Springer. Permissions for use may be obtained through RightsLink at the Copyright Clearance Center. Violations are liable to prosecution under the respective Copyright Law.
The use of general descriptive names, registered names, trademarks, service marks, etc. in this publication does not imply, even in the absence of a specific statement, that such names are exempt from the relevant protective laws and regulations and therefore free for general use.
While the advice and information in this book are believed to be true and accurate at the date of publication, neither the authors nor the editors nor the publisher can accept any legal responsibility for any errors or omissions that may be made. The publisher makes no warranty, express or implied, with respect to the material contained herein.

Printed on acid-free paper

Springer is part of Springer Science+Business Media (www.springer.com)

Preface

This book focuses on nanotechnology in electrocatalysis for energy applications. In particular it covers nanostructured electrocatalysts for low temperature fuel cells, low temperature electrolyzers, and electrochemical valorization of carbon dioxide.

In recent years a variety of papers have been published on this subject. Nevertheless, the availability of introductory monographs on such a hot topic is still limited.

Researchers and professionals new to this field often find it difficult to navigate through the huge amount of information being constantly produced in such a quickly growing area. For this reason we have tried to design a book whose function is to provide an introduction to the basic principles of electrocatalysis, together with a review of the main classes of materials and electrode architectures. We feel that this approach has the potential to illustrate the basic ideas behind material design, providing also an introductory sketch of the current research focuses. The book is conceived to be as self-contained as possible. Here and there, especially in the chapters concerning basic thermodynamic and kinetic principles, we advise the reader to refer to the many excellent textbooks that already cover these areas. We hope we have succeeded in making this book readable enough to allow a graduate in technical and scientific disciplines with a fair background in chemistry (i.e. physicists, engineers, chemists, electrochemists, etc.) to understand the basic concepts. A reader with such a background will experience a gentle introduction allowing him to grasp the main design criteria driving the development of new nanomaterials for electrocatalysis. We also hope that the material presented in the book will help the reader to seek more specialized literature, developing his or her own opinion about the pros and cons of the very many existing approaches (at times, a nontrivial task).

The subject has been limited to low temperature electrocatalysis (below 120 °C). We are conscious that this is a limit. But on the other hand, extension to high temperature systems would have required much more space and the illustration of a variety of complex principles, something we believe does not match the original objectives we had for this book.

Discussion focuses on the three main fields where nanostructured and molecular electrocatalysts play a major role: (i) polymer electrolyte membrane fuel cells, (ii) electrolytic hydrogen production, and (iii) CO_2 electroreduction.

The book consists of three parts. After a short introduction (Chap. 1) that reports the general framework and outlines the concept of the book. Part I, entitled *Fundamentals,* then begins. This is aimed at giving an introduction to the basic concepts of electrocatalysis (Chap. 2), also describing the main devices where nanomaterials are exploited (Chap. 3). The text has been organized in such a way that no complex derivations or lengthy descriptions are given. Only the major formulas and concepts are reported in a simple fashion, to help the reader to understand the philosophy behind electrocatalytic material development. This part closes with a discussion of the factors affecting the design of electrocatalysts (Chap. 4), describing the main issues and also stressing the constraints which have to be necessarily accounted for. After such a discussion the role of nanotechnology in addressing the targets for effective electrocatalyst development is considered.

Building upon sound foundations, the description of the various materials begins. Each chapter regarding materials begins with a key concepts paragraph, giving the essential background that lies behind the development of research in each area.

Part II, entitled *Support Materials*, is devoted to catalyst support materials. The part starts (Chap. 5) with a discussion of carbon blacks, the ubiquitous porous carbons widespread in commercial electrocatalyst technology. Then carbon nanomaterials are reviewed, with a special emphasis on the "rising stars," such as carbon nanotubes and graphene. Chapter 6 deals with other support materials. Titania nanotubes and other conducting oxides are considered. These are especially important for fuel cells fed with liquid fuels. The use as innocent support and promoter of the kinetics of a variety of other nanomaterials is also described, completing the scenario.

Part III is entitled *Active Materials*. Chapter 7 describes the main approaches to metal nanoparticle synthesis and the main commercial electrocatalysts. A variety of nanostructured metals with shape and structure control (Chap. 8) are considered. A special emphasis is laid on control of the surface structure, with a discussion of the recent discovery of new synthetic routes to high index faceting for activity enhancement. Chapter 9 reports classes of nanoparticles engineered for the reduction of noble metal loading. The focus is on "hollow" and "core" shell nanoparticles. Chapter 10 reports a "molecular" approach to electrocatalysis. The use of macrocycles and heat treated macrocycles in electrocatalysis is extensively reviewed with a special emphasis on the most recent findings. The description of the breakthrough discovery of organometallic complexes employed in electrocatalysis is also given. The objective is to provide examples of single site processes leading to a completely new approach which could be considered to go "beyond nanotechnology." A short conclusion summarizing the main aspects of each single material category is then reported in Chap. 11.

Acknowledgments

The authors are grateful to holders of copyright who have kindly consented to the use of their illustrations. Should any omissions have inadvertently occurred, sincere apologies are offered. The authors are indebted to Dr. Jonathan Filippi (ICCOM-CNR), Dr. Manuela Bevilacqua (ICCOM-CNR), and Dr. Andrea Marchionni (ICCOM-CNR) who have generously provided scientific information. Their dedication and skills in assisting with the reproduction of the illustrations are also much appreciated.

Finally the authors wish to acknowledge with gratitude the patience of their wives, Alberta Bacchelli, Daniela Cacioli and Enza Larosa who provided encouragement during the preparation of this book.

Contents

Part I
Fundamentals

Chapter 1
Introduction

1.1 Key Concepts

The present chapter is intended as a somewhat "non-technical" introduction to the field covered in the book. We have structured it in such a way that the reader may find a short overview of the fundamental issues for which nanotechnology is relied upon to provide solutions. It starts with a short review of the current world energy and resources situation. This is important because the finiteness of resources is not only the reason why we look to renewable energy sources, but it also defines an important constraint to material design. Indeed many of the best known metals which can be used as electrocatalysts in electrochemical energy conversion devices are rare and expensive. For this reason they should be preserved or recycled and used only in negligible amounts in order for these devices to help to obtain a "sustainable" future. Part 2 focuses on environmental issues, stressing the need for a transition to a completely renewable energy system that does not poison our planet. The concept of renewable resources is also defined and some of the most relevant renewable energy resources are reviewed, with a special emphasis on the connection between their application in energy harvesting and electrochemical energy conversion technologies. The concepts of EROEI and net energy are also quickly discussed, together with the approach of Life Cycle Analysis. This is especially relevant for the book, as we believe that both researchers and professionals operating in the field should be aware that the materials they develop are just a small part of a complex system. The purpose of this system is to deliver energy in a clean and efficient way. It is never unworthy to stress that efficiency has to be provided at the level of the whole system. Basically the point is that not only is the performance important, but also the energetic and environmental impact of the materials and the processes used to obtain them. All of which must be carefully considered.

Next we introduce the concept of the hydrogen economy and the use of energy vectors as a part of a sustainable energy paradigm. A short introduction to fuel cells and electrolyzers follows, thus defining the devices where the materials focused on in this book will be potentially exploited. Next we introduce the

A. Lavacchi et al., *Nanotechnology in Electrocatalysis for Energy*,
Nanostructure Science and Technology 170, DOI: 10.1007/978-1-4899-8059-5_1,
© Springer Science+Business Media New York 2013

electroreduction of CO_2. This field is much less developed than fuel cells and electrolysis and no commercial devices are at present available. Nevertheless, the valorization of CO_2 is an area of outstanding importance and promises to be one where nanotechnology may give a truly relevant contribution. We will limit our discussion to low temperature devices. A wider discussion is beyond the scope of the book. The need for nanotechnology in energy-related electrocatalysis is then discussed at the end of the chapter where a short presentation of the concept of the book is also given.

1.2 Energy and Resources

The world over the last couple of centuries has experienced an unprecedented global economic development. This has undoubtedly been attributed to the rising availability of fossil energy resources, such as oil, coal, and natural gas. Among these resources the most notable has definitely been oil, as its transformation into liquid fuel provided an impressive energy vector, easily stored and transported and with a large energy density.

Nowadays, almost every aspect of human life, at least in most developed countries, strongly depends on the availability of a large amount of energy and ultimately fossil fuels (Scheme 1.1). The role of fossil fuels has also been

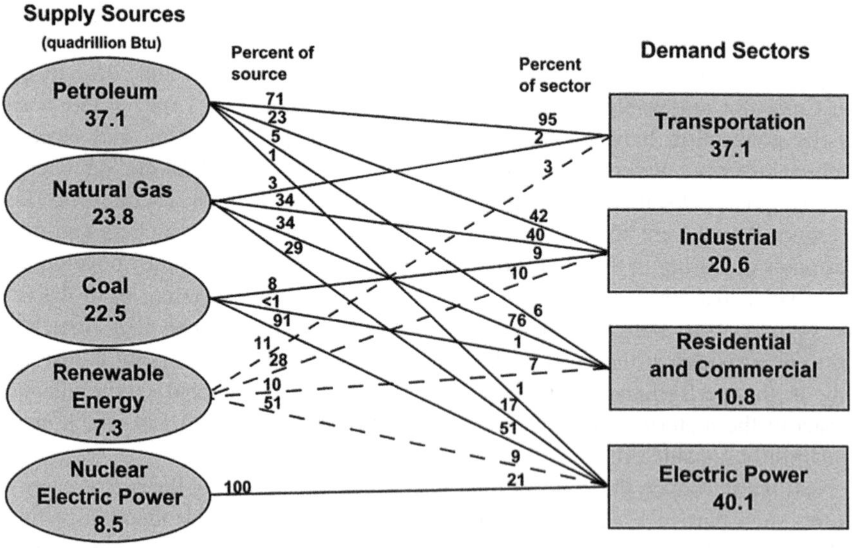

Scheme 1.1 Diagram illustrating the global dependency on fossil fuels for the USA within the overall energy supply sector and where each supply source is primarily consumed. Reproduced from Ref. [6] permission from Elsevier

fundamental for the population boom of the twentieth century. They made possible the diffusion of the energy intensive atmospheric nitrogen fixation into artificial fertilizers, contributing to the rise of intensive agriculture. Commodities such as wheat, corn, rice, and meat became progressively less expensive. At the same time agriculture itself became more and more demanding in terms of resources consumption.

Fossil energy sources are finite. This fact poses severe limitations to future economic development. The expanding world population, together with the rising standards of living, continuously pushes up the demand for energy. Energy hungry developing countries are putting increasing pressure on the continuous diminishing fossil fuel resources, making them even more costly. In 2005, oil consumption was approximately 1,000 barrels per second [1]. Just to give a rough idea this corresponds to 2 liters per person/day if we average over the population living on the Earth [2]. The current global power consumption sits around 13 terawatts (TW) and projections of energy consumption indicate that the demand could rise dramatically even in the very near future.

The problem of the exhaustion of fossil resources has been an intense subject of investigation starting from the second half of the twentieth century. Marion King Hubbert [3] proposed in the 1950s that the production of resources, particularly oil, would follow a "bell shaped," symmetric curve. The curve shows clearly a peak (Fig. 1.1).

He came to this conclusion by the analysis of the prototypical case of crude oil in the United States and, specifically, in the lower 48 states (US 48). By extrapolation, Hubbert was able to predict that the peak year for oil production in the US 48 would have been 1970. Indeed that was the case. Today, Hubbert's model is well known and has been applied to the whole world's oil production, (e.g., see [5]. According to the estimation, the global production peak is expected to occur

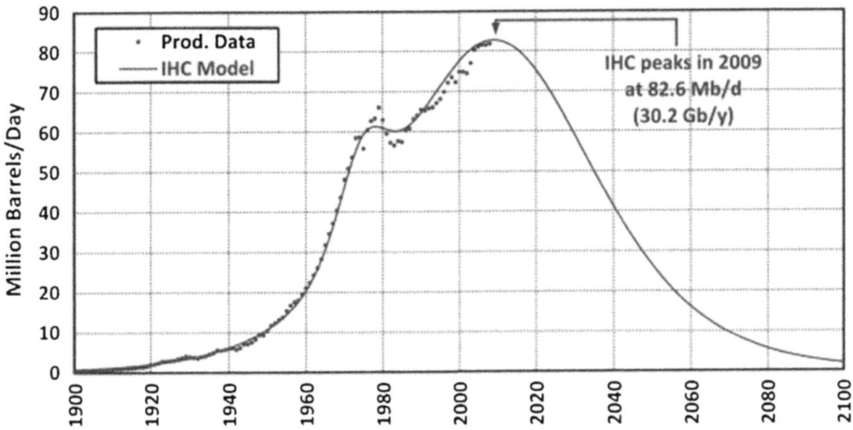

Fig. 1.1 A case of Hubbert "bell shaped" curve. Global oil annual production and its extrapolation to 2100. Reproduced from Ref. [4] with permission from Elsevier

within the first two decades of the twenty-first century (Fig. 1.1). While there may be uncertainties about the exact peak time, it is clear that the fossil fuels resources will, sooner or later, become scarce.

A Hubbert behavior has been found also in the production of coal and minerals. Some biological resources such as whale oil in the eighteenth century and Caspian caviar in the twentieth century have been shown to follow the same behavior see e.g. [7]. These findings suggest that Hubbert's model can be applied to all those resources for which the production rate is much faster than regeneration. This concept gives us the opportunity to introduce a definition of a renewable resource. Indeed we could define renewable resources as those resources able to regenerate at a rate larger than their consumption.

The general applicability and occurrence of bell-shaped production curves for the production of all sorts of nonrenewable resources has profound implications. From the Hubbert lesson we may learn that there is no way to plan a long-term "sustainable" economy if we focus on fossil resources alone. Production peak of fossil resources marks a critical moment in a economic system which is geared to maintain its growth forever and we may expect it to seriously affect the whole world's economy. It is hard to underestimate the importance of these concepts.

Not only is oil finite but also minerals and among minerals, those containing noble metals are the most likely to show scarcity in the near future. This is a matter of concern. Indeed these elements are essential in our technology and any effort that can be made to reduce their use is definitely worthy. A prototypical example is platinum which is essential in catalysis and, at present, is still the most relevant electrocatalyst for low temperature fuel cells. Platinum is a precious metal, hence it is scarce, hence has to be used with attention and if possible alternatives have to be developed. There will be no future for fuel cells and the hydrogen economy if we will not be able to reduce the platinum content in such devices. Nanotechnology is expected to play a key role in addressing this task, offering the chance to lower the noble metal loading also leading, in some cases, to its complete elimination (Fig. 1.2).

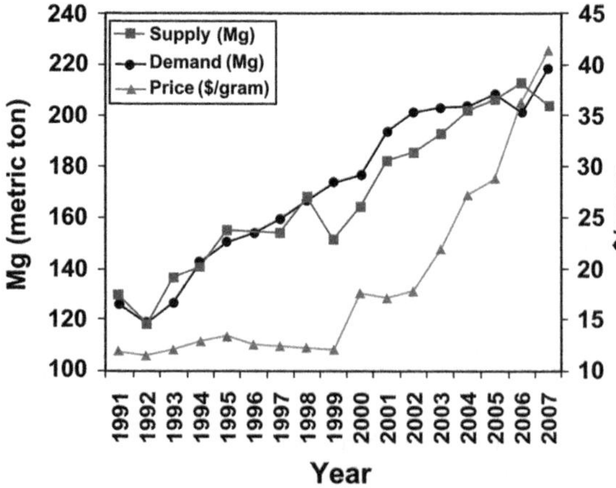

Fig. 1.2 Platinum supply, demand, and price. (Compiled from annual reports on platinum supply, demand, and prices published by Johnson Matthey Plc., and Kito.com.) reproduced from Ref. [8] with permission from Elsevier

1.3 Environmental Concerns

Finiteness is just one of the "bad sides" of fossil resources. The potentially damaging environmental effect of continuous carbon, natural gas, and oil usage has also to be seriously considered. CO_2 emission resulting from fossil fuel combustion has been shown to be the most relevant cause of the anthropogenic "greenhouse effect." For this reason the burning of fossil fuels has been recognized as a primary cause for global warming. Global warming may result in significant changes to ecosystems. Its implications are still difficult to predict accurately and its consequences on humanity and the entire ecosystem could be, potentially, catastrophic. The global climate change issue is indeed on the agenda of both national and international institutions.

The exploitation of fossil fuel resources is a source of further relevant health and environmental hazards. Such risks are linked to virtually any stage of the life cycle of the resource, from extraction, to transportation and storage. Coal, oil, together with its derivatives, and methane burning results in the release of serious pollutants as effluents in the atmosphere, on the land and in water. Among the most notable and ubiquitous pollutants we may cite: CO, CH_4 (also relevant as a greenhouse gas), NO_x, SO_x, volatile organic compounds (VOCs), heavy metals, particulate matter (PM) and, as previously stated, very large quantities of carbon dioxide (CO_2). A vast variety of epidemiology investigations show that environmental pollution significantly increases the risk of contracting cancer and other pollution-related pathologies. This risk is nowadays considered unacceptable by the population. The rise of awareness in the public is pushing governments to deliver year after year more stringent regulations on pollutants emission. Ultimately new regulations might even result in the impossibility of applying certain technologies, imposing the transition to alternatives.

The development of electrochemical technology for energy conversion and storage offers us the chance to address directly some of these issues. Among other topics the electrochemical conversion of CO_2 into organic compounds, which in turn can be used as fuels, is of primary importance. This is an outstanding opportunity to mitigate the impact of fossil fuels on "greenhouse emissions." At the same time this is indeed an extraordinary difficult challenge for electrochemical science and technology. Nanotechnology does offer the chance of designing materials on which devices can be built capable of addressing the targets for the proposed technology.

1.4 Renewable Energy Resources

Planning a sustainable future is now a priority. The development of new energy supply able to meet the demands for consumption in sectors such as household, commerce, industry, and transportation is a challenge that cannot be eluded. This must be done without impacting on the environment also assuring a long-term stability of the supply. There is no magic formula allowing the transition to such an energy sustainable system. Rather a mix of renewable energy sources is probably the best solution. What has to be common to all the possible power sources is the fact of being renewable.

Common applications of renewable energies are electricity generation and production of fuels for transportation. Nowadays, it is recognized that renewable energy has the potential to replace fossil fuels. Nevertheless such a change implies a system transition, requiring major technological changes and ultimately huge capital investment. Indeed and interestingly the variety of potential renewable energy resources offers the possibility to tailor the energy production according to local conditions (e.g. Iceland is not ideal for using photovoltaics due to its latitude, but it is exceptional in terms of geothermal energy for the geological nature of the island). The diversification of renewable energy resources [9, 10] hence provides a unique opportunity to create an energy system where the supply is not concentrated. In history concentration of essential resources has been proved to be risky due to geopolitical instability of the countries owning the resources.

Just to prove that renewable energy could, in principle, "power the world" we may analyze the case of solar energy. The radiation coming from the Sun and reaching the earth's surface is indeed the most abundant renewable resource available. In one year, the Sun delivers energy exceeding by a factor of 10,000 the energy consumed by humanity in the same time span. The problem arises in how we may efficiently collect it.

Let us go back to oil for a moment. One of the fortunes of oil is that it is found in oilfields, where it is (or better was) abundant and easy to extract. At the very beginning oil collection was an extraordinary straightforward task, providing large amounts of energy easy to transport and store with little effort. In terms of economy that was a huge opportunity for investment. While solar energy is and will definitely be more abundant than the energy from oil and other fossil resources, nevertheless, it suffers from the drawback of being diffused (we have seen that this has some advantage in terms of stability of the supply). This fact implies the need of a considerable amount of land to be used to recover it. Furthermore, the energy produced has to be stored and transported. That was easy with oil, but will be somewhat more troublesome with solar. This implies a complete rethinking of the energy system introducing new and clean technologies not only for energy harvesting, but also for storage transportation and conversion.

A large variety of renewable resources can be used for energy production. Here there is a short and somewhat arbitrary list, reporting some of the most popular approaches:

(1) Sunlight (converted into heat or electricity respectively by solar thermal or photovoltaics).
(2) Wind (to be converted into electricity by wind turbines or other mechanical devices).
(3) Hydroelectric (the mechanical energy of water can be converted into electrical energy by water turbines).

(4) Tides (mechanical energy of tides can be converted into electric energy).
(5) Geothermal heat (the heat contained in the earth crust can be employed in the production of vapor. The energy of the vapor can be then converted into electrical energy with turbines and alternators).
(6) Biomasses (photosynthesis provides CO_2 fixation in plants, which in turn can be burnt to render energy in the form of heat. The heat can be converted into electrical energy. Recently, direct conversion of biomass-derived compounds in fuel cells has been proposed).

Apart from tides (originated by gravitational interaction) and geothermal (originated by earth′s internal heat), all the other energy resources are the result of the impact of solar radiation onto the atmosphere. Wind comes from atmospheric thermal convection, biomasses result from the radiation promoted CO_2 fixation, and hydroelectric from the rain which is due to water evaporation provoked in lakes, seas, and rivers by the heating induced by the absorption of solar radiation.

Solar radiation can be converted into electricity, either directly using photovoltaics (PV), or indirectly using concentrated solar power (CSP) [11, 12]. PV has experienced an outstanding commercial growth in the last decade, providing the opportunity of realizing a distributed energy generation system. Even large PV fields have been established. PV is now a solid reality in the renewable energy panorama and its growth is expected to continue in the future.

A significant issue connected with most of the renewable energy technologies is the intermittent nature of energy generation. While a power plant fuelled with fossil resources can produce more or less on demand, this is not true for, e.g., of PV or wind energy. Energy production occurs as a result of local environmental conditions. Sometimes the production is not easy to forecast, producing difficulties in the electric grid management. As a result of this there is a huge demand for energy storage systems for buffering the electrical energy production from renewables. Coupling PV with electrolysis for the production of H_2 is in principle a solution. This would support the development of the so-called hydrogen economy that will be discussed later on. H_2 can then be exploited in fuel cells for transportation or even for producing electrical energy on demand in stationary systems. This is straightforward to understand. But it has to be pointed out that each energy conversion process leads to energy loss. Again, nanotechnology may lead to the synthesis of materials with improved electrocatalytic performance, leading to highly energy efficient processes potentially capable of improving the energy efficiency of a renewable energy-based economy.

An interesting alternative to solar energy storage is the CO_2 fixation into organic compounds through the photosynthetic biomass production (Scheme 1.2). Biofuels such as ethanol, biodiesel, and biogas [13] can be considered as the carriers of solar energy. This point is considered in this book. Indeed a variety of the nanostructured materials presented are aimed at the direct conversion of biomass derived compounds into electrical energy. Among such compounds a very special place is occupied by ethanol, which, when derived from biomasses is often referred to as "bioethanol." It is obtained mostly from sugars which are contained in corn [14], sugarcane [15], or sugar beet [16]. Recently the possibility of obtaining ethanol and other bio-alcohols from cellulose has been deeply investigated. This has led to the possibility of producing hydrogen by steam reforming of bioethanol obtained using portions of land not previously devoted to food production [17]. This is an essential issue, as a sustainable future cannot accept a competition between energy production and the human food energy chain. Subtraction of land from food production would especially damage less developed countries, with scarcity in essential commodities.

Biofuels can be directly burned in internal combustion engines. This is true for bioethanol and for biodiesel especially [18]. Nevertheless, combustion, even if performed with biomass derived compounds, is a source of pollution even if not a "greenhouse" contribution anymore. Whenever possible a direct, low temperature, conversion of the energy contained in biomass-derived compounds into electrical energy (e.g., through alcohol oxidation in direct alcohol fuel cell [19–22]) would definitely be advantageous for the environment.

At present the common problem of renewable resources, apart probably from hydroelectric, is that they are still not competitive in terms of produced energy prices with the nonrenewable resources. Technology is by the way quickly evolving and as many of the illustrated approaches have reached mass production, the cost of production is lowering (Scheme 1.2).

The relatively high prices of renewable energy requires a governance directed by major political institutions, with the target of setting up long-term strategies aimed at moving to a fully energetically renewable system. This has been done in a

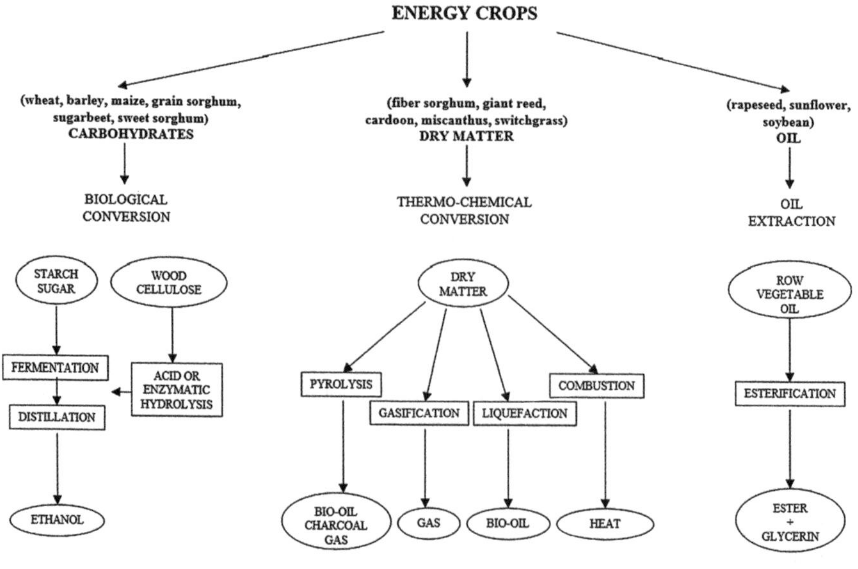

Scheme 1.2 An example of biomass exploitation. Main energy crops, conversion processes, and available products for energy uses. Reproduced from Ref. [23] with permission from Elsevier

variety of cases. PV has been recently extensively supported by economic incentives to its installation and consequent energy production.

1.4.1 The EROEI and the Life Cycle Analysis

In the previous chapters we have defined the concept of renewable. While this concept is pretty straightforward to understand, it is worthy to dedicate some room to the way that is currently in use to establish and quantify the energy impact of such technologies.

Thermodynamics states that the exploitation of an energy resource can never be 100 % efficient. For instance, before recovering the chemical energy stored in oil, we must spend some energy in a variety of operations which include: (i) prospecting; (ii) drilling; (iii) extracting; (iv) processing, and (vi) transporting. The concept of EROEI (energy return for energy invested) allows us to quantify all of these contributions and can be used to understand if the exploitation of a resource is convenient or not. Furthermore, it can tell us the amount of energy that the resource can return as compared to the energy we spend to recover it. In the end the EROEI can be defined as "the ratio of the energy obtained from the resource to the energy expended in production" [24–26]. Alternatively, one could consider the concept of net energy (in practice the energy gain) which is defined as the energy produced minus the energy expended by the resource. The relation between the two quantities is the following: if EROEI is equal to 1 or lower, the net energy is 0 or lower, while when the EROEI is larger than 1 the net energy gain is larger than 0. EROEI values up to 40 and more have been reported for oil [27].

The larger the EROEI the more preferable the resource exploitation. This is true in principle. However, in practice, some processes have been carried out even at low EROEIs, sometimes even smaller than 1. This has happened as a result of specific choices of political and economic systems. The most notable example is probably the production of biofuels from corn and in particular ethanol. Detailed analysis has shown that this process has a very low EROEI [28, 29]. Interestingly, according to the Pimentel and Patzek analysis bioethanol production from corn may lead to a use of energy from fossil fuels larger than the energy contained in the resulting biofuel. Nevertheless, the US government decided to proceed providing substantial financial aid to support the activity. Production from switchgrass, lignocelluloses, rapeseed, and sugarcane has been proved to be much more efficient with EROEI in some cases larger than 10 and those are, in the opinion of these authors, the future of biomasses.

The calculation of EROEI may be a difficult task. Many factors and boundaries need to be accounted for. These aspects may be considered according to the "Life Cycle Analysis" (LCA) concept. LCA is a well-established approach. Standards defining protocols for performing the analysis have been elaborated (e.g., ASTM E1991-05 [30]).

It is now worth looking at how the life cycle of a product is performed. LCA attempts to include all stages of a product's life in an evaluation. In doing this it considers that all these aspects are interdependent, leading one operation to the next. Hence LCA provides a comprehensive view of the environmental aspects of all the stages involved in the product realization. Next a list of the main aspect usually considered in the LCA is reported:

- collection of all the significant energy and material inputs and the associated emissions to the environment;
- evaluation of the potential environmental impacts associated with all inputs and emissions;
- results, analysis, and reporting for decision making.

The collection of relevant information requires both input and output. The following elements are usually accounted for:

- Input
  - Raw materials;
  - Manufacturing;
  - Use/reuse/maintenance;
  - Recycle/waste management.
- Output
  - The products;
  - Atmospheric emissions;
  - Waterborne wastes;
  - Solid wastes;
  - Co-products;
  - Other releases.

LCA can be extended to the analysis of energy and to the definition of the EROEI of energy resources or for the evaluation of the energy efficiency of energy conversion systems. These considerations lead to the life cycle energy analysis (LCEA). LCEA accounts for all energy inputs as well, not only the direct energy inputs during manufacture, but also those needed to produce components, materials, and services needed. With LCEA, the *total life cycle energy input* is established. This approach is now spreading amongst the researcher community. Extensive evaluations have recently been undertaken in Europe, China, US, and especially the UK to determine the life cycle energy (alongside full LCA) impacts of a number of renewable technologies [31].

We believe that being aware of the existence of the concepts of EROEI and net energy as well as of the LCA and LCEA is fundamental for both researchers and professionals involved in the field of synthesis and the application of electrocatalytic materials for electrochemical energy conversion and storage. Such evaluations are usually meaningless, if not impossible, at the research lab stage. But it is important to stress that a well performing material is not the whole story. Even the nature of the material, its manufacturing, the use that it makes of resources can play a determining role in defining whether its exploitation in technology is profitable or not. When considering material design criteria it could be a good exercise not to think of the material functionality alone but also to consider resource-related aspects. This is important in a modern context where research is often asked to follow directions indicated by institutions and formulated to give answers to specific needs of society.

1.4.2 The Role of Hydrogen and Energy Vectors

Energy alone means nothing. Society needs energy in the right place at the right time. This is especially true for transportation. The automotive sector needs reliable energy storage systems capable of delivering the required amount of power on demand. Hydrogen has been proposed as the energy carrier of the future as, in principle, it is applicable as a fuel for transportation and an intermediate in the conversion of renewable energy sources.

Hydrogen that is produced mainly by steam reforming of methane is used primarily to produce NH_3 which in turn is transformed into urea, and then in fertilizers. Hydrogen can be exploited in an internal combustion engine, but this is not the best solution. Internal combustion engines suffer from the thermodynamic efficiency limitation of thermo-mechanical cycles; furthermore the high temperature produced by the combustion of hydrogen with air, apart from heat and water, may release nitrogen oxides which have been recognized as extremely hazardous pollutants.

The combination of molecular hydrogen and oxygen in a fuel cell is a cleaner opportunity to generate electricity only resulting in the release of heat and water into the environment. Coupling hydrogen with low temperature fuel cells gives the opportunity to make the transportation sector energetically and environmentally sustainable. By the way, there are a variety of challenges that need to be considered before hydrogen can become a commercial reality for energy storage. First, of all hydrogen does not form spontaneously, at least in the large amounts potentially required by our society, it has to be produced. Doing this in a clean and efficient way from chemical compounds requires energy [32–34]. Indeed this is the reason why hydrogen is not an energy resource, but just an energy carrier.

At present most of hydrogen production employs fossil fuels both as energy and as hydrogen sources. Such methods require high temperatures to be effective. Reforming processes from fossil fuels result in what is usually called "syngas" which is a blend of CO and H_2. This is a drawback for fuel cell applications, as CO is a serious poison for platinum electrocatalysts. Hence, after hydrogen synthesis a further purification of the syngas from CO has to be performed. Lastly, reforming is not carbon dioxide neutral, contributing to the rise of CO_2 concentration in the atmosphere and ultimately to the increase of the "greenhouse effect."

To fulfill the requirements for a sustainable energy carrier, hydrogen has to be produced from water using renewable energies (such as solar energy). A scheme summarizing the possible ways to sustainably produce hydrogen is reported in Scheme 1.3 [35]. Hydrogen production from renewables is a concrete chance for the future and it has been hypothesized that the optimal endpoint would be the setup of a "Hydrogen Economy" Hydrogen economy should replace fossil fuels with hydrogen produced from renewables. Many environmental advantages will result from the Hydrogen Economy, and as such, it can be referred to as the Hydrogen Environmental Economy.

Criticism has also been expressed toward the concept of hydrogen economy based on its overall energy efficiency. It has been pointed out that hydrogen has to be made from renewable electricity by the electrolysis of water and then its chemical energy has to be converted back into electricity with fuel cells. Fuel cells efficiency maximum ranges around 50 %. Moreover, there are problems related to the storage technology [36] and to the creation of a safe distribution and transport network for this new energy carrier. When delivering hydrogen, whether by truck or pipeline, the energy costs are several times that for established energy carriers like natural gas or gasoline.

Biomasses conversion into fuel could be an interesting alternative to hydrogen as an energy carrier, even if more demanding in terms of catalysis. Bioethanol and other biofuels are liquid, with high energy density and easy to store and transport. In such a sense they are appealing, at least in principle, for powering automobiles. The efficiency of energy conversion in direct alcohol fuel cells is still not sufficient for powering cars or trucks. Nevertheless, it is the opinion of the authors that there is huge room for the development of such devices. Electrocatalysis is the main issue here and we expect that the ability to manipulate matter at the nanoscale holds the key to increasing the energy efficiency of such devices.

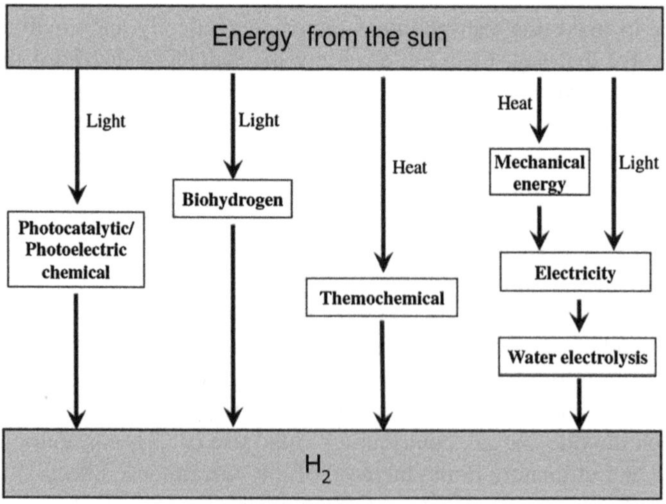

Scheme 1.3 Hydrogen from solar energy; production processes

1.5 Fuel Cells as Power Sources

A fuel cell is a device that converts the chemical energy from a fuel into electricity through a chemical reaction with oxygen [37]. Other oxidizing agents could in principle be used but their application is limited to very specialized niches. Hydrogen is the most commonly employed fuel in fuel cells, but hydrocarbons such as natural gas and alcohols like methanol and ethanol are also being used. Direct Alcohol Fuel Cells (DAFCs) have attracted increasing interest over the past decade, with a special emphasis on alkaline devices [38]. Easy storage and handling, high energy density, and wide availability are features that make alcohols attractive liquid fuels for the most promising alternative power sources for transportation, portable electronics, and stationary applications.

What makes fuel cells so appealing is the fact that they generate electricity through electrochemical processes, rather than combustion. Typical fuel cells consist of an anode (negative side), a cathode (positive side), and an electrolyte that allows ions to move between the two sides of the fuel cell (a detailed description of the fuel cell structure and components is deferred to Chap. 3). The energy efficiency of fuel cells is generally between 40 and 60 %. Values up to 85 % may be obtained if heat recovery systems are used. Figure 1.3 reports the main classes of fuel cells. They include: alkaline fuel cells (AFC), proton exchange membrane (PEM) fuel cells, direct alcohol fuel cells (DAFC), molten carbonate fuel cells (MCFC), phosphoric acid fuel cells (PAFC), and solid oxide fuel cells (SOFC). While these technologies are not yet mass technologies, some of them have been exploited at the commercial level. Each of these technologies has its own characteristics, such as different operating temperatures, catalysts, and electrolytes. The operating conditions of fuel cells define its range of application.

In this book we will limit our discussion to low temperature fuel cells and among them to the polymer electrolyte membrane fuel cells. Electrocatalysis in this category of fuel cells is particularly demanding and nanotechnology has the potential to dramatically improve the energy performance and feasibility of such devices.

Why direct chemical energy conversion into electricity in fuel cells is potentially so relevant? First, we know they are functional to the hydrogen economy. But there is more. Fuel cells have many potential benefits against competing technologies. Among them we cite the usually high efficiency, their modular nature which make fuel cell power units suitable for scale-up. A somewhat detailed list of major fuel cell advantages is reported below:

(1) Low-to-Zero Emissions and High Efficiency:
Fuel cells especially in low temperature technologies such as those based on polymer electrolyte membranes only emit neglegible amounts of hazardous effluents. Furthermore, the fact they don't use thermomechanical cycles is an advantage in terms of thermodynamic efficiency.
(2) Fuel flexibility and connection to sustainable development:
Fuel cells are a key element in the hydrogen economy and sustainable

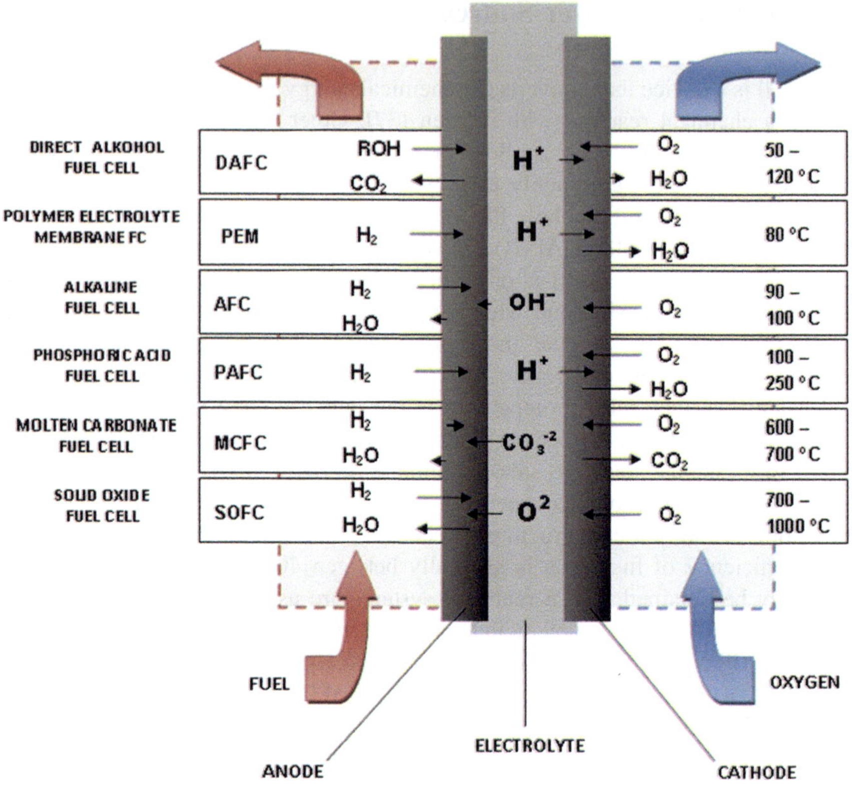

Fig. 1.3 Typical fuel cells consist of an anode (negative side), a cathode (positive side), and an electrolyte that allows charges to move between the two sides of the fuel cell. Adapted from Ref. [39] with permission from Elsevier

development based on renewable energy resources. As stated in the previous sections they may employ a variety of fuels such as hydrogen but also alcohols and organic compounds derived from biomasses. Solid oxide fuel cells may also directly employ hydrocarbons, without the need for using precious metal electrocatalysts.

(3) Reliability and Energy Security:
Practical applications need to rely on constant power to maintain operations. Buildings require power that is available practically without discontinuing service operations (e.g., hospitals). In some areas the electrical grid cannot guarantee such continuity of the service. Fuel cells can supply power independently of the grid. They can act as backup power to a grid-connected building. Fuel cells can also be configured to be a building's primary source of power (e.g., SOFC). Fuel cells can also be located in extreme climates, and rural areas where the grid may not be present and transportation is difficult due to the lack of infrastructure [39].

(4) Durability:
Without any important mechanical parts fuel cells may in principle be suitable candidates for power generation with low maintenance and durable operations. By the way for demanding applications such as automotive, fuel cells still do not meet the durability targets, but there are chances that this will happen in the near future as indicated by the US Department of Energy. Fuel cells are in principle also suitable for portable power electronic devices such as laptops and cell phones. For these applications fuel cells may exhibit much longer service life as compared to batteries, and since fuel cells have a higher energy density, they offer the chance to realize higher power sources. Furthermore, no electric grid is required as only the replacement of the fuel load is required. The recharging operation is also much shorter than that of batteries.

1.6 Electrolytic Hydrogen Production

The "hydrogen economy" calls for efficient processes for the production of hydrogen from renewables. Electrolytic water splitting (water electrolysis) is, at present, the technology that best matches the requirements for hydrogen production from renewable energy sources. While fuel cells have not yet been fully exploited commercially, electrolysis is in some sense a more mature technology and has been widely applied to the production of pure hydrogen. Electrolyzers in a wide variety of sizes are available and it is easy to find them in research labs around the world utilized for in situ hydrogen production. Electrolytic hydrogen production accounts for approximately 1 % of the overall hydrogen production in the world.

What makes electrolysis particularly appealing in terms of sustainability is its ability to directly convert electric energy into molecular hydrogen. So, any possible renewable source of electricity may be used to drive electrolytic hydrogen production, leading to a valuable system for storing energy from intermittent sources. The energy consumption for hydrogen production with state-of-the-art technologies is around 50 kWh kg^{-1} of molecular hydrogen of which 33.6 comes from thermodynamics (1.23 V is the standard thermodynamic potential). The rest comes from a variety of contributions, including the activation potential for the hydrogen and oxygen evolution reactions. Such contributions may be lowered by using better electrocatalysts whose performance can be tuned by using nanotechnology.

As an alternative to traditional water electrolysis the introduction of sacrificial agents in the anode compartment may provide large energy savings [40–42]. Water can be substituted at the oxygen evolution electrode (anode) with an easily oxidizable species such as, for example, ammonia or ethanol, at which point oxygen is no longer evolved at the anode. The uses of such easily oxidizable species as sacrificial agents allows us to reduce the thermodynamic contribution to values close to 0 V. Under these conditions it has been shown that electrolysis may occur at potentials lower than 1 V [41], leading to large energy savings. If the sacrificial agents are produced from biomasses with processes having large EROEI such a process may be advantageous with respect to conventional electrolytic technologies. It has been shown that from electrolytes containing ethanol hydrogen can be produced with electrical energy consumptions lower than 20 kWh kg^{-1}. As by-products organic compounds with high added value may also be obtained and this could contribute to the profitability of the process.

1.7 CO_2 Electroreduction

The steady increase in atmospheric CO_2 concentration is a pressing global environmental issue; as a consequence the electroreduction of carbon dioxide is currently investigated by many scientists as a one of the most promising ways to convert waste CO_2 into useful organic compounds which can be used as fuels or raw chemicals [43]. The field of electroreduction of CO_2 is by the way far less mature than fuel cells or electrolysis. No commercial application is actually known, but there is truly a huge interest in the subject. The task is by the way extraordinarily difficult. The main reason is due to the exceptional stability of CO_2 which requires a lot of energy for its activation. CO_2, according to thermodynamics, is more stable than any other organic compound in our atmosphere. Furthermore, the reduction mechanism requires the formation of a radical species whose activation potential is −1.9 V against the standard hydrogen electrode (SHE) [44]. Hence, CO_2 reduction in electrolytic devices requires a very high electrical energy input which is mostly related to the cathodic reduction of CO_2. There is hence great potential to improve the energetic performance of devices for CO_2 reduction to fuels concentrating on the cathode electrocatalyst. Material science and nanotechnology are expected to contribute massively to this subject resulting in improvements in the energy efficiency and compound selectivity through the design and realization of nanostructured electrode architectures.

1.8 Electrocatalysis and the Need for Nanotechnology

All the devices and processes covered by this book make use of electrocatalysts to enhance energy efficiency. Electrocatalysis is the branch of electrochemistry devoted to understanding and modifying reaction mechanisms through the use of catalytic materials. Electrocatalysis is a very old science. According to Jaksic et al. "…electrocatalysis and the search for promising electrocatalysts effectively started its development after two distinct core discoveries in the science: (i) Sir William Grove's inventive discovery and theoretical definition of (H_2/O_2) fuel cells and their fundamental structure in 1842 and (ii) Tafel plots in the year 1905, when various metals were distributed and ranged on the $\eta = a - b \log j$ coordination chart, with clear distinction amongst good and bad, or, on more or less polarizable, mostly transition elements or their composite electrode materials." [45] The development of new electrode architectures with enhanced electrocatalytic properties is a subject strictly connected to material science. The electrocatalytic properties of electrodes can be tuned by acting on a variety of material characteristics; the most relevant aspects connected to the electrocatalytic activity are:

(1) Composition;
(2) Surface structure;
(3) Morphology.

All of these elements together determine the ability of a given material to accomplish a given task. In electrocatalysis a smart design of a material can have the huge payback of improving the rate of a given electrochemical reaction with positive impact on the energy efficiency of the processes. We will not give here a complete description of how these aspects relate to the electrocatalytic activity as they will be extensively considered in Part 1 chapters.

There is still one important element which has not been explicitly considered yet and that may be considered the true reason for advocating nanotechnology in electrocatalysis. This is the surface area of the catalyst. The importance of this aspect resides in the nature of electrochemical reactions. Electrochemical reactions typically occur at the surface of an electronically conductive material (the electrocatalyst) which is in contact with an ionically conductive medium containing the electroactive species. According to this consideration and from the fundamentals of electrochemistry it is known that the energy absorbed by a given electrochemical reaction to proceed at a certain rate decreases as the extent of the electrode–electrolyte interface increases. Hence the exploitation of spontaneous electrochemical reactions (e.g., hydrogen or alcohols oxidation or oxygen reduction) in devices requires the electrode electrolyte interface to be as large as possible. Basically, the surface of the electrocatalyst has to be as large as possible and this is possible introducing in the material features with a scale length of the order of a nanometer. Indeed the success of polymer electrolyte fuel cells in the conversion of hydrogen chemical energy into electrical power is largely due to the availability of high surface area platinum electrocatalysts. The carbon supported platinum nanoparticles of the catalyst layer shows diameters ranging well below 10 nm and are employed both as anode and cathode. To go a bit more in detail, in PEMFC systems electrocatalysts with a metal loading of a fraction of mg cm^{-2} are employed. This is because platinum is rare and we do not want to waste it. The platinum specific surface area of such catalysts may range even over 100 m^2 g^{-1} [46]. Just to have a rough idea this means that with 0.4 g we would be able to cover a 40 m^2 floor with platinum. What would be the thickness of the floor? Very small indeed, just in the nanometer range. In turn the metal loading in a fuel cell electrode could be in the range of 0.1 mg cm^{-2}. This means that an electrode with a section area of 1 cm^{-2} would show a real catalyst area of 100 cm^2. With just

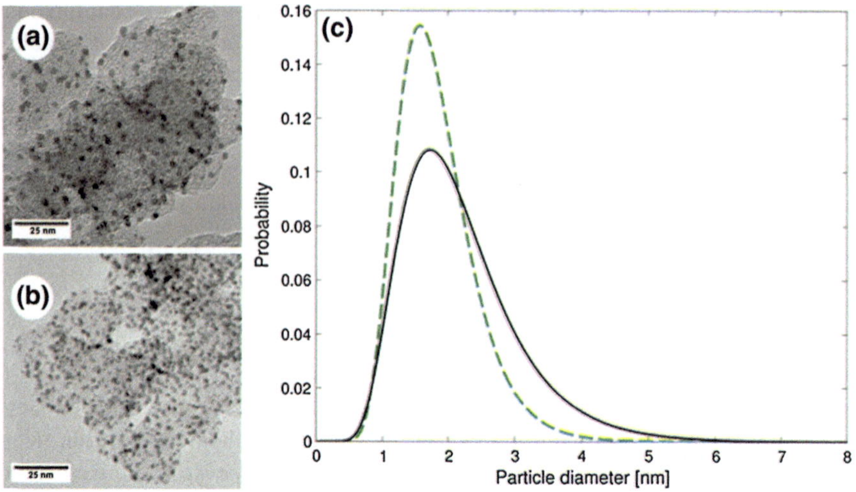

Fig. 1.4 Small supported platinum nanoparticles for oxygen reaction reduction in fuel cells. TEM micrographs of a 30 wt% Pt/Ketjenblack (A) and 30 % wt. Pt/Vulcan (B) catalysts. Particle size distribution as determined by Small Angle X-ray Scattering (SAXS) (C): 30 wt% Pt/ Ketjenblack (*solid*) and 30 % Pt/Vulcan (*dashed*). Reproduced from Ref. [46] with permission from Elsevier

0.1 mg cm^{-2} of a very dense metal we can obtain a ratio between the real and geometric surface area of 100. This is impressive. This is the power of nanotechnology (Fig. 1.4).

This is truly the issue. This is why in order to have efficient electrocatalysts we do need nanotechnology. Only a nanotechnology approach may deliver materials with a complete control of structure and composition. The ability to manipulate matter on the nanoscale to reach these targets is the true core of nanotechnology. This cannot be obtained by using conventional micro-machining methods as no mechanical method is capable of providing control on such small scale.

1.9 This Book's Approach

A huge variety of methods have been developed as well as many thousands of nanostructured materials and materials architecture have been produced. This is still an expanding research field with an incredibly large amount of scientific information produced each day. It is not easy to move through this sea of information, even for people who have been working in the field already for some time. Those approaching the subject for the first time may find it very difficult to find his or her bearings.

This book aims at providing a guide where we provide an *as gentle* as possible introduction to the basic principles followed by a review of a variety of materials currently at the state of the art. Such a review is necessarily not complete. It would not have been possible to cover all the relevant contributions to the subject. Rather than providing a comprehensive treatise, for which many excellent reviews and books exist, we have preferred to describe some of the main material classes with the relative application to particular electrocatalytic reactions. Sometimes such a selection may have been arbitrary and affected by the authors' personal view and knowledge. By the way we feel that the purpose of our work is not to be complete, rather to illustrate materials continuously recalling the principle behind their development. In some sense the book may be intended as a kind of "advanced" introduction to the subject.

We are also conscious that most of the works regarding electrocatalytic materials are generally logically arranged following certain reactions. Most of the books report sections on Hydrogen Evolution Reaction, Hydrogen Oxidation Reaction, Small Organic Molecules Oxidation, Oxygen Evolution Reaction, and Oxygen Reduction Reaction. For the nature of this book and for the sake of providing a presentation aimed at illustrating principles for electrocatalytic material design, we have preferred to organize the book following materials classes. So, we have a first Part devoted to the basic principles and then a Part focused on substrate materials for the high surface area catalyst supports. This is followed by the Part devoted to the active phases where we present some of the possible approaches to obtain enhancements in electrocatalytic activity and stability.

References

1. P. Tertzakian, *A Thousand Barrels a Second : The Coming Oil Break Point and The Challenges Facing An Energy Dependent World* (McGraw-Hill, New York, 2007), pp. xvi, 272 pp
2. R. A. Kerr, Do we have the energy for the next transition? Science **329**, 780 (2010)
3. M. K. Hubbert, *Energy Resources*. (National Academy of Sciences-National Research Council, Washington, 1962), p. 141
4. B. Gallagher, Peak oil analyzed with a logistic function and idealized Hubbert curve. Energy Policy **39**, 790 (2011)
5. C. J. Campbell, J. H. Laherrere, The end of cheap oil. Sci. Am. **78**, 3 (1998)
6. J. Van Hoesen, S. Letendre, Evaluating potential renewable energy resources in Poultney, Vermont: A GIS-based approach to supporting rural community energy planning. Renewable Energy **35**, 2114 (2010)
7. U. Bardi, Energy prices and resource depletion: lessons from the case of whaling in the nineteenth century. Energy Sources Part B: Econ. Plan. Policy **2**, 297 (2007)
8. C.-J. Yang, An impending platinum crisis and its implications for the future of the automobile. Energy Policy **37**, 1805 (2009)
9. B. Everett, Open University. *Energy Systems and Sustainability : Power for a Sustainable Future*, 2nd edn. (Oxford University Press, Oxford, 2012), pp. xiii, 654 p
10. R.U. Ayres, H. Turton, T. Casten, Energy efficiency, sustainability and economic growth. Energy **32**, 634 (2007)
11. D. Buchan, *The rough guide to the energy crisis* (Rough Guides, London, 2010)

12. S. J. Wagner, E. S. Rubin, Economic implications of thermal energy storage for concentrated solar thermal power. Renewable Energy
13. B.N. Divakara, H.D. Upadhyaya, S.P. Wani, C.L.L. Gowda, Biology and genetic improvement of *Jatropha curcas* L.: a review. Appl. Energy **87**, 732 (2010)
14. K.L. Kadam, J.D. McMillan, Availability of corn stover as a sustainable feedstock for bioethanol production. Bioresour. Technol. **88**, 17 (2003)
15. M.R.L.V. Leal et al., Sugarcane straw availability, quality, recovery and energy use: a literature review. Biomass Bioenergy **53**, 11 (2013)
16. C.M. Hoffmann, Root quality of sugarbeet. Sugar Tech **12**, 276 (2010)
17. Y. Sun, J. Cheng, Hydrolysis of lignocellulosic materials for ethanol production: a review. Bioresour. Technol. **83**, 1 (2002)
18. M. Balat, Biomass energy and biochemical conversion processing for fuels and chemicals. Energy Sources Part A: Recovery Util. Environ. Effects **28**, 517 (2006)
19. C. Lamy et al., Recent advances in the development of direct alcohol fuel cells (DAFC). J. Power Sources **105**, 283 (2002)
20. C. Lamy, E.M. Belgsir, J.M. Leger, Electrocatalytic oxidation of aliphatic alcohols: application to the direct alcohol fuel cell (DAFC). J. Appl. Electrochem. **31**, 799 (2001)
21. A. Marchionni et al., Electrooxidation of ethylene glycol and glycerol on Pd-(Ni-Zn)/C Anodes in direct alcohol fuel cells. Chemsuschem **6**, 518 (2013)
22. V. Bambagioni et al., Direct alcohol fuel cells as chemical reactors for the sustainable production of energy and chemicals Energy and chemicals from renewables by electrocatalysis. Chim Oggi **28**, Vii (2010)
23. P. Venturi, G. Venturi, Analysis of energy comparison for crops in European agricultural systems. Biomass Bioenergy **25**, 235 (2003)
24. C.A.S. Hall, An assessment of several of the historically most influential theoretical models used in ecology and of the data provided in their support. Ecol. Model. **43**, 5 (1988)
25. C.A.S. Hall, S. Balogh, D.J.R. Murphy, What is the minimum EROI that a sustainable society must have? Energies **2**, 25 (2009)
26. C. A. S. Hall, R. Powers, W. Schoenberg, Peak Oil, EROI, Investments and the Economy in an Uncertain Future. *In Renewable Energy Systems: Environmental and Energetic Issues*, 113 (2008)
27. M.K. Heun, M. de Wit, Energy return on (energy) invested (EROI), oil prices, and energy transitions. Energy Policy **40**, 147 (2012)
28. D. Pimentel, T.W. Patzek, Ethanol production using corn, switchgrass, and wood; Biodiesel production using soybean and sunflower. Nat. Resour. Res. **14**, 65 (2005)
29. I.C. Macedo, J.E.A. Seabra, J.E.A.R. Silva, Green house gases emissions in the production and use of ethanol from sugarcane in Brazil: The 2005/2006 averages and a prediction for 2020. Biomass Bioenergy **32**, 582 (2008)
30. ASTM E Standard guide for environmental life cycle assessment (LCA) of Building Materials/Products (1991-2005)
31. M.C. McManus, Life cycle impacts of waste wood biomass heating systems: a case study of three UK based systems. Energy **35**, 4064 (2010)
32. M. Momirlan, T. Vezirolu, Recent directions of world hydrogen production. Renew. Sustain. Energy Rev. **3**, 219 (1999)
33. M. Momirlan, T.N. Veziroglu, The properties of hydrogen as fuel tomorrow in sustainable energy system for a cleaner planet. Int. J. Hydrogen Energy **30**, 795 (2005)
34. R. Kothari, D. Buddhi, R.L. Sawhney, Comparison of environmental and economic aspects of various hydrogen production methods. Renew. Sustain. Energy Rev. **12**, 553 (2008)
35. Y. Lu, L. Zhao, L. Guo, Technical and economic evaluation of solar hydrogen production by supercritical water gasification of biomass in China. Int. J. Hydrogen Energy **36**, 14349 (2011)
36. L. Zhou, Progress and problems in hydrogen storage methods. Renew. Sustain. Energy Rev. **9**, 395 (2005)

37. R. P. O'Hayre, *Fuel Cell Fundamentals*, 2nd edn. (Wiley, Hoboken, N.J, 2009), pp. xxv, 546 p., 4 p. of plates
38. E. Antolini, E. R. Gonzalez, Alkaline direct alcohol fuel cells. J. Power Sources **195**, 3431 (2010)
39. http://www.fuelcells.org/fuel-cells-and-hydrogen/types/
40. C. Lamy, T. Jaubert, S. Baranton, C. Coutanceau, Clean hydrogen generation through the electrocatalytic oxidation of ethanol in a Proton Exchange Membrane Electrolysis Cell (PEMEC): Effect of the nature and structure of the catalytic anode. J. Power Sources **245**, 927 (2014)
41. V. Bambagioni et al., Self-sustainable production of hydrogen, chemicals, and energy from renewable alcohols by electrocatalysis. Chemsuschem **3**, 851 (2010)
42. F. Vitse, M. Cooper, G.G. Botte, On the use of ammonia electrolysis for hydrogen production. J. Power Sources **142**, 18 (2005)
43. K.P. Kuhl, E.R. Cave, D.N. Abram, T.F. Jaramillo, New insights into the electrochemical reduction of carbon dioxide on metallic copper surfaces. Energy & Environmental Science *5*, 7050 (2012)
44. M. Jitaru, Electrochemical carbon dioxide reduction—fundamental and applied topics. J. Univ. Chem. Technol. Metall. **42**, 333 (2007)
45. M.M. Jaksic, W. Schmickleer, G. Botton, Advances in electrocatalysis. Adv. Phys. Chem. **2012**, 4 (2012)
46. J. Speder et al., On the influence of the Pt to carbon ratio on the degradation of high surface area carbon supported PEM fuel cell electrocatalysts. Electrochem. Commun. **34**, 153 (2013)

Chapter 2
A Bird's Eye View of Energy-Related Electrochemistry

2.1 Key Concepts

An electrochemical cell is a device capable of either [1] obtaining electrical energy directly from a chemical reaction or [2] of converting electrical energy into chemical transformations. Electrochemical devices where the conversion of the chemical energy (the free energy of a spontaneous chemical reaction) into electrical energy (e.g., combination of molecular hydrogen and oxygen to form water) occurs in fuel cells and batteries. The second type of device, known as an electrolyzer is the class of electrochemical cell where an electrical energy input is supplied to drive an uphill chemical reaction (e.g., water splitting into elementary hydrogen and oxygen).

Now, two questions arise naturally. First, we wonder if a given electrochemical cell will supply or absorb electric energy under a given set of conditions (pressure, concentration, temperature, etc.). The answer to this question has been found by applying the principles of thermodynamics to electrochemical cells. The application of thermodynamic principles to electrochemistry has led to the world famous Nernst equation which, among its many applications, allows the prediction of the maximum energy which may be delivered by a fuel cell or a battery or on the opposite side the minimum energy supply required by an electrolyzer. Second, we may want to know at what rate an electrochemical reaction will proceed and what energy price we will have to pay to have it going at such a given rate. Electrochemical kinetics (sometimes referred to as dynamic electrochemistry) provides the framework for such an understanding. Electrochemical kinetics is truly a "lighthouse" in electrocatalysis, for its deep implications in the design of electrode materials suitable for the exploitation of electrochemical reactions in devices. Indeed, the role of nanotechnology in electrocatalysis comes as a direct consequence of the application of the fundamental laws of electrochemical kinetics. It will become clear throughout the book chapters devoted to electrocatalytic materials [3–8], that for efficient electrochemical processes materials need to be nanostructured in order to have a surface area large enough to allow fast

A. Lavacchi et al., *Nanotechnology in Electrocatalysis for Energy*,
Nanostructure Science and Technology 170, DOI: 10.1007/978-1-4899-8059-5_2,
© Springer Science+Business Media New York 2013

kinetics. The first two sections of this chapter illustrate the basic principle of electrochemical thermodynamics and kinetics. No derivation for the presented formulas is given, rather we have preferred to discuss their major implications, in the form we feel more appropriate for pursuing the aims of this book.

The discussion is accessible to graduates who have received classes in general chemistry, physical chemistry, or thermodynamics. If the reader feels a need to refresh some fundamental concepts of chemical thermodynamics, they may refer to a variety of textbooks in chemical physics and chemical thermodynamics (see, e.g., [1]). The reader interested in a deeper understanding of both the thermodynamics of the electrochemical cell and electrochemical kinetics is instead referred to excellent specialized monographs [2, 9].

This chapter concludes with a third section describing in detail the electrochemical reactions that are relevant to electrochemical energy conversion, i.e., Hydrogen Evolution Reaction (HER), Hydrogen Oxidation Reaction (HOR), Oxygen Evolution Reaction (OER), and Oxygen Reduction Reaction (ORR). Alcohol oxidation reactions and CO_2 reduction will also be covered. Each of these reactions has been extensively investigated in the literature, leading to an enormous volume of articles and reviews. Making a selection has been necessary. Only major facts and findings have been considered, as the purpose of this section is the illustration of general concepts functional for the understanding of principles and for the design of nanostructured electrode materials. However, here and there across the proceeding chapters of the book which is essentially devoted to materials, a more in-depth discussion essential for a correct understanding of the concepts underlying some specific architectures will also be given.

2.2 Thermodynamics

2.2.1 The Electrochemical Cell

In some sense, we may refer to electrochemistry as the art of splitting reactions. To better understand this concept, let us consider the simple chemical reaction:

$$H_2 + \frac{1}{2}O_2 \rightarrow H_2O \tag{2.1}$$

Equation (2.1) is the water formation starting from the elements. It is also the overall reaction occurring in fuel cells employing hydrogen as the fuel. It is known that such a reaction is spontaneous ($\Delta G = -237.1$ kJ mol^{-1}). According to the thermodynamic definition, ΔG is also the maximum amount of energy delivered by the reaction which can be converted into mechanical work. From general chemistry it is known that both oxygen and hydrogen are in oxidation state 0 when in elementary form. In water they combine in such a way that hydrogen delivers electrons to oxygen. In molecular water, hydrogen is said to be oxidized to +1 having lost one electron in the reaction, while oxygen is reduced to −2 having acquired two electrons. Charge neutrality is fulfilled. Such reactions are called red-ox as one species loses electrons (hydrogen), and the other acquires the electron being reduced (oxygen). The trick of electrochemistry is to separate the oxidation and reduction reactions making them occur in physically distinct regions.

Such a separation is possible thanks to the architecture shown in Fig. 2.1.

The cell consists of two electrodes, an ionic conductor and an electronic conductor. An electrode is denoted as the anode if the oxidation occurs there. If a reduction reaction occurs the electrode is defined as the cathode.

Now let us consider again the reaction of Eq. (2.1). It can be formally split in the following way:

$$H_2 \rightarrow 2H^+ + 2e^- \tag{2.2}$$

$$\frac{1}{2}O_2 + 2e^- \rightarrow O^{2-} \tag{2.3}$$

Equation (2.2) is the HOR, hence, according to our above considerations this takes place at the anode, while Eq. (2.3) is the ORR and takes place at the cathode. There is a substantial difference between this set of equations and Eq. (2.1). Charged species are explicitly considered here. Particularly at the anode we observe the formation of hydrogen ions and electrons, while at the cathode oxygen

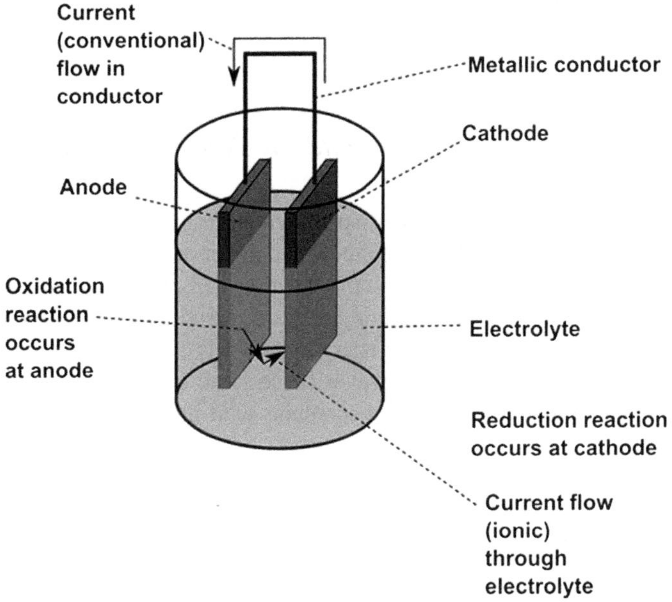

Fig. 2.1 Sketch of a simple electrochemical cell reporting the four essential components, the anode (oxidation electrode), the cathode (reduction electrode), the electrolyte (ion conductor), and the metal connection between the electrodes (electronic conductor)

is reduced to negative ions and electrons are consumed. Since the reaction occurs in physically separated regions there is the need to allow the transport of such charged species. Furthermore, we have to avoid the same path for ions and electrons. Fortunately, it is known that the transport of ions and electrons need different kinds of conductors. Generally what conducts electrons does not conduct ions and vice versa. Electrons need the so-called first type conductors (e.g., metals, a variety of conducting oxides and compounds), while ions need second type conductors, namely electrolytes (e.g., liquid solutions containing significant amounts of dissolved ions or ceramic materials at high temperature). Therefore, in order to close the circuit and start the reaction, there is the need to have an electron conductor connecting the anode and the cathode and an electrolyte allowing ions to freely move across the cell. If in the middle of the electron conductor an electrical load (e.g., a lamp, a radio, or any other electrical device) is applied, it consumes part of the energy of the electrons which uses this for its functioning. This is the way that chemical energy can be transformed into electrical energy without passing through thermo-mechanical cycles with their known efficiency limitations.

In principle all this could seem a bit artificial, but the implications of this approach are incredibly important. Indeed, as just seen, it is this art of splitting reactions which allows the existence of fuel cells and batteries, the only devices where chemical energy is directly transformed into electrical energy without the need of heat and mechanical work. The same is the basis for the electrolytic generation of hydrogen, electro-synthesis of chemical compounds, and electrolytic metal production and refining.

2.2.2 *Electrochemical Reaction and the Nernst Equation*

It is now clear how electrochemistry splits redox reactions into half reactions. What is still not defined is how to predict if under a given set of conditions (temperature, pressure, concentrations) the reaction will produce energy or energy will be required to keep it going. For chemical reactions such a prediction is possible thanks to the concept of free energy. Basically a chemical transformation is, in principle (the kinetics may however be so slow to completely hamper the course of the reaction), said possible or spontaneous if its associated free energy is negative.

Furthermore, it is essential to have the chance of a priori prediction of the potential of an electrochemical reaction. The last task is accomplished calculating the cell potential according to Eq. (2.4).

$$E^0 = -\frac{\Delta G^0}{nF} \tag{2.4}$$

Let us consider again the reaction of Eq. (2.1). We have seen that the associated standard Gibbs free energy is -237.1 kJ mol^{-1}. The application of Eq. (2.4) leads to a cell standard potential of 1.23 V. Hence, a cell with HOR as anode reaction and ORR as cathode reaction is capable of delivering at maximum 1.23 V when hydrogen and oxygen are fed in standard conditions (1 atm and 298 °C). This is the theoretical cell potential of a hydrogen fuel cell. From Eq. (2.4), we also notice that as the potential gets more positive the reaction will become more spontaneous.

The prediction of the cell potential in nonstandard conditions is possible through the application of the Nernst equation (Eq. 2.5). The Nernst equation explicitly accounts for the dependence of the cell potential on parameters such as partial products and reactants activity (in a first approximation and for diluted solutions or gas activity may be replaced by concentrations or partial pressure) and temperature.

$$E = E^0 - \frac{RT}{nF} \ln \frac{\prod a_{\text{products}}^{v_i}}{\prod a_{\text{reactants}}^{v_j}} \tag{2.5}$$

With the Nernst equation, we are able to predict the thermodynamic potential of an electrochemical cell under a given set of operating conditions. For the water formation reaction reported in Eq. (2.1) the Nernst equation takes the form of Eq. (2.6).

$$E = E^0 - \frac{RT}{2F} \ln \frac{a_{H_2O}}{a_{H_2} a_{O_2}^{1/2}} \tag{2.6}$$

As water is liquid under 100 °C its activity is conventionally fixed as 1. In a first approximation, we may also consider the activity of hydrogen and oxygen equal to their partial pressures (for practical purposes the activity is replaced by concentrations for liquid solutions and partial pressures for gases), leading to Eq. (2.7) which may be used in practice for rough estimations:

$$E = E^0 - \frac{RT}{2F} \ln \frac{1}{p_{H_2} p_{O_2}^{1/2}} \tag{2.7}$$

From Eq. (2.7), we notice that as we increase the partial pressure of both hydrogen and oxygen the potential gets more positive indicating that the reaction is more energetically favored.

It is worthy to spend some more words on the dependence of the cell potential on the temperature. Such a dependence is more subtle than what is suggested by Eq. (2.5). Basically this is because even the term E^0 is affected by the temperature. The dependence of E^0 on the temperature may be accounted for by applying Eq. (2.8).

$$E = E^0 + \frac{\Delta S}{nF}(T - T^0) - \frac{RT}{nF} \ln \frac{\prod a_{\text{products}}^{v_i}}{\prod a_{\text{reactants}}^{v_j}} \tag{2.8}$$

Nevertheless the temperature dependence may often be neglected, at least in practical technologies, as such dependence is usually small. Values of the order of

0.1–1 mV/°C are common. For low temperature systems (operating between 20 and 100 °C), such variations are not relevant as compared to the overall cell potential and for rough calculations this may be neglected.

Very often in aqueous systems chemical reactions include the production or the consumption of H^+ or OH^- species. The concentration of such species, hence, the pH of the electrolyte, affect the cell potential. Again such a dependence can be calculated using the Nernst equation.

As an example, let us consider the partial oxidation of ethanol in alkaline environments, a reaction exploited for energy production by Direct Ethanol Fuel Cells (DEFC).

$$CH_3CH_2OH + O_2 + OH^- \rightarrow CH_3COO^- + 2H_2O \tag{2.9}$$

The reaction Eq. (2.9) occurs in alkaline environments and the OH^- species is a reactant. In a first approximation, we can substitute the activity with the concentration, furthermore considering that four electrons are exchanged in Eq. (2.9), we obtain the following Nernst relationship:

$$E = E^0 - \frac{RT}{4F} \ln \frac{c_{CH_3COO^-}}{c_{CH_3CH_2OH} c_{OH^-} p_{O_2}} \tag{2.10}$$

Then considering the ionic water product and the pH definition this equation together with some algebraic workout results in Eq. (2.11):

$$E = E^0 + \frac{RT}{4F} pH - \frac{RT}{4F} \ln \frac{c_{CH_3COO^-}}{c_{CH_3CH_2OH} p_{O_2}} \tag{2.11}$$

Hence, the higher will be the pH, the higher will be the potential and the higher the reaction will be spontaneous.

The use of the Gibbs free energy for the calculation of the standard cell potential can be avoided introducing the concept of the standard reduction potential for half reactions. As previously calculated water formation in an electrochemical cell under standard conditions produces a potential difference of 1.23 V. The cell potential may hence be considered as the contribution of the sum of a potential contribution coming from the half cell anode (Eq. 2.2) and the half cell cathode reactions (Eq. 2.3), respectively. As in physical systems only potential differences matter, there is no chance to define an absolute potential scale for half reactions. Nevertheless for practical purposes the problem can be solved tabulating all half cell potentials with respect to the same standard electrode providing a practical working framework. The role of reference for the potential scale is assigned to the so-called standard hydrogen electrode (SHE) to which a conventional value of 0 V is assigned. A further convention is that each half reaction in the standard potential table is reported in such a way that the direct reaction is the reduction one. Table 2.1 lists values for a variety of reduction half reactions. According to Table 2.1, the standard electrochemical cell potential for water formation can be calculated as follows:

Table 2.1 Table of standard reduction potential for selected half cell reactions

Reduction half reaction	Standard reduction potential E° (V)
$Li^{+}(aq) + e^{-} \text{ - > } Li(s)$	−3.04
$K^{+}(aq) + e^{-} \text{ - > } K(s)$	−2.92
$Na^{+}(aq) + e^{-} \text{ - > } Na(s)$	−2.71
$Mg^{2+}(aq) + 2e^{-} \text{ - > } Mg(s)$	−2.38
$Al^{3+}(aq) + 3e^{-} \text{ - > } Al(s)$	−1.66
$2H_2O(l) + 2e^{-} \text{ - > } H_2(g) + 2OH^{-}(aq)$	−0.83
$Zn^{2+}(aq) + 2e^{-} \text{ - > } Zn(s)$	−0.76
$Cr^{3+}(aq) + 3e^{-} \text{ - > } Cr(s)$	−0.74
$Fe^{2+}(aq) + 2e^{-} \text{ - > } Fe(s)$	−0.41
$Ni^{2+}(aq) + 2e^{-} \text{ - > } Ni(s)$	−0.23
$Sn^{2+}(aq) + 2e^{-} \text{ - > } Sn(s)$	−0.14
$Pb^{2+}(aq) + 2e^{-} \text{ - > } Pb(s)$	−0.13
$Fe^{3+}(aq) + 3e^{-} \text{ - > } Fe(s)$	−0.04
$2H^{+}(aq) + 2e^{-} \text{ - > } H_2(g)$	0.00
$Sn^{4+}(aq) + 2e^{-} \text{ - > } Sn^{2+}(aq)$	0.15
$Cu^{2+}(aq) + e^{-} \text{ - > } Cu^{+}(aq)$	0.16
$ClO_4^{-}(aq) + H_2O(l) + 2e^{-} \text{ - > } ClO_3^{-}(aq) + 2OH^{-}(aq)$	0.17
$AgCl(s) + e^{-} \text{ - > } Ag(s) + Cl^{-}(aq)$	0.22
$Cu^{2+}(aq) + 2e^{-} \text{ - > } Cu(s)$	0.34
$ClO_3^{-}(aq) + H_2O(l) + 2e^{-} \text{ - > } ClO_2^{-}(aq) + 2OH^{-}(aq)$	0.35
$Cu^{+}(aq) + e^{-} \text{ - > } Cu(s)$	0.52
$I_2(s) + 2e^{-} \text{ - > } 2I^{-}(aq)$	0.54
$Fe^{3+}(aq) + e^{-} \text{ - > } Fe^{2+}(aq)$	0.77
$Hg_2^{2+}(aq) + 2e^{-} \text{ - > } 2Hg(l)$	0.80
$Ag^{+}(aq) + e^{-} \text{ - > } Ag(s)$	0.80
$Hg^{2+}(aq) + 2e^{-} \text{ - > } Hg(l)$	0.85
$ClO^{-}(aq) + H_2O(l) + 2e^{-} \text{ - > } Cl^{-}(aq) + 2OH^{-}(aq)$	0.90
$2Hg^{2+}(aq) + 2e^{-} \text{ - > } Hg_2^{2+}(aq)$	0.90
$NO_3^{-}(aq) + 4H^{+}(aq) + 3e^{-} \text{ - > } NO(g) + 2H_2O(l)$	0.96
$O_2(g) + 4H^{+}(aq) + 4e^{-} \text{ - > } 2H_2O(l)$	1.23
$Cr_2O_7^{2-}(aq) + 14H^{+}(aq) + 6e^{-} \text{ - > } 2Cr^{3+}(aq) + 7H_2O(l)$	1.33
$Cl_2(g) + 2e^{-} \text{ - > } 2Cl^{-}(aq)$	1.36
$Ce^{4+}(aq) + e^{-} \text{ - > } Ce^{3+}(aq)$	1.44
$MnO_4^{-}(aq) + 8H^{+}(aq) + 5e^{-} \text{ - > } Mn^{2+}(aq) + 4H_2O(l)$	1.49
$H_2O_2(aq) + 2H^{+}(aq) + 2e^{-} \text{ - > } 2H_2O(l)$	1.78
$Co^{3+}(aq) + e^{-} \text{ - > } Co^{2+}(aq)$	1.82
$S_2O_8^{2-}(aq) + 2e^{-} \text{ - > } 2SO_4^{2-}(aq)$	2.01
$F_2(g) + 2e^{-} \text{ - > } 2F^{-}(aq)$	2.87

$$E^0_{H_2O} = E^0_{\frac{O_2}{H_2O}} - E^0_{\frac{H^+}{H_2}} = 1.23\,V - 0.00\,V = 1.23\,V \quad (2.12)$$

The term related to hydrogen is negative in sign as hydrogen is actually oxidized. Furthermore, the term concerning hydrogen is 0 V as the hydrogen oxidation in standard conditions is our actual reference.

In summary, we have seen how thermodynamics considerations allow us to make predictions of the maximum energy we may extract from spontaneous chemical reactions. In turn, we can also predict the minimum potential we have to supply to a given system to drive an uphill reaction (this is the case in electrolysis).

2.3 Electrochemical Kinetics

Electrochemical kinetics is a complex subject and its comprehensive description is out of the scope of the present book. Nevertheless as pointed out before a review of its fundamental concepts is essential for the understanding of the major implications that the theory of the rate of electrochemical reactions has in the design of efficient electrocatalysts.

In order to understand electrochemical kinetics we have to rely on the actual picture of what an electrochemical interface is. Conventionally, we refer to the electrochemical interface as the spatial region including the electrode and the portion of the electrolyte where the "most important" things for determining the rate of an electrochemical reaction occur. This commonly accepted vision of the electrochemical interface is actually sketched in Fig. 2.2.

Three distinct sections are identified in the electrolyte. The first one, directly in contact with the electrode surface, is the so-called Helmholtz layer. In such a portion, positive ions accumulate as a result of the electrostatic attraction with the negative charge accumulated at the surface of the electrode. The Helmoltz layer may divide in the so called inner Helmoltz layer and outer Helmoltz layer. The inner Helmoltz layer consists of the molecules or ions directly adsorbed at the electrode surface. The outer Helmoltz layer is formed by the ion electrostatically attracted by the electrode

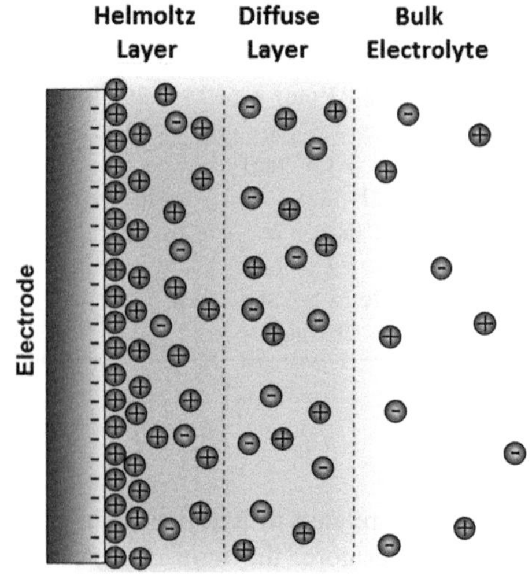

Fig. 2.2 Pictorial view of a negatively polarized electrochemical interface

but still being coordinated by a shell of solvent molecules. Specific adsorption of chemical species may also occur. Specific adsorption, also called chemisorption, is the result of strong covalent interactions between chemical species and the electrode surface. It will be clear when describing in detail some electrochemical reactions that adsorption may play a very important role in the determination of the rate of an electrochemical reaction. Sometimes, as in the case of CO for platinum adsorption, this may also result in the most relevant poisoning pathway for electrocatalysts.

There is also another essential step in electrochemical reactions which occurs here and that has to be considered. As we know electrons cannot directly flow through the electrolyte. In order to guarantee a current flow through the cell, we need the electrons to *hop on* or *hop off* from chemical species such as ions. The stage where the electrons *hop on* or *hop off* the ions is the so-called charge transfer. The order of magnitude for the spatial size of the Helmholtz layer is less than 1 nm as it is identified as the order of magnitude of an atomic distance. Charge transfer mainly occurs here because it has been demonstrated to be the result of a quantum tunneling effect whose probability quickly drops as the distance from the electrode surface increases. The second section of the electrolyte consists in the diffuse layer. In such a portion of the electrolyte, only coulombic interactions and polarization effects occur. In the bulk or third section of the electrolyte, we may have what is called bulk diffusion which may also produce limitations to the rate, especially for low reactant concentrations or high polarizations of the interface. A mathematical description of the model reported here has been attained so the dependence of the value of the potential across the interface can be calculated.

2.3.1 Charge Transfer

A key quantity for electrochemical kinetics is the overpotential which is defined as the potential difference between the reduction potential, as determined by thermodynamic considerations, and the actual half cell potential.

$$\eta = E - E^0 \tag{2.13}$$

The overpotential is the true driving force for a half cell reaction. The definition of the overpotential is the first step in the determination of the rate of an electrochemical reaction. The equation linking the rate (current density) of an electrochemical reaction where only charge transfer limits the rate is the so-called Butler–Volmer equation which takes the name from the scientists who independently derived it.

$$j = j^0\left(e^{\alpha_a nF\eta/RT} - e^{-\alpha_c nF\eta/RT}\right) \tag{2.14}$$

In Eq. (2.14) j^0 is the exchange current density and it is the common absolute value of the anodic and cathodic current densities when the system is at the equilibrium (no net current flowing through the electrochemical cell). This is

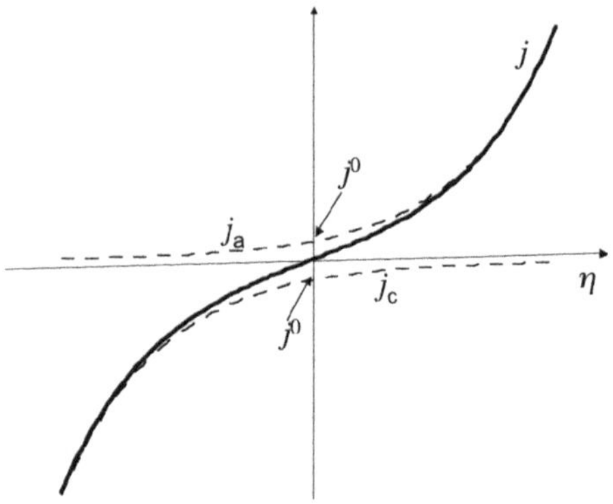

Fig. 2.3 Plot for the Butler–Volmer equation showing anodic current density j_a (*upper dashed curve*), cathodic current density j_c (*lower dashed curve*), and net current density j (*solid curve*) dependence on the overpotential

probably the most determining quantity in electrochemical kinetic, as the current for overpotential away from zero is directly proportional to it. Basically the larger is the exchange current density the faster the reaction will occur. In other words, large values of j^0 will require small polarization, absorbing only a limited fraction of energy as compared to systems with smaller j^0. It is also worth noticing that such a density is usually referred to the geometric area only and does not account for the real surface area of the electrode. According to this we can have an immediate taste of why nanotechnology is so important in electrocatalysis. Indeed, nanotechnology provides an approach to increase j^0 through the increase of the real surface area while keeping the geometric area constant.

Other essential quantities for electrocatalysis are the coefficients α_a and α_c. Such coefficients define how large is the increment in the rate of the reaction for a given variation of the overpotential shows a typical shape for the Bulter-Volmer curve (Fig. 2.3). The curve may be divided into two distinct regions. The first one is the closest to the equilibrium ($\eta = 0$). Here the shape is approximately linear. This region does not invest much interest in electrocatalysis as reaction rates are usually very small and unpractical for devices. On the contrary this is very important in the investigation of metal corrosion. The second region is where the potential significantly departs from 0 showing an exponential dependence of j on η. In this region, only the cathodic or anodic terms of the Butler–Volmer equation may be considered, as the entity of the reverse reaction can be neglected. At large overpotential deviation from the Butler–Vomer behavior occurs. This is because mass transfer limitation due to the diffusion of the electroactive species occurs. When mass transport controls the reaction rate nothing else matters. Even if the overpotential rises we will never see our reaction going faster. A current density increase may be eventually observed, but that is generally due to the onset of other parasitic electrochemical reactions. The second region is worthy of more consideration as it allows, in principle, the determination of the cathodic and anodic coefficients. Equations (2.15) and (2.16) show the dependence of α_c and α_a on the current density and the overpotential when $|\eta| \gg 0$ (absolute value is used to include in the relationship both anodic and cathodic region).

$$\alpha_c = -\left(\frac{RT}{nF}\right)\left(\frac{\mathrm{d}lnj_c}{d\eta}\right) \tag{2.15}$$

$$\alpha_a = \left(\frac{RT}{nF}\right)\left(\frac{\mathrm{d}lnj_a}{d\eta}\right) \tag{2.16}$$

At first glance one could also be tempted to use Eqs. (2.15) and (2.16) to estimate the number of exchanged electrons before the rate determining step of an electrochemical reaction. By the way as it is going to be pointed out by a specifically addressed IUPAC commission, drawing mechanistic conclusion from this is not allowed. Indeed IUPAC is proposing the following definitions for α_c and

$$\alpha_c = -\left(\frac{RT}{F}\right)\left(\frac{\mathrm{d}lnj_c}{d\eta}\right) \tag{2.17}$$

$$\alpha_a = \left(\frac{RT}{F}\right)\left(\frac{\mathrm{d}lnj_a}{d\eta}\right) \tag{2.18}$$

This simply requires the removal of the factor n from the denominator of Eqs. (2.15) and (2.16). Such a definition is based directly on a experimental quantity and avoids any possible attempt to draw mechanistic conclusions. Hence the two following relationships can be derived

$$j_a = j^0 \exp[\alpha_a F\eta/RT] \tag{2.19}$$

$$|j_c| = j^0 \exp[-\alpha_c F\eta/RT] \tag{2.20}$$

The above relationships may be passed to logarithms returning the following equations:

$$\ln j_a = \ln j^0 + \frac{\alpha_a F}{RT}\eta \tag{2.21}$$

and

$$\ln |j_c| = \ln j^0 - \frac{\alpha_c F}{RT}\eta \tag{2.22}$$

From Eqs. (2.21) and (2.22) it is clear that plotting η versus ln j returns a straight line whose slope is proportional to the anodic or cathodic coefficients. The intercept provides a tool for determining the value of j_0. Such plots are referred to as Tafel plots (Fig. 2.4). Deviation from linearity of the Tafel plot at low overpotentials is due to the rising contribution of the inverse reaction. It is important to mention that a linear Tafel plot might eventually not be encountered in practice. This may be the consequence of a variety of reasons, such as the onset of other electrochemical reactions, variation in the mechanism as a result of the variation of the overpotential, adsorption of poisoning species, etc.

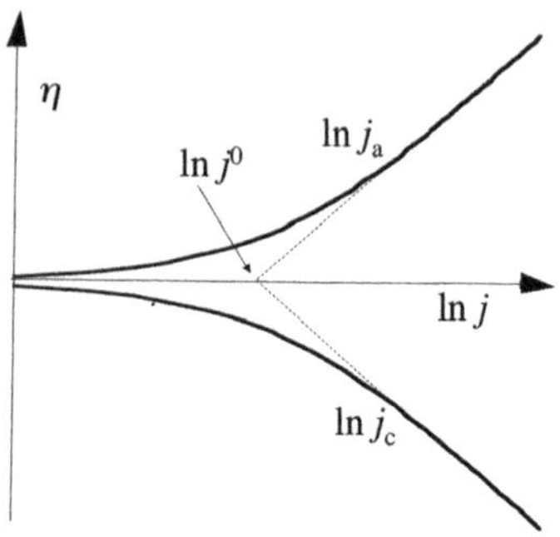

Fig. 2.4 Tafel plot. A line fit for the high overpotential region allows the estimation of the anodic and cathode coefficients, while the intercept provides a tool for the determination of the exchange current density j_0

By the way, for practical purposes only, a fully empirical knowledge of the relationship between the overpotential and the current density may be enough, at least for defining the energy of a given electrochemical reaction.

2.3.2 Mass Transfer

When electrochemical reactions are driven far from the equilibrium depletion of reactants at the electrochemical interface may occur. On the other hand, difficulties in the removal of the reaction products may lead to accumulation in close proximity of the electrode surface. In still electrolyte the mass transfer usually only occurs for diffusion even if local variation of the concentration of chemical species may drive variation in the density leading to buoyant convection. Mass transfer may affect both thermodynamics and kinetic reactions. The way diffusion affects thermodynamics is intuitive as in the Nernst equation only depends on the actual value of the concentration at the electrode surface. So reactant depletion and product pile up may slow down the reactions rates departing from the value expected from bulk concentration. Kinetics also is significantly altered by variation in the concentration. Indeed the Butler–Volmer equation can be modified to account for the actual concentration value at the electrode–electrolyte interface leading to

$$j = j^0 \left[\frac{c_R}{c_R^O} \exp(\alpha_a nF\eta/RT) - \frac{c_P}{c_P^O} \exp(-\alpha_c nF\eta/RT) \right] \tag{2.23}$$

Under complete diffusion control the Butler–Volmer equation loses its validity and the current density becomes independent of the overpotential, in the sense that further polarization of the interface does not provide any further increment in the current, which in turn will diminish with the increasing time. The effect of diffusion may be limited by applying convection. Under convection it is possible to replenish (or remove) chemical species from the electrode–electrolyte interface. Controlled convection conditions are sometimes employed for expanding the possibility of electrochemical characterization as, e.g., in the case of the rotating disk experiments which will be discussed later.

It is worthy to mention that diffusion phenomena may significantly be hampered in certain nanoarchitectures. This will be discussed in some detail in Chap. 4.

2.3.3 Adsorption

We have seen that adsorption occurs within the Helmoltz layer. Such a stage is often determining for the rate of electrochemical reactions. Conventionally, we may discriminate between various kinds of adsorption interaction between the electrode surface and the molecules or ions in the solutions. The first one is the previously encountered chemisorptions. It occurs when between the electrode surface and the molecules or ions in the solution a covalent interaction subsists. We will see later that this is truly an issue of outstanding importance, as the extent of the specific adsorption interaction may heavily determine the effectiveness of a material as an electrocatalyst. Particularly it is usually necessary to find a good compromise between the strength of the adsorption which is accompanied by a high adsorption rate and the ability of the adsorbed species to be removed from the surface. Practically the adsorption interaction has to be "just right." Right enough to guarantee high adsorption rated and low enough to gurantee a quick desorption. If the adsorbed specie sticks strongly to the surface it will hamper the catalytic activity of the site. This is the basis of the Sabatier principle [10] whose pictorial view are the volcano plots which will be discussed in some detail for the hydrogen and oxygen oxidation and reduction. A case of very strong adsorption interaction is that of CO with platinum [3–6]. The adsorption of carbon monoxide may determine the deactivation of the platinum catalyst, leading, e.g., to overall efficiency losses of the polymer electrolyte fuel cells fuelled with methanol.

Adsorption may also be aspecific when due to weak dispersive interaction. Under such circumstances adsorption is usually denoted as physisorption. The interactions are much weaker than in chemisorptions and such phenomena is not as relevant as chemisorptions is. A third kind of adsorption may also result from the interaction of charged surface with ions of the opposite sign. In some cases adsorption may be the rds of a electrochemical reaction. As an example, we may cite the case of the alcohols oxidation on palladium in an alkaline environment. It is known and it will be discussed later that in such cases the formation $Pd\text{-}(OH)_{ads}$ is the rate determining step for the reaction at least at low overpotential [7]. To circumvent the problem palladium may be mixed with materials (e.g., CeO_2, NiO, etc.) [8, 11] which have a stronger tendency to form hydroxide species at the surface. In such oxides, such species are also very mobile and can be transferred to the metal affecting the rate of the rate determining step. Such a phenomenon is known as "spillover of the primary oxide" [12, 13] and involves the adsorption of hydroxide at both an oxide and the palladium metals. It has been demonstrated that the primary oxide spillover can be exploited enhancing the efficiency of the DEFCs. Analogous phenomena may also occur with hydrogen. If palladium particles are deposited onto a gold support they can collect hydrogen at their surface and transfer it to the surface of the gold support. Such a phenomenon has been investigated in electrochemistry and it is a well-known case of "hydrogen spillover." Spillover phenomena are sometimes considered when designing catalysts, as they may significantly enhance the rate of electrochemical reactions.

2.4 Electrochemical Techniques

Phenomena occurring at the electrochemical interface may be investigated by means of the so-called three-electrode cell or half cell. This electrochemical cell arrangement allows the identification of phenomena occurring at a single interface. The experimental setup consists in a cell containing a working electrode, a counter electrode, and a reference electrode. The working electrode is the one where the interface subject of the investigation is located. The counter is where reactions necessary to close the circuit occur. The reference is indeed a kind of "spectator" which allows only a negligible current flow, keeping its potential constant during the measurement (usually this condition is referred to as nonpolarized). The three electrodes are connected to a potentiostat. With the potentiostat the operator is allowed to directly control the potential difference between the reference and the working electrode. Once this potential has been set, the instrument adjusts the potential between the working and the counter electrode so that the potential difference between the reference electrode and the working electrode sticks to the value set by the operator. Historically, the potentiostat is the instrument which produced a true "quantum leap" in electrochemical characterization. Most of the present understanding of electrochemical kinetics we owe to the potentiostat.

2.4.1 Voltammetry

Voltammetry is one of the major experiments in dynamic electrochemistry. In a voltammetric experiment, the potential at the working electrode interface is scanned with the time using a three-electrode cell arrangement and a potentiostat. If the scan is performed with the potential varying linearly with the time and just in one direction, while recording the current passing through the cell, the experiment takes the name of Linear Sweep Voltammetry (LSV). If the same scan is performed at very low scan rate (e.g., 1 mV s^{-1} this condition is usually referred to as potentiostatic), it may allow the acquisition of Tafel plots, described in the section devoted to mass transport. More commonly scans are performed both in the forward and backward directions and eventually repeated many times, leading to the so-called cyclic voltammetry (CV) (Fig. 2.5). Cyclic voltammetry is possibly the most widespread experiment for evaluating the effectiveness of an electrocatalyst. At least is the first performed by electrochemists when they want to have a

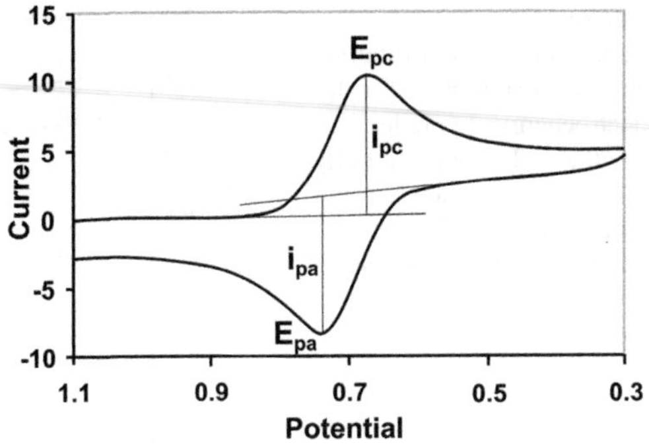

Fig. 2.5 A cyclic voltammetry curve. Amongst the information which can be obtained from the CV, there are the degree of reversibility of electrochemical reactions, the on-set potential, and the peak current densities for anodic and cathodic processes

preliminary understanding of the electrochemical behavior of systems. CV has been the subject of entire books and its deep understanding would require much more space that this book can accommodate. We limit the treatment to the set of information which can be obtained by a cyclic voltammogram, inviting the reader to refer to the excellent book from Bard and Faulkner for a comprehensive review of CV basic principles.

Cyclic voltammetry provides a wide variety of relevant information about the electrocatalytic properties of materials, the following list reports some of them:

(1) Onset potential (it is the potential at which the reaction current density significantly departs from the background) for a given electrochemical reaction.
(2) Determination of the diffusion coefficient of the reactants.
(3) Oxidation, dissolution, and electrochemical behavior of the active catalyst phase.
(4) Determination of the electrochemically active surface area of the catalyst.
(5) Direct comparison of the electrocatalytic activity of various materials.

The obtainment of this set of information may require the acquisition of CV varying parameters such as the potential span (minimum and maximum potential), the scan rate, and eventually the nature of the electrolyte.

2.4.2 Rotating Disk and Rotating Ring-Disk Methods

Controlled mass transport conditions may be employed to design electrochemical experiments helpful to elucidate electrochemical reaction mechanisms. The rotating disk electrode consists of a disk electrode embedded in an inert non-conductive cylinder. The cylinder is then connected to the shaft of an electric motor with a precise control of the rotation rate (usually between 100 and 3,000 r.p.m.). The disk rotation provides the instauration of a laminar flow which is the result of a centrifugal force pushing the electrolyte away from the center of the electrode in the radial direction and a liquid movement from the bulk of the electrolyte to the electrode surface in the axial reaction. For large overpotentials complete mass transport control of the electrochemical reaction occurs and

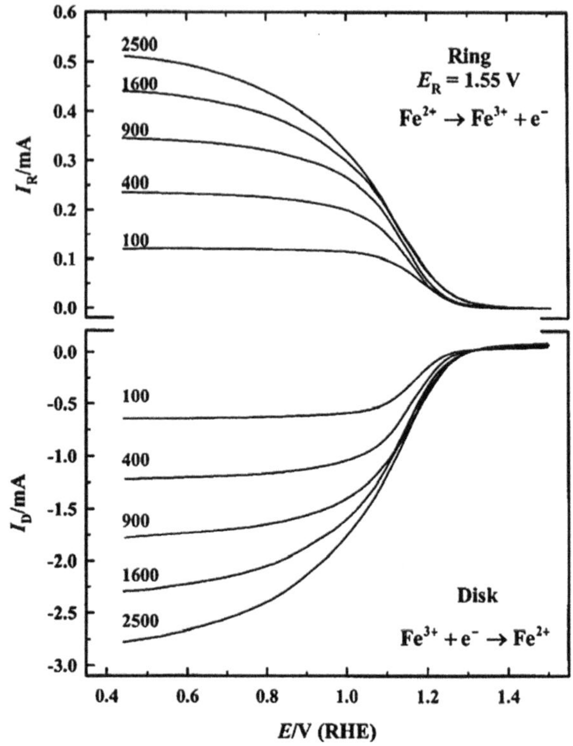

Fig. 2.6 Classical representation for a rotating ring disk experiment for the electrochemical couple Fe^{3+}/Fe^{2+}. The ring partially collects the products generated at the disk as a result of the rotation-induced convection

formulas describing the resulting convection–diffusion problem can be obtained. The resulting Levich equation (Eq. 2.24) [14] provides a reliable method for estimating the number of exchange electrons in a electrochemical reaction.

$$j_l = 0.62\ n\ FD^{2/3}\omega^{1/2}\upsilon^{-1/6}C. \tag{2.24}$$

When the control of the reaction is mixed kinetic and convective–diffusive the Koutecky–Levich (Eq. 2.25) equation may be used [2]. Such an equation allows the determination of the kinetic current density for a given overpotential and has been widely applied for extracting electrocatalytic parameters for reactions relevant to electrochemical energy conversion. By the way, its application is not straightforward and questions regarding the quality of the information that can be obtained may arise [14, 15].

$$1/j = 1/j_l + 1/j_k \tag{2.25}$$

If a second working electrode in the form of a ring is embedded in the nonconductive cylinder in such a way that it encircles the disk, a rotating ring-disk electrode is obtained. As two working electrodes are present, to operate the RRDE a special potentiostat capable of controlling four electrodes is necessary (bi-potentiostat).

The current at both electrodes is recorded and the most common data representation correspond to that is shown in Fig. 2.6.

The RRDE is a fundamental tool for determining properties of electrocatalysts for fuel cells. As an example, it has been widely employed in the characterization of the ORR as cathode reaction for PEMFC. This is mainly because partial ORR may result in the production of hydrogen peroxide a specie harmful in fuel cell for its significant contribution to carbon support corrosion. Experiments are performed coating the disk with a thin layer of electrocatalyst and then driving a potential scan to the reduction of oxygen. Any product generated at the disk electrode is removed from the disk in virtue of the flow regime above described and reaches the ring electrode. In the case of the ORR, the potential of the ring electrode is poised to detect any hydrogen peroxide that may have been generated at the disk leading to quantitative information about the amount of partially reduced oxygen species.

2.5 Major Energy-Related Electrochemical Reactions

2.5.1 Hydrogen Oxidation and Evolution Reactions

As hydrogen has been recognized as a possible energy vector for the future, much attention has been devoted to both its production through water electrolysis and oxidation in fuel cells.

As a consequence the major energy-related hydrogen reactions are the HER and the molecular HOR.

As previously seen in the section devoted to thermodynamics, hydrogen reduction is taken as a reference for the standard potential table and its reduction potential is by convention set to 0 V at standard conditions (pH = 0, T = 25 °C, H_2 partial pressure 101,325 Pa). To drive the reaction far from equilibrium and having hydrogen oxidation or reduction proceeding at a given rate we have hence to force the potential away from 0. Namely positive potentials provide oxidation and while negative evolution. At present hydrogen evolution and oxidation are not the main concern in electrocatalysis. Indeed, their kinetics is fast as compared to the more sluggish OER and ORR. The kinetic of hydrogen oxidation and reduction is known to be relatively fast on noble metals. Indeed, at present, best candidate electrocatalysts for such reactions for application in devices are metals such as, e.g., Pt, Pd, and Ru in pure form or in alloys. In our description, all the considered mechanisms refers to platinum without loss of generality. Indeed, analogous mechanisms apply to a wide variety of metal electrocatalysts.

Hydrogen evolution reaction

HER is reported in Eq. (2.26). Equation (2.26) shows the way that hydrogen ions are oxidized to produce H_2 in acidic environment.

$$H^+ + e^- \rightarrow \frac{1}{2}H_2 \tag{2.26}$$

In alkaline environment the HER takes the form of Eq. (2.27).

$$H_2O + e^- \rightarrow \frac{1}{2}H_2 + OH^- \tag{2.27}$$

A glance at the HER reactions immediately reveals that the reduction potential of hydrogen reduction depends on the pH. Indeed, as previously pointed out the potential for pH $= 0$ is 0 V, while the Nernst equation shows that at pH 14 the potential is -0.83 V.

HER has been thoroughly investigated, leading to a generally accepted mechanistic scheme. Nowadays, it is widely recognized that the HER occurs through two main pathways, the Vomer–Tafel and the Volmer–Heyvrosky. For both pathways the first step is the adsorption of hydrogen on platinum through the so-called Volmer reactions which for acidic and alkaline conditions take the form of Eqs. (2.28) and (2.29), respectively [16]

$$Pt + H^+ + e^- \rightarrow Pt - H \tag{2.28}$$

$$Pt + H_2O + e^- \rightarrow Pt - H + OH^- \tag{2.29}$$

The Volmer reaction is essentially a combined charge transfer and adsorption step. Then adsorbed hydrogens atoms may react together according to the Tafel (Eq. 2.30).

$$2(Pt - H) \rightarrow 2Pt + H_2 \tag{2.30}$$

Alternatively the Heyvrosky reactions for acidic and alkaline environment (Eqs. 2.31 and 2.32) may occur again leading to H_2 evolution

$$Pt - H + H^+ + e^- \rightarrow Pt + H_2 \tag{2.31}$$

$$Pt - H + H_2O + e^- \rightarrow Pt + H_2 + OH^- \tag{2.32}$$

It has been demonstrated that the Volmer–Heyvrosky mechanism dominates for large overpotentials, while at low overpotentials the Volmer–Tafel is the preferred pathway [16].

Both hydrogen evolution pathways show charge transfer, adsorption and desorption steps. The search for good catalysts for such reactions need to account for all these elements, but adsorption and desorption definitely play the major role. The interaction of adsorbed hydrogen with the catalyst has to be strong enough to guarantee a sufficient adsorption rate, while on the other side, has not to be too strong to allow fast desorption of the product. The most widely accepted figure to elucidate the relationships between material fundamental properties and its activity is the so-called volcano plot (Fig. 2.7). The volcano plot provides a graphical representation of the dependence of the exchange current density on the strength of the bond between the catalyst surface and hydrogen. Indeed, volcano plots are a pictorial view of the Sabaltier principle previously discussed in the section dedicated to adsorption.

The plot peaks just around platinum (Fig. 2.7). This is because platinum has a good tendency to adsorb hydrogen at high rate, but at the same time doesn't form a

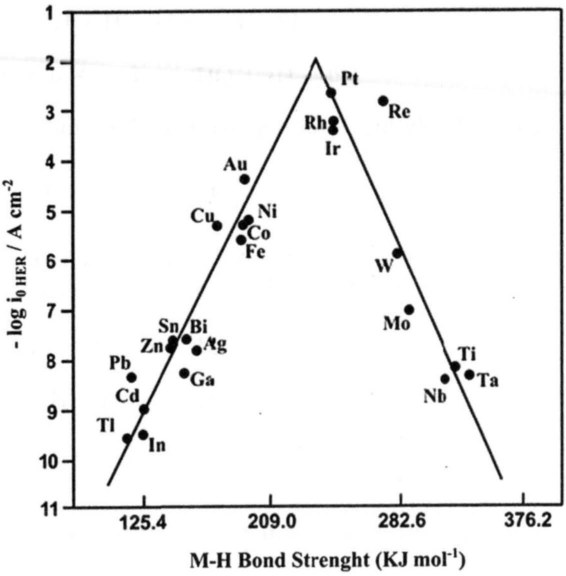

Fig. 2.7 Volcano plot for hydrogen evolution exchange current density

bond whose strength hamper desorption and mobility of atoms. The volcano plot justifies why platinum is the most performing, at least among all the pure metals, for both hydrogen evolution and oxidation reaction. By the way other metals even not noble, may be used to reduce or oxidize hydrogen. This is the case of Ni a transition metal scoring pretty well in the plot. In virtue of its position in the volcano plot and its stability in alkaline environment, Ni has found application as a catalyst for hydrogen evolution in electrolytic water splitting.

Hydrogen oxidation reaction

HOR is the anode reaction occurring in hydrogen fuel cells. It's existence is one of the foundations of the hydrogen economy paradigm. Most of the consideration for the HER also holds for the HOR [16, 17]. The mechanism is the same, but the reaction has to be viewed as the reverse. Next, the HOR for acidic (Eq. 2.33) and alkaline (Eq. 2.34) environments are reported.

$$\frac{1}{2}H_2 \rightarrow H^+ + e^- \tag{2.33}$$

$$\frac{1}{2}H_2 + OH^- \rightarrow H_2O + e^- \tag{2.34}$$

Hydrogen oxidation reaction takes place in similar way indeed the oxidation of hydrogen to protons may occur through the acidic and (Eq. 2.35) alkaline (Eq. 2.36) Heyrovsky reactions

$$Pt + H_2 \rightarrow Pt - H + H^+ + e^- \quad (2.35)$$

$$Pt + H_2 + OH^- \rightarrow Pt - H + H_2O + e^- \quad (2.36)$$

Alternatively the Tafel reaction (Eq. 2.37) may occur

$$\frac{1}{2}H_2 + Pt \rightarrow Pt - H \quad (2.37)$$

Then hydrogen adsorbed on Pt can release produced hydrogen ions in acidic environment or water in alkaline through the Volmer reactions (Eqs. 2.38 and 2.39)

$$Pt - H \rightarrow Pt + H^+ + e^- \quad (2.38)$$

$$Pt - H + OH^- \rightarrow Pt + H_2O + e^- \quad (2.39)$$

2.5.2 Oxygen Evolution and Reduction Reaction

Oxygen Reduction Reaction (ORR) and Oxygen Evolution Reaction (OER) are the cathode and the anode reactions in fuel cells and electrolyzers, respectively. Despite the fact that oxygen is a powerful oxidizing agent, its complete reduction is still a major issue in electrocatalysis. Actually ORR limits the energy efficiency of fuel cells.

A list of the main reactions involved in ORR and OER is shown in Table 2.2 for both the acidic and the alkaline environments. The list also reports the value for the thermodynamic potentials reported against the Normal Hydrogen Electrode (NHE).

Oxygen reduction reaction

At present, the best catalyst for ORR is platinum. Because these Pt-based catalysts are too expensive for making commercially viable fuel cells, extensive research over the past several decades has focused on developing alternative catalysts, including non-noble metal catalysts. Damjanovic in [18] summarizes the following:

(1) Few materials (only noble metals and alloys) can withstand the highly positive potential associated with the oxygen reduction (or evolution) without undergoing dissolution themselves and hence contributing to the overall current.

Table 2.2 Table of the reactions related to the ORR with the relative reduction potentials

$O_2 + 4H^+ + 4e^- \rightarrow 2H_2O$	1.229 V (NHE)
$O_2 + 2H^+ + 2e^- \rightarrow H_2O_2$	0.682 V (NHE)
$H_2O_2 + 2H^+ + 2e^- \rightarrow 2H_2O$	1.77 (NHE)
$O_2 + 4H_2O + 4e^- \rightarrow 4OH^-$	0.401 (NHE)
$O_2 + H_2O + 2e^- \rightarrow OH^- + HO_2^-$	− 0.427 (NHE)
$HO_2^- + H_2O + 2e^- \rightarrow 3OH^-$	0.942 (NHE)

(2) The mechanistic analysis of oxygen–reduction reactions is relatively complex owing to the numerous reaction steps and reaction intermediates with the energy of adsorption varying with the electrode potential and coverage. Furthermore, the reduction of O_2 may proceed either by a two-electron process to hydrogen peroxide or by a four-electron process leading to water. Hydrogen peroxide may, at least partially, reduce to water, or it may catalytically decompose.
(3) In the potential range in which oxygen dissolution occurs, the electrode surface may be covered with oxide or relatively bare, so meaningful conclusions on the catalysis may be drawn.
(4) Due to the low-exchange current densities, the electrode reactions may become potential controlling, particularly at low current densities. So even traces of impurities can profoundly affect the overall kinetic.

It is known that ORR in aqueous electrolytes may occur via four or two electron reaction pathways. The four electron pathway results in the formation of water, while the two electron pathway leads to the formation of hydrogen peroxide. There are a variety of reasons why the four electron pathway is preferred to the two electron one. First, the production of H_2O_2 in a fuel cell delivers less energy according to the reduction potential table. Second, the formation of H_2O_2 may be dangerous for the materials of the fuel cells. Indeed, hydrogen peroxide is a powerful oxidizing agent (this is again shown by the potential table) and may lead to the oxidation of the carbonaceous materials on the cathode side. That is one of the major reason for the performance degradation in PEMFC. Hence, the search for electrocatalysts for the O_2 reduction need to account for such an aspect and the number of exchanged electrons is a major criterium for the definition of the quality of a suitable electrocatalyst for ORR.

A sketch of the ORR mechanism on metal (Pt) catalyst is reported [wroblòowa 76] as follows:

As shown in Scheme 2.1 the first stage for the ORR is the O_2 adsorption onto the surface of the metal. Such an adsorbed oxygen may directly react to form water. Alternatively oxygen may be oxidized to the adsorbed hydrogen peroxide which either may be further oxidized to H_2O or may be released from the electrode surface. As for HER and HOR adsorption plays a fundamental role. A volcano plot reporting the activity of a variety of metals against the oxygen adsorption energy is shown in Fig. 2.8. Again just as the case of HER and HOR platinum is on top of the diagram indicating that among the transition metals platinum is the most active.

The effectiveness of a variety of other catalysts for the ORR has been demonstrated. Indeed, macrocycles of iron and cobalt (Chap. 10) as well as metal oxides have been proved to be effective for the ORR, providing a cheap alternative route to platinum and noble metals. Nevertheless, apart from a few examples reported in the literature, are stable in alkaline conditions only. It is worth remembering that PEMFCs for automotive and general high power applications operate in acidic conditions preventing the use of such materials. The state of the art for electrocatalysis in commercial PEMFCs is still platinum. The required

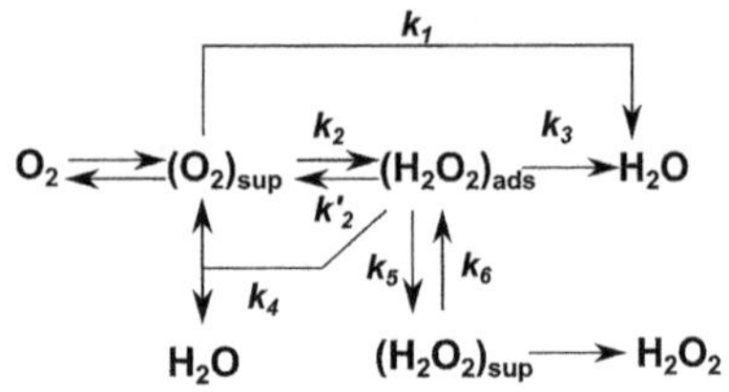

Scheme 2.1 Scheme of the possible pathways for the ORR

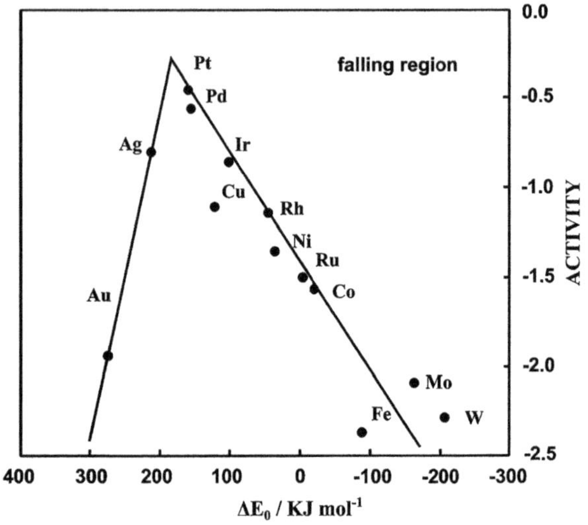

Fig. 2.8 Volcano plot for ORR

platinum loading at the cathode is larger that required at the anode as the kinetic of the HOR is much faster than that of ORR.

Oxygen evolution reaction
The OER occurs as the anode reaction in water electrolysis. The mechanism of such a reaction is indeed very complex and involves many steps with high energy intermediates explaining the need for large overpotentials. Consequently, the OER is the reaction absorbing most of the overpotentials required for water electrolysis [19]. The first step of the ORR for the acidic and the alkaline environment is as follows:

$$H_2O \rightarrow OH_{ads} + H^+ + e^- \tag{2.40}$$

$$OH^- \rightarrow OH_{ads} + e^- \tag{2.41}$$

These reactions are the rate determining steps for the OER on materials where the affinity toward the hydroxyl adsorption is weak. Materials with this characteristic are unpractical for application in electrolysis. On the other hand, materials strongly adsorbing OH, the rate determining step may become OH_{ads} desorption, suggesting again that the behavior of materials may be explained with the volcano plot reported in Fig. 2.8.

From the volcano plot is clear that, among all the pure metals platinum is the most active metal. After OH has been adsorbed according to Eqs. (2.40) or (2.41) for the acid and alkaline conditions, respectively, the following cascade of reactions occurs (Eqs. 2.42, 2.43 and 2.44).

$$OH_{ads} + OH_{ads} \rightarrow O_{ads} + H_2O \tag{2.42}$$

$$OH_{ads} + OH_{ads} \rightarrow O_{ads} + e^- \tag{2.43}$$

$$O_{ads} + O_{ads} \rightarrow O_2 \tag{2.44}$$

This is the so-called chemical oxide pathway which in principle should result in Tafel slopes of the order of 30 mV. Alternatively, the electrochemical pathway may occur, consisting in the hydroxyl adsorption (Eq. 2.40) followed by Eq. (2.45) :

$$OH_{ads} \rightarrow O_{ads} + H^+ + e^- \tag{2.45}$$

According to this scheme the formation of the surface oxide occurs electrochemically. A cascade of two successive electron transfer is observed with an estimated Tafel slop of 40 mV for low OH coverage. In some cases, larger slope of around 60 mV may be observed leading to the so-called acid base equilibrium path. The Path's first step is again the hydroxyl adsorption (Eq. 2.40). Adsorbed hydroxyl gets then deprotonated according to Eq. (2.46).

$$OH_{ads} \rightarrow O^-_{ads} + e^- \tag{2.46}$$

Deprotonated adsorbed oxygen can then release one electron according to Eq. (2.47):

$$O^-_{ads} \rightarrow O_{ads} + e^- \tag{2.47}$$

Once O_{ads} has been produced it may react producing molecular oxygen.

Whatever the mechanism is of a given material OER is a complex and essential task. Since the formation of oxides is involved in the mechanism this imply that not only bare metals may be active but also the corresponding oxide. Indeed this is the case. It is known and even commercially exploited that in alkaline environment NiO is a good material for the OER also in virtue of its stability in alkaline environment.

2.5.3 Methanol Oxidation

Since 1990 methanol has been investigated as a possible fuel for direct fuel cells. Methanol shows some principle advantages over hydrogen, above all the fact of being a liquid is definitely the most relevant. Nevertheless a full exploitation of methanol oxidation in fuel cells has not yet been achieved. As compared to HOR, MOR electrocatalysis is still an challenge and the existence of poisoning pathways strongly hampers the diffusion of such devices. Again the most effective catalysts for the MOR are noble metals, with Pt and its alloys (e.g., Pt–Ru) being the best. Next a short survey of the mechanism for the MOR is reported. The treatment only refers to acidic conditions as at present this is the state of the art for the DMFC.

The complete oxidation of methanol to CO_2 proceeds according to Eq. (2.48):

$$CH_3OH + H_2O \rightarrow CO_2 + 6H^+ + 6e^- \tag{2.48}$$

Despite the fact that only simple molecules are involved in the reaction, the mechanism is rather complex and it involves eleven possible steps. We present here the mechanism by Leger [20]. According to Leger. the first step of the reaction is the adsorption of a methanol molecule according to Eq. (2.49).

$$Pt + (CH_3OH)_{sol} \rightarrow Pt - (CH_3OH)_{ads} \tag{2.49}$$

Adsorption is then followed by dissociation of several species which may lead to a variety of reaction pathways. The first step after adsorption is the formation of radical species with the release of 1e$^-$ according to Eqs. (2.50) or (2.51).

$$Pt - (CH_3OH)_{ads} \rightarrow Pt - (CH_2OH)_{ads} + H^+ + e^- \tag{2.50}$$

$$Pt - (CH_3OH)_{ads} \rightarrow Pt - (CH_3O)_{ads} + H^+ + e^- \tag{2.51}$$

Further monoelectronic oxidation occurs, leading to the Eqs. (2.52) or (2.53)

$$Pt - (CH_2OH)_{ads} \rightarrow Pt - (CHOH)_{ads} + H^+ + e^- \tag{2.52}$$

$$Pt - (CH_3O)_{ads} \rightarrow Pt - (CH_2O)_{ads} + H^+ + e^- \tag{2.53}$$

Both the products of Eqs. (2.52) or (2.53) are then oxidized to adsorbed reactive intermediate $Pt - (CHO)_{ads}$ through Eqs. (2.54) or (2.55).

$$Pt - (CHOH)_{ads} \rightarrow Pt - (CHO)_{ads} + H^+ + e^- \tag{2.54}$$

$$Pt - (CH_2O)_{ads} \rightarrow Pt - (CHO)_{ads} + H^+ + e^- \tag{2.55}$$

$Pt - (CHO)_{ads}$ may then be oxidized to $Pt - (CO)_{ads}$ according to Eq. (2.56). Such a reaction is responsible for the poisoning of Pt-based catalysts in DMFCs as CO is so strongly adsorbed that it may hamper further oxidation.

$$Pt - (CHO)_{ads} \rightarrow Pt - (CO)_{ads} + H^+ + e^- \tag{2.56}$$

In parallel with the formation of the formyl like species and CO adsorption of OH at the Pt surface also occurs according to Eq (2.57).

$$Pt + H_2O \rightarrow Pt - (OH)_{ads} + H^+ + e^- \tag{2.57}$$

Adsorbed OH species may then react with the adsorbed formyl to directly render CO_2 (Eq. 2.58) or to form adsorbed COOH groups (Eq. 2.59) which according to Eq. (2.60) are successively oxidized to CO_2.

$$\mathrm{Pt-(CHO)_{ads}+Pt-(OH)_{ads}\rightarrow 2Pt-CO_2+2H^++2e^-} \tag{2.58}$$

$$\mathrm{Pt-(CHO)_{ads}+Pt-(OH)_{ads}\rightarrow Pt+Pt-(COOH)_{ads}+H^++e^-} \tag{2.59}$$

$$\mathrm{Pt-(COOH)_{ads}\rightarrow Pt+CO_2+H^++e^-} \tag{2.60}$$

To oxidize CO larger overpotentials are required. Under such conditions, production of CO_2 occurs directly via Eq. (2.61) or indirectly via Eq. (2.62) followed by Eq. (2.60).

$$\mathrm{Pt-(CO)_{ads}+Pt-}(OH)_{\mathrm{ads}}\rightarrow \mathrm{2Pt+CO_2+H^++e^-} \tag{2.61}$$

$$\mathrm{Pt-(CO)_{ads}+Pt-}(OH)_{\mathrm{ads}}\rightarrow \mathrm{Pt+Pt(COOH)_{ads}} \tag{2.62}$$

Discussion on the mechanism is limited to the acidic environment, as in these conditions platinum and ruthenium alloys exhibit the lowest overpotentials for the MOR. Low overpotential operation prevents catalyst poisoning prolonging the life of the catalyst. Platinum ruthenium effectiveness in the MOR reaction is due to a synergistic effect between the two metals. Indeed, platinum is very active for the dissociative chemisorptions of methanol, while the oxidation of the carbonaceous adsorbate to CO_2 is favored by the presence of oxidized form of ruthenium [21]. At present platinum ruthenium alloys are the state of the art catalysts for the MOR.

2.5.4 Ethanol Electroxidation

Ethanol is receiving much attention for exploitation in DEFCs, largely for its renewable nature, its well-established distribution infrastructure and lower toxicity as compared to methanol. Scheme 2.2 reports the variety of oxidation products in principle attainable through electrocatalysis. Complete oxidation to CO_2 would render 12 e^- but it is hard to obtain requiring a breakage of the C–C bond. Alternative products can be acetic acid and acetaldehyde delivering $4e^-$ and $2e^-$, respectively [22].

Scheme 2.2 Possible oxidation reaction products for ethanol electroxidation

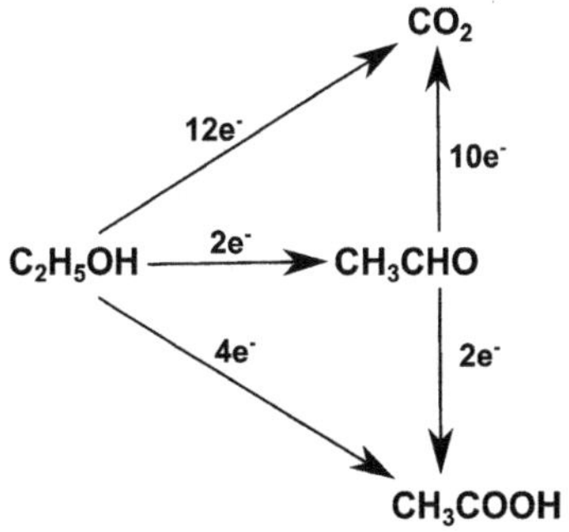

As pointed out complete oxidation is difficult to obtain. Next a description of the complex mechanism leading to the formation of the above-mentioned products is reported.

Aldehydes may react through the following pathway to render acetic acid:

$$CH_3CHO + Pt \rightarrow Pt - (CH_3CHO)_{ads} \text{ and } (CH_3CHO)_{ads} + Pt - (OH)_{ads} \rightarrow CH_3COOH + H^+ + e^- \tag{2.63}$$

Obtaining CO_2 requires a difficult pathway which also may lead to the formation of methane.

$$Pt + Pt(CO - CH_3) \rightarrow Pt - (CO)_{ads} + Pt - (CH_3)_{ads} \tag{2.64}$$

Equation 2.64 illustrates the dissociation of the adsorbed aldehyde into adsorbed CO and CH_3.

$$2Pt + H_2O \rightarrow Pt - H_{ads} + Pt - (OH)_{ads} \tag{2.65}$$

Hence, the adsorbed methyl may recombine with adsorbed hydrogen to produce methane and refresh catalytic metal sites.

$$Pt - (CH_3)_{ads} + Pt - H_{ads} \rightarrow CH_4 + 2Pt \tag{2.66}$$

On the other side, CO may react with the hydroxyl adsorbed at the platinum surface to produce CO_2.

$$Pt - (CO)_{ads} + Pt - (OH)_{ads} \rightarrow CO_2 + H^+ + e^- + 2Pt \tag{2.67}$$

It is worthy to spend a few words on the role that the adsorbed hydroxyl species play in the oxidation of ethanol. As we have learnt for the oxidation of methanol, the presence of adsorbed CO species at the platinum surface may hamper the catalytic activity. Nevertheless its occurrence is essential to produce a full oxidation to CO_2. Indeed, it is the presence of adsorbed hydroxyl which allows CO to be oxidized to CO_2. Coupling materials capable of increasing the rate of formation of the adsorbed hydroxyl at the platinum surface is indeed a key for increasing the effectiveness of the ethanol electroxidation.

Lately, ethanol electroxidation has been widely explored in alkaline media mainly because Pt can be effectively substituted by Pd indeed leading to even better performance. Nevertheless the C–C bond cleavage in alkaline environments has been proved to not occur for ethanol at pHs larger than 13 [23], acetate being the only oxidation product according to Eq. (2.68)

$$C_2H_5OH + 5OH^- \rightarrow CH3COO^- + 4H_2O + 4e^- \tag{2.68}$$

which proceeds according to the mechanism illustrated in Scheme 2.3.

From the mechanistic scheme. it is clear that in order to produce full oxidation to acetate the adsorption of hydroxyl is essential. Indeed, it has been demonstrated

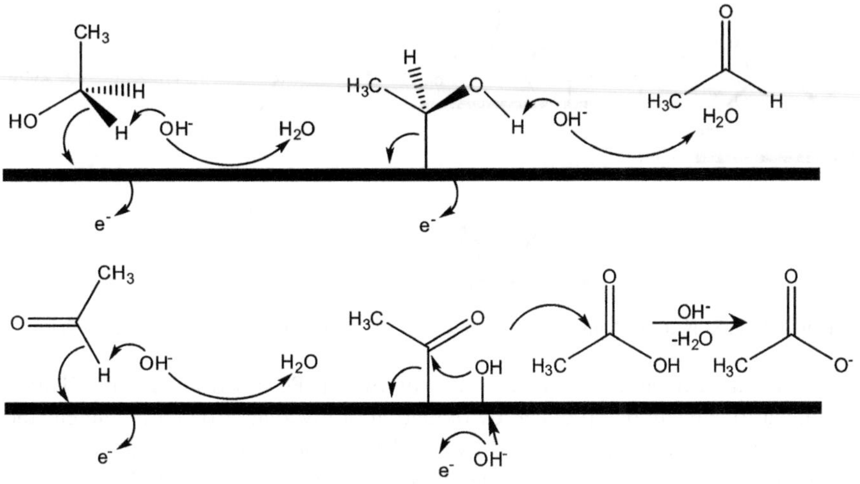

Scheme 2.3 Ethanol electroxidation mechanism for alkaline environment

that with palladium hydroxyl adsorption is the rate determining step [7], at least at low over potentials. The addition of materials which may increase the hydroxyl adsorption rate on palladium has been proved to be effective in enhancing ethanol electroxidation. This is the case of Ceria, which is capable of improving kinetics via spillover of the primary oxide [8].

2.5.5 Other Alcohols

Alcohols with a higher molecular weight than ethanol are now arousing great interest as fuels in direct alcohol fuel cells (DAFC). That is for a variety of reasons, including relatively low toxicity, low vapor pressure, high energy density, and the fact, at least for some of them, of being renewable. Included in this group are ethylene glycol (EG) and glycerol (G) [24]. Again, as in the case of ethanol and even more so here, the complete oxidation of these species has not yet been achieved and appears to be a very challenging task. To the best of our knowledge, all known Pd- or Pt-based electrocatalysts cannot promote the complete oxidation of EG or G to CO_2 both in acidic and alkaline media. EG or G may eventually undergo C–C bond scission with formation of CO_2 or carbonate. By the way this is a minor reaction pathway as compared to the oxidation to various carboxylic acids and carboxylates. Such reactions are, in turn, not completely selective. The subject is extensive and for the sake of illustrating the peculiar feature of electrocatalysis for such complex process we describe the oxidation of EG and G only in alkaline environments. The selection has been made as there are many more reports of alkaline fuel cells employing these fuels as compared to the acidic.

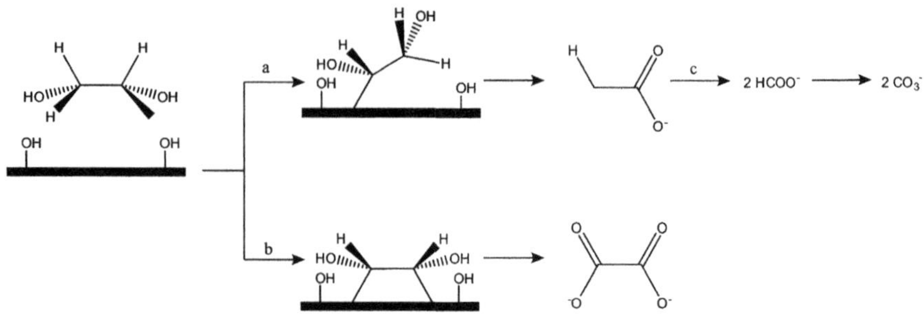

Scheme 2.4 Proposed mechanism for the electrooxidation of EG on Pd in alkaline media

As shown in Scheme 2.4, in the case of EG on Pd-based anodes in alkaline media, carbonate can be produced by oxidation of glycolate (route c), whereas the major product, oxalate, is produced by the direct oxidation of EG (path b).

The oxidation of G is more complex than that of EG. Scheme 2.5 illustrates the possible reaction mechanisms proposed for G oxidation on Pd-, Pt-, and Au-based anode electrocatalysts.

Glycerol on Pd-based anodes is first oxidized to aldehyde (path a), which in turn is quickly oxidized to glycerate in a subsequent two-electron transfer step. Glycerate is further oxidized to tartronate and, by cleavage of C–C bond, into glycolate and formate. The latter species lead to carbonate and glycolate is oxidized to oxalate [25–28]. G is also oxidized directly to tartronate (path b) by chelating adsorption on catalytic surfaces. However, the oxidation of G yields significant amounts of carbonate. Since oxalate is very slowly oxidized on Pd-based electrodes [25, 26, 29, 30] CO_2 is prevalently a by-product of the oxidation of either glycerate or glycolate.

2.5.6 Formic Acid

Formic acid has been proposed as a liquid fuel for powering fuel cells [31]. It shares the advantage of high energy density with alcohols. Further its oxidation in acidic environment is favorable ($E^0 = -0.25$) and combined to oxygen reduction ($E^0 = 1.23$ V) leads to a cell potential of 1.48 V.

$$HCOOH \rightarrow CO_2 + 2H^+ + 2e^- \qquad (2.69)$$

On the other hand, complete oxidation produces only two electrons which is much less as compared to methanol ($6e^-$) or ethanol ($12e^-$).

The so-called "parallel or dual pathway" is the recognized mechanism for formic acid oxidation. Direct oxidation (pathway 1) occurs via a dehydrogenation reaction, without forming CO as a reaction intermediate Eq. (2.69) [32] Alternatively, adsorbed carbon monoxide (CO) may form as a reaction intermediate by dehydration leading to Eq. (2.70):

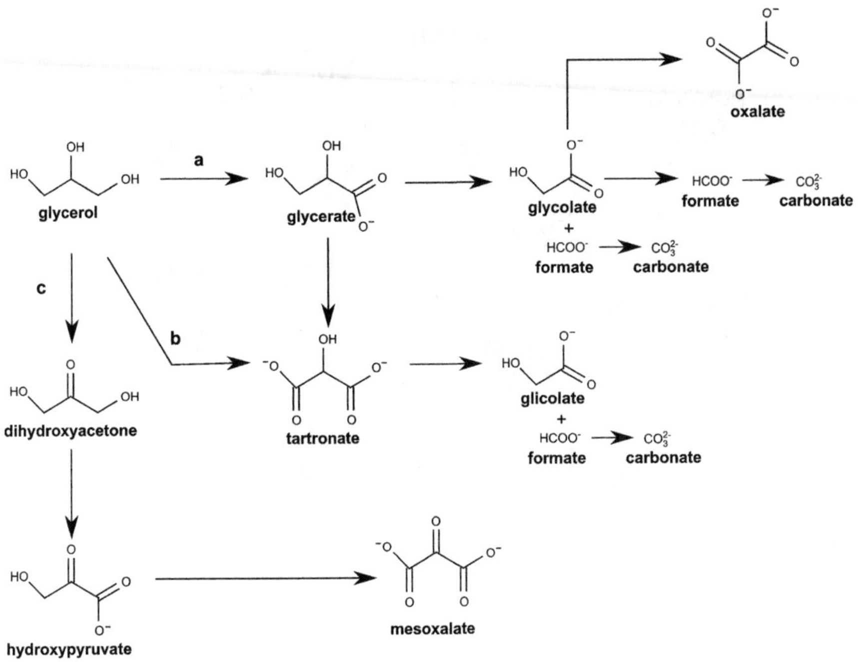

Scheme 2.5 Main oxidation products of glycerol oxidation

$$HCOOH \rightarrow CO_{ads} + H_2O \rightarrow CO_2 + 2H^+ + 2e^- \quad (2.70)$$

For direct formic acid fuel cells, dehydrogenation is the desired reaction pathway, to enhance overall cell efficiency and avoid poisoning of the catalyst. Anode catalyst selection is pivotal in directing formic acid oxidation to proceed via Eq. (2.69).

Recently, oxidation of formate (the deprotonated form of formic acid) has also been investigated in alkaline environment. Under such conditions, the oxidation leads to carbonate as major oxidation product and catalysts such as palladium have been proved to be effective [33].

2.5.7 CO_2 Electroreduction reaction

The steady increase in the atmospheric CO_2 concentration is a pressing global environmental issue; as a consequence, the reduction of carbon dioxide is currently investigated by many scientists as one of the most promising ways to convert waste CO_2 to useful materials and products (Scheme 2.6).

CO_2 is very stable (standard free energy of formation $\Delta G° = -394{,}359$ kJ mol^{-1}) and is the most oxidized form of carbon; therefore, the only chemical transformation

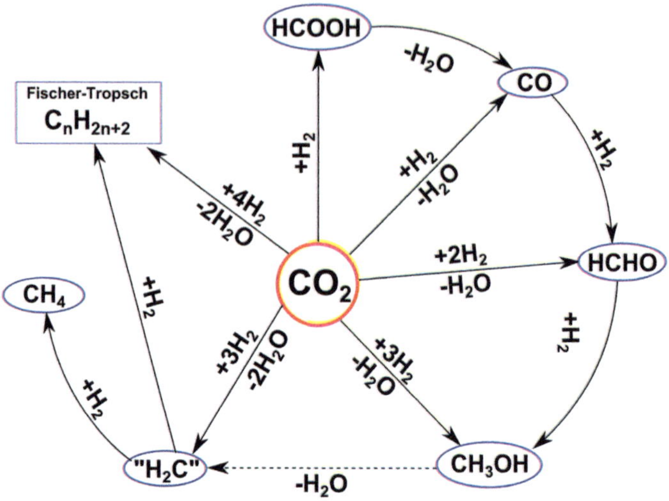

Scheme 2.6 Reduction of CO_2 to industrial products

possible at normal energies should be its reduction which can be performed using different approaches such as: (a) radiochemical methods [34, 35]; (b) chemical reduction by metals [36]; (c) Thermo-chemically [37]; (d) Bio-chemically [38]; (e) Bio-photochemically [39]; (f) Photo-electrochemically [40]; and (g) Bio-photo-electrochemically [41];

A bibliometric analysis of the scientific literature concerning CO_2 reduction published in the Science Citation Index (listed periodicals from 1999 to 2009) was presented by C.-M. Shu in 2011 (see Table 2.3) to highlight the emerging nature of this research issue worldwide [42]. The subject categories which deal with CO_2 reduction include Material Science, Electrochemistry, and Environmental Science.

Regarding the electrochemical reduction of CO_2, a status report on metal-based cathodes was published in 2004 and various factors which influence the process efficiency and selectivity were critically evaluated [43].

In order to exploit the electroreduction of CO_2 an understanding of the chemistry of CO_2 activation is fundamental. Carbon dioxide electroreduction to useful fuels is the reverse electrochemical process with respect to the anode reactions which occur in direct fuel cells. From the thermodynamic point of view, the Gibbs free energy of this reduction process is always positive at medium and high pHs and the theoretical potentials are negative; therefore, CO_2 reduction is an electrolysis process that requires an electrical energy input. Concerning the kinetic aspect, the over-potential needed to electrochemically reduce CO_2 is always >1.0 V (in order to obtain a high yield of products, i.e., methane, ethylene, etc.). In addition, when using an aqueous electrolyte, at these potentials water reduction will always occur and as a consequence H_2 is a major by-product of the process.

The reactions that occur on the electrode in an aqueous solution at pH 7 at 25 V (E^0 vs. SHE) are shown in Table 2.4 [44, 45]:

Table 2.3 Journals that publish articles on CO_2 reduction

Journal	Articles in the field (%)	Subject category
Abstracts of Papers of the American Chemical Society	33 (3.9)	–
Journal of the American Chemical Society	19 (2.2)	Chemistry, multidisciplinary
Journal of Electroanalytical Chemistry	18 (2.1)	Chemistry, analytical electrochemistry
Energy Policy	16 (1.87)	Energy and fuels, Environmental sciences
Inorganic Chemistry	12 (1.4)	Chemistry, inorganic and nuclear
Catalysis today	11 (1.28)	Chemistry, Applied Chemistry, Physical, Engineering, Chemical
Electrochimica Acta	11 (1.28)	Electrochemistry
Industrial and Engineering Chemistry Research	11 (1.28)	Engineering, Chemical
Tetrahedron Letters	11 (1.28)	Chemistry, organic
Carbon Dioxide Utilization for Global Sustainability	10 (1.17)	–
Chemical Communications	10 (1.17)	Chemistry, multidisciplinary
European Journal of Inorganic Chemistry	10 (1.17)	Chemistry, inorganic and nuclear
Journal of the Electrochemical Society	10 (1.17)	Electrochemistry
Studies in Surface Science and Catalysis	10 (1.17)	–
Green Chemistry	9 (1.05)	Chemistry, multidisciplinary
Applied Catalysis B—Environmental	8 (0.94)	Chemistry, Physical, Engineering, Environmental Engineering, Chemical
Chemistry Letters	8 (0.94)	Chemistry, multidisciplinary
Journal of Applied Electrochemistry	8 (0.94)	Electrochemistry
Angewandte Chemie-International Edition	7 (0.82)	Chemistry, multidisciplinary
Applied Catalysis A-General	7 (0.82)	Chemistry, Physical, Environmental science

Table 2.4 Reactions related to carbon dioxide reduction with relative reduction potentials

$CO_2 + 4e^- + 3H_2O \rightarrow HCHO + 4OH^- \quad E^0 = -0.48\,V$

$CO_2 + 2e^- + H_2O \rightarrow CO + 2OH^- \quad E^0 = -0.52\,V$

$CO_2 + 2e^- + H_2O \rightarrow HCOO^- + OH^- \quad E^0 = -0.43\,V$

$CO_2 + 8e^- + 6H_2O \rightarrow CH_4 + 8OH^- \quad E^0 = -0.25\,V$

$2CO_2 + 12e^- + 8H_2O \rightarrow C_2H_4 + 12OH^- \quad E^0 = -0.34\,V$

$3CO_2 + 12e^- + 9H_2O \rightarrow C_2H_5OH + 12OH^- \quad E^0 = -0.33\,V$

$3CO_2 + 18e^- + 13H_2O \rightarrow C_3H_7OH + 18OH^- \quad E^0 = -0.32\,V$

$2H_2O + 2e^- \rightarrow 2OH^- \quad E^0 = -0.41\,V$

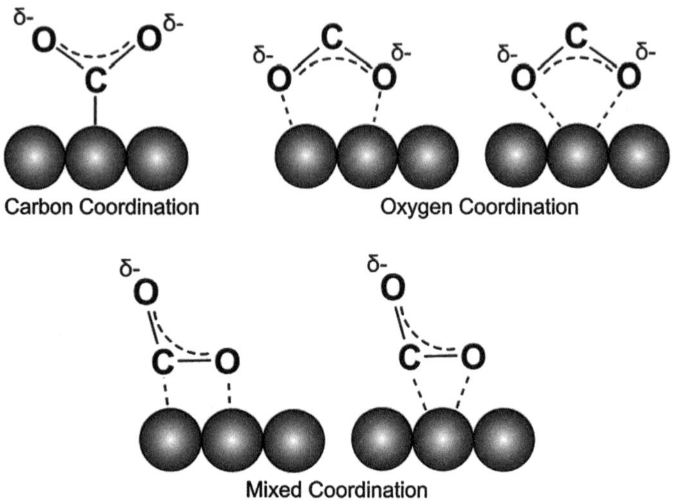

Scheme 2.7 Proposed structures for adsorbed $CO_2^{\delta-}$ on metals

CO_2 reduction on various metal-based electrodes has been extensively investigated, even if the reaction mechanism is only partially understood. The rate determining step is represented by the single electron reduction of CO_2 to CO_2^- occurring at the standard potential of −1.90 V through $CO_2^{\delta-}$ adsorption. The five possible molecular structures for adsorbed $CO_2^{\delta-}$ are shown in Scheme 2.7 It is not clear which is the actual geometry as the expected molecular vibrations can be easily affected by the surface geometry (we know that $CO_2^{\delta-}$ is most easily produced at surface defects).

In addition, the adsorbed state of $CO_2^{\delta-}$ is also influenced by the type of metal on the electrode surface.

Scheme 2.8 shows the main reaction pathways for the electroreduction of CO_2.

In case (a): (i.e., Hg, Cd, Pb, Tl, In, and Sn), high hydrogen overvoltage electrodes with weak adsorption of CO, there is also a high overvoltage for CO_2 reduction to CO_{2ads}^- and as a consequence these metals generally produce predominantly formate (or oxalate in nonaqueous CO_2 electrolysis); in case (b) (i.e., Pt, Ni, and Fe) these low hydrogen overvoltage metals also reduce CO_2 at low overpotentials to form tightly adsorbed CO; and in case (c) (i.e., Au, Ag, Zn, and Cu) metals with a low hydrogen overvoltage that reduce CO_2 at medium overpotentials form more loosely held CO. Scheme 2.9 shows a schematic overview of CO_2 reduction on a Cu electrode. Two pathways are shown, one leading to formate and the other to CO and hydrocarbon products. Both CO_2 reduction and hydrogen evolution occur on unblocked areas of the electrode surface, with CO being the primary cause of surface inhibition.

The development of Cu-based or Cu-coated metallic electrodes for the preparation of hydrocarbons and/or alcohols from CO_2 in aqueous media has been a major breakthrough in this area.

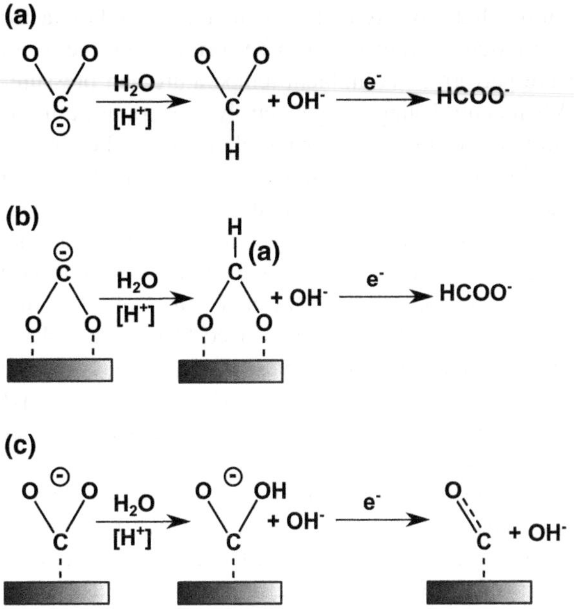

Scheme 2.8 Proposed reaction pathways for the reduction of adsorbed CO_2^- on different metals

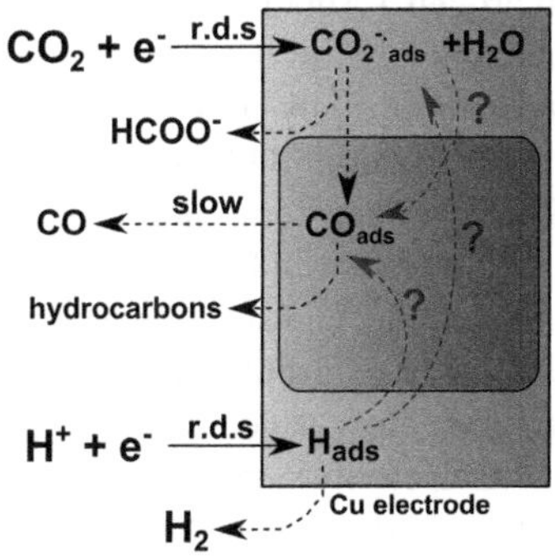

Scheme 2.9 CO_2 reduction reaction pathways on a Cu electrode surface

The electrocatalytic activity of Cu–based alloys, such as Cu-Sn and Cu–Zn in addition to elemental Cu, Sn, and Zn cathodes, at low temperatures has also been studied (275 K) [46].

The highly selective reduction of CO_2 to C2 compounds such as CH_3CHO, C_2H_5OH, and C_2H_4 on Cu–Hg alloy electrodes by employing pulsed electroreduction has been reported [47]. The Faradic efficiencies (rf) for the production of C2 compounds produced on Cu–Ag alloy electrodes were found to vary with the atomic ratio of Cu and Ag. The total rf for C2 compounds was 54.2 % for the pulsed reduction on a Cu–Ag alloy electrode (Cu/Ag = 28/72) with an anodic bias of Va = −0.4 V and a cathodic bias of Vc = −2.0 V versus Ag/AgCl. It was found that the formation of an oxide layer on Cu and the desorption of intermediates on Ag under anodic bias were key factors for the product selectivity. The products were found to be formic acid, methanol, and CO. These reductions were performed at ambient temperature and pressure and at high current density. When neutral supporting electrolytes were used the rf was quantitative. The major limitation of these copper-based electrodes lies in their quick deactivation (within 20–30 min. of electrolysis). A periodic anodic activation procedure can be used and this has allowed high hydrocarbon yields over prolonged electrolysis, which has been confirmed by Cook et al. [48] who found that the electrocatalytic activity of Cu depends on the renewal of the electrode surface. The clean Cu surface allowed for the achievement of rf values of 73 % for CH_4 and 25 % for C_2H_4. Li and Prentice 1997 [47], synthesized MeOH (rf = 40 %) via the electroreduction of CO_2 under aqueous high pressure conditions with a LiCl electrolyte using a Cu electrode, at a current density of 9 mA/cm^2 at -1.1 V versus Ag/AgCl. These researchers used a rotating disk electrode in order to achieve hydrodynamic conditions.

The mechanism of CO_2 electroreduction on Cu remains not yet fully understood [44]. Generally, for the electrochemical reduction of CO_2 in water, hydrogen formation competes with the CO_2 reduction reaction. Therefore, the depression of hydrogen formation is very important because the applied energy is wasted on hydrogen evolution instead of being used for the reduction of CO_2. In a recent example using a Na^+/methanol-based electrolyte, the faradic efficiency for hydrogen formation on a Cu electrode at 243 K was suppressed to 17.9 %. An efficiency below 1 % was achieved when sodium hydroxide and thiocyanate salts were used. Only in a CH_3COONa/electrolyte hydrogen-formation efficiency was relatively high (34.0 %), as with other acetates, also, the efficiency was very high. The hydrogen formation roughly decreased with a decreasing cation size [49].

Numerous studies have been carried out with the objective of elucidating the reaction mechanisms. However, liquid phase CO_2 electrocatalytic reduction suffers from serious problems which are difficult to overcome, such as sluggish reaction kinetics, low selectivity of CO_2 reduction formation of various by-products (nonselectivity), the low solubility of CO_2 in aqueous solutions, and the deactivation of the electrode catalysts.

For these reasons, an alternative to CO_2 reduction in aqueous electrolytes has been developed and is represented by the use of GDEs (Gas Diffusion Electrodes) and solid polymer electrolytes such as cation exchange membranes as well as anion exchange membranes (AEM) for CO_2 reduction to formic acid or methane as well as ethylene in devices similar to that developed for fuel cell technology, usually composed of Teflon-bonded catalyst particles and carbon blacks [50–68].

The electroreduction of CO_2 promoted by molecular complexes will be described in Chap. 10.

References

1. P.W. Atkins, J. De Paula, *Physical Chemistry*, 9th edn. (W.H. Freeman, New York, 2010), pp. xxi, 1139 p
2. A.J. Bard, L.R. Faulkner, *Electrochemical Methods: Fundamentals and Applications*, 2nd edn (Wiley, New York, 2001), pp. xxi, 833 p
3. C.A. Lucas, N.M. Markovic, B.N. Grgur, P.N. Ross, Structural effects during CO adsorption on Pt-bimetallic surfaces I. The Pt(100) electrode. Surf Sci **448**, 65 (2000)
4. C.A. Lucas, N.M. Markovic, P.N. Ross, Structural effects during CO adsorption on Pt-bimetallic surfaces II. The Pt(111) electrode. Surf. Sci. **448**, 77 (2000)
5. C.A. Lucas, N.M. Markovic, P.N. Ross, Structural effects induced by CO adsorption on Pt-bimetallic surfaces. Surf. Rev. Lett. **6**, 917 (1999)
6. M. Hachkar, T. Napporn, J.M. Leger, B. Beden, C. Lamy, An electrochemical quartz crystal microbalance investigation of the adsorption and oxidation of CO on a platinum electrode. Electrochim. Acta **41**, 2721 (1996)
7. Z.X. Liang, T.S. Zhao, J.B. Xu, L D. Zhu, Mechanism study of the ethanol oxidation reaction on palladium in alkaline media. Electrochim. Acta **54**, 2203 (2009)
8. M. Bevilacqua et al., Improvement in the efficiency of an OrganoMetallic Fuel Cell by tuning the molecular architecture of the anode electrocatalyst and the nature of the carbon support. Energy Environ. Sci. **5**, 8608 (2012)
9. J.O.M. Bockris, A.K.N. Reddy, M.E. Gamboa-Aldeco, *Modern Electrochemistry*, 2nd edn. (Plenum Press, New York, 1998)
10. A.B. Laursen et al., Electrochemical hydrogen evolution: Sabatier's principle and the Volcano plot. J. Chem. Educ. **89**, 1595 (2012)
11. C.W. Xu, P.K. Shen, Y.L. Liu, Ethanol electrooxidation on Pt/C and Pd/C catalysts promoted with oxide. J. Power Sour. **164**, 527 (2007)
12. J.A. Jaksic et al., Spillover of primary oxides as a dynamic catalytic effect of interactive hypo-d-oxide supports. Electrochim. Acta **53**, 349 (2007)
13. J. M. Jaksic, D. Labou, G. D. Papakonstantinou, phenomena and significance of intermediate spillover in electrocatalysis of oxygen and hydrogen electrode reactions. Chem. Ind. **66**, 425 (2012)
14. S. Treimer, A. Tang, D.C. Johnson, A consideration of the application of Koutecky-Levich plots in the diagnoses of charge-transfer mechanisms at rotated disk electrodes. Electroanalytical **14**, 165 (2002)
15. F.J. Vidal-Iglesias, J. Solla-Gullon, V. Montiel, A. Aldaz, Errors in the use of the Koutecky-Levich plots. Electrochem. Commun. **15**, 42 (2012)
16. W.C. Sheng, H.A. Gasteiger, Y. Shao-Horn, Hydrogen oxidation and evolution reaction kinetics on platinum: acid vs alkaline electrolytes. J. Electrochem. Soc. **157**, B1529 (2010)
17. B.E. Conway, B.V. Tilak, Interfacial processes involving electrocatalytic evolution and oxidation of H2, and the role of chemisorbed H. Electrochim. Acta **47**, 3571 (2002)
18. A. Damjanovic, in *Modern Aspects of Electrochmistry*, vol. 5, ed. by B.E. Conway, J.O'M. Bockris (Plenum, New York, 1965)
19. E. Guerrini, S. Trasatti, *Electrocatalysis in water electrolysis* ed. by C. Bianchini, P. Barbaro. Catalysis for Sustainable Energy Production (John Wiley and Sons, 2009), p. 235 (2009)
20. J.M. Leger, Mechanistic aspects of methanol oxidation on platinum-based electrocatalysts. J. Appl. Electrochem. **31**, 767 (2001)
21. H.A. Gasteiger, N. Markovic, P.N. Ross, E.J. Cairns, Methanol electrooxidation on well-characterized Pt-Rn alloys. J. Phys. Chem.-Us **97**, 12020 (1993)
22. C. Lamy et al., Recent advances in the development of direct alcohol fuel cells (DAFC). J. Power Sources **105**, 283 (2002)
23. X. Fang, L.Q. Wang, P.K. Shen, G.F. Cui, C. Bianchini, An in situ fourier transform infrared spectroelectrochemical study on ethanol electrooxidation on Pd in alkaline solution. J. Power Sources **195**, 1375 (2010)

24. A. Marchionni et al., Electrooxidation of ethylene glycol and glycerol on Pd-(Ni-Zn)/C anodes in direct alcohol fuel cells. Chemsuschem **6**, 518 (2013)
25. V. Bambagioni et al., Ethylene glycol electrooxidation on smooth and nanostructured pd electrodes in alkaline media. Fuel Cells **10**, 582 (2010)
26. V. Bambagioni et al., Ethanol oxidation on electrocatalysts obtained by spontaneous deposition of palladium onto nickel-zinc materials. Chemsuschem **2**, 99 (2009)
27. Y. Kwon, K.J.P. Schouten, M.T.M. Koper, Mechanism of the catalytic oxidation of glycerol on polycrystalline gold and platinum electrodes. Chem. Cat. Chem. **3**, 1176 (2011)
28. C. Bianchini, P.K. Shen, Palladium-Based electrocatalysts for alcohol oxidation in half cells and in direct alcohol fuel cells. Chem. Rev. **109**, 4183 (2009)
29. P.A. Christensen, A. Hamnett, The oxidation of ethylene glycol at a platinum electrode in acid and base: an in situ FTIR study. J. Electroanal. Chem. Interfacial Electrochem. **260**, 347 (1989)
30. F. Hahn, B. Beden, F. Kadirgan, C. Lamy, Electrocatalytic oxidation of ethylene glycol: Part III. In-situ infrared reflectance spectroscopic study of the strongly bound species resulting from its chemisorption at a platinum electrode in aqueous medium. J. Electroanal. Chem. Interfacial Electrochem. **216**, 169 (1987)
31. X. Yu, P.G. Pickup, Recent advances in direct formic acid fuel cells (DFAFC). J. Power Sources **182**, 124 (2008)
32. J.M. Feliu, E. Herrero, Handb. Fuel Cells **2**, 679 (2003)
33. A.M. Bartrom, J.L. Haan, The direct formate fuel cell with an alkaline anion exchange membrane. J. Power Sources **214**, 68 (2012)
34. A. Voss, The future-role of nuclear and renewable forms of energy. Brennst-Warme-Kraft **42**, 579 (1990)
35. C. Willis, A.W. Boyd, Excitation in the radiation chemistry of inorganic gases. Int. J. Radiat. Phys. Chem. **8**, 71 (1976)
36. E.L. Quinn, C.L. Jones, *Carbon Dioxide*. American Chemical Society Monograph series (Reinhold Publishing Corporation, New York, 1936), pp. 294, 6 p. incl. illus., tables, diagrs
37. C.E. Bamberger, P.R. Robinson, Thermochemical splitting of water and carbon dioxide with cerium compounds. Inorg. Chim. Acta **42**, 133 (1980)
38. J.M. Lehn, R. Ziessel, Photochemical generation of carbon monoxide and hydrogen by reduction of carbon dioxide and water under visible light irradiation. Proc. Natl. Acad Sci. USA **79**, 701 (1982)
39. D. Mandler, I. Willner, Photochemical fixation of carbon dioxide: enzymic photosynthesis of malic, aspartic, isocitric, and formic acids in artificial media. J. Chem. Soc. Perkin Trans. 2 **0**, 997 (1988)
40. I. Taniguchi, B. Aurian-Blajeni, J.O.M. Bockris, The reduction of carbon dioxide at illuminated p-type semiconductor electrodes in nonaqueous media. Electrochim. Acta **29**, 923 (1984)
41. B.A. Parkinson, P.F. Weaver, Photoelectrochemical pumping of enzymatic CO2 reduction. Nature **309**, 148 (1984)
42. T.J. Wan, S.M. Shen, A. Bandyopadhyay, C.M. Shu, Bibliometric analysis of carbon dioxide reduction research trends during 1999–2009. Sep. Purif. Technol. **94**, 87 (2012)
43. M.A. Scibioh, B. Viswanathan., Electrochemical reduction of carbon dioxide: a status report. Proc. Indian Natn. Sci. Acad **70**, 65 (2004)
44. M. Gattrell, N. Gupta, A. Co, A review of the aqueous electrochemical reduction of CO2 to hydrocarbons at copper. J. Electroanal. Chem. **594**, 1 (2006)
45. J.A. Dean, N.A. Lange, *Handbook of Chemistry* (McGraw-Hill, New York, 1973), p. v
46. M. Watanabe, M. Shibata, A. Katoh, H. Uchida, M. Tomkiewietz, H. Yoneyami, Y. Hori, R. Haynes, Eds. (1993)
47. J.W. Li, G. Prentice, Electrochemical synthesis of methanol from CO2 in high-pressure electrolyte. J. Electrochem. Soc. **144**, 4284 (1997)
48. R.L. Cook, R.C. MacDuff, A.F. Sammells, Electrochemical reduction of carbon dioxide to methane at high current densities. J. Electrochem. Soc. **134**, 1873 (1987)

49. S. Kaneco, H. Katsumata, T. Suzuki, K. Ohta, Electrochemical reduction of CO2 to methane at the Cu electrode in methanol with sodium supporting salts and its comparison with other alkaline salts. Energ. Fuel **20**, 409 (2006)
50. K. Hara, A. Kudo, T. Sakata, M. Watanabe, High-Efficiency electrochemical reduction of carbon-dioxide under high-pressure on a gas-diffusion electrode containing Pt catalysts. J. Electrochem. Soc. **142**, L57 (1995)
51. K. Hara, A. Kudo, T. Sakata, Electrochemical reduction of carbon-dioxide under high-pressure on various electrodes in an aqueous-electrolyte. J. Electroanal. Chem. **391**, 141 (1995)
52. N. Furuya, T. Yamazaki, M. Shibata, High performance Ru-Pd catalysts for CO2 reduction at gas-diffusion electrodes. J. Electroanal. Chem. **431**, 39 (1997)
53. J. Lee, Y. Kwon, R.L. Machunda, H.J. Lee, Electrocatalytic recycling of CO2 and small organic molecules. Chem-Asian J **4**, 1516 (2009)
54. T.Yamamoto, D.A. Tryk, A. Fujishima, H. Ohata, Production of syngas plus oxygen from CO2 in a gas-diffusion electrode-based electrolytic cell. Electrochim. Acta **47**, 3327 (2002)
55. C.M. Sanchez–Sanchez, V. Montiel, D.A. Tryk, A. Aldaz, A. Fujishima, Electrochemical approaches to alleviation of the problem of carbon dioxide accumulation. Pure Appl. Chem. **73**, 1917 (2001)
56. D.A. Tryk et al., Recent developments in electrochemical and photoelectrochemical CO2 reduction: involvement of the (CO2)(2)(.-) dimer radical anion. Appl. Organomet. Chem. **15**, 113 (2001)
57. T. Yamamoto, D.A. Tryk, K. Hashimoto, A. Fujishima, M. Okawa, Electrochemical reduction of CO2 in the micropores of activated carbon fibers. J. Electrochem. Soc. **147**, 3393 (2000)
58. Y. Hori et al., "Deactivation of copper electrode" in electrochemical reduction of CO2. Electrochim. Acta **50**, 5354 (2005)
59. S. Komatsu, M. Tanaka, A. Okumura, A. Kungi, Preparation of Cu-solid polymer electrolyte composite electrodes and application to gas-phase electrochemical reduction of Co2. Electrochim. Acta **40**, 745 (1995)
60. C. Delacourt, P.L. Ridgway, J.B. Kerr, J. Newman, Design of an electrochemical cell making syngas (CO + H-2) from CO2 and H2O reduction at room temperature. J. Electrochem. Soc. **155**, B42 (2008)
61. D. Dewulf, A. Bard, The electrochemical reduction of CO2 to CH4 and C2H4 at Cu/Nafion electrodes (solid polymer electrolyte structures). Catal. Lett. **1**, 73 (1988)
62. R.L. Cook, R.C. Macduff, A.F. Sammells, High-rate gas-phase Co2 reduction to ethylene and methane using gas-diffusion electrodes. J. Electrochem. Soc. **137**, 607 (1990)
63. R.L. Cook, R.C. Macduff, A.F. Sammells, Gas-phase Co2 reduction to hydrocarbons at metal solid polymer electrolyte interface. J. Electrochem. Soc. **137**, 187 (1990)
64. M. Shibata, N. Furuya, Electrochemical synthesis of urea at gas-diffusion electrodes Part VI. Simultaneous reduction of carbon dioxide and nitrite ions with various metallophthalocyanine catalysts. J. Electroanal. Chem. **507**, 177 (2001)
65. M. Shibata, N. Furuya, Simultaneous reduction of carbon dioxide and nitrate ions at gas-diffusion electrodes with various metallophthalocyanine catalysts. Electrochim. Acta **48**, 3953 (2003)
66. Y. Hori, A. Murata, S.-Y. Ito, Y. Yoshinami, O. Koga, Nickel and Iron Modified Copper Electrode for Electroreduction of CO_2 by In-situ Electrodeposition. Chem. Lett. **18**, 1567 (1989)
67. Y. Hori, H. Ito, K. Okano, K. Nagasu, S. Sato, Silver-coated ion exchange membrane electrode applied to electrochemical reduction of carbon dioxide. Electrochim. Acta **48**, 2651 (2003)
68. S. Ikeda, T. Ito, K. Azuma, K. Ito, H. Noda, Electrochemical mass reduction of carbon-dioxide using Cu-loaded gas-diffusion electrodes .1. Preparation of electrode and reduction products. Denki Kagaku **63**, 303 (1995)

Chapter 3
Electrochemical Devices for Energy Conversion and Storage

3.1 Key Concepts

After having reviewed the electrochemical concepts behind electrochemical energy conversion in Chap. 2, here we describe in some detail the architectures for the most commonly used devices. This chapter is divided into two main sections, Fuel Cells and Electrolyzers. Concerning Fuel Cells first we give a short review of the technological background, introducing the main constitutive elements and the key parameters for performance assessment. Then a somewhat detailed description of fuel cell operation with a variety of fuels is given, reporting relevant data such as stability and maximum power output. As fuels we focus on hydrogen, methanol, and other alcohols.

Next we will give a general background on electrolysis for hydrogen production. While not yet competitive with fossil fuel-based technologies, electrolysis is widely recognized as the technology of the future. This is because it directly converts electrical energy into chemical energy and can be linked to renewable power sources such as photovoltaics or wind turbines. As these power sources are by their nature not continuous and often not completely predictable in their power output, electrolysis may be considered as a system for buffering their production through the storage of energy as hydrogen. As in the case of fuel cells, the section devoted to electrolysis starts with a general background review, where the current state-of-the-art devices are described. A short review on the most relevant key parameters for performance comparison is also reported. Then the main technologies for water splitting are considered with a special emphasis on alkaline electrolyzers and polymer electrolyte membrane electrolyzers. The chapter ends with a new class of electrolyzers that have recently appeared on the electrolysis panorama, namely devices using sacrificial agents. We believe this class of devices may play an important role in the future of electrolytic technology. By employing easily oxidizable species such as ammonia or ethanol at the anode, oxygen evolution is suppressed. Water electrolysis requires a minimum voltage of 1.23 V. This indeed is the potential required by thermodynamics to split water into hydrogen and oxygen. In practice the potential for water splitting is much higher

A. Lavacchi et al., *Nanotechnology in Electrocatalysis for Energy*,
Nanostructure Science and Technology 170, DOI: 10.1007/978-1-4899-8059-5_3,
© Springer Science+Business Media New York 2013

and may range between 1.6 and 2 V due to kinetic and ohmic resistance factors. The use of easily oxidizable species reduces dramatically the thermodynamic contribution to values close to 0 V, making, electrode kinetics and resistance the only significant contributions to the overall cell voltage. Thus electrolysis may occur at potentials lower than 1 V, leading to large energy savings. In the case of the oxidation of alcohols interesting byproducts may also be obtained, indeed this approach has been proposed as an innovative, green, and energetically self-sustainable class of processes for the production of raw chemicals. It has to be pointed out that the energy saving occurs at the expense of the sacrificial agent which has its own energy costs. The overall convenience of these processes also depends on the availability of energy efficient ways to produce the sacrificial agent.

3.2 Fuel Cells: General Background

As stated in Chap. 2 a fuel cell is an electrochemical device which converts the free energy of a spontaneous chemical reaction into electrical energy. A fuel cell consists of an electrolyte that separates an anode (negative electrode, oxidation site) and a cathode (positive electrode, reduction site). The fuel is fed to the anode and can be either a liquid or a gas. In the course of the oxidation reaction, the fuel releases electrons at the anode catalyst layer. Electrons are then allowed to flow through an electronic conductor to the cathode side where they produce the reduction reaction. A contemporaneous flow of ionic species occurs in the electrolyte and these contribute to closing the circuit [1].

As an example we will describe the oxidation of hydrogen in a PEMFC in acidic environment (Fig. 3.1). Protons are produced at the anode and consumed at the cathode. Hence, in order to close the electrical circuit they have to flow through the cell following a different path than the electrons. The flow of the ions occurs through the electrolyte.

Fuel cells may operate under continuous flow of fuel or by just consuming the fuel contained in a tank directly connected to the anode. In the first case devices are defined as "active," while in the second case they are denoted as "passive." Active devices are usually meant for applications where a large amount of energy is required, for example automotive. Passive devices, usually operating with liquid fuels, are meant for low power generation, e.g., for power supply of portable electronics . Furthermore, especially for high power applications fuel cells are not single units but they are connected in stacks. A single fuel cell is capable of delivering power with a maximum potential of around 1 V. Such voltages are often impractical for power application, so fuel cell designers have linked together many individual cells to form a "stack" in series thus delivering higher voltages. Furthermore, a stack can be arranged with many groups of cells which can also be arranged in parallel to deliver the desired voltage, current, and power. Often a stack includes more than 50 or more unit cells (Fig. 3.2).

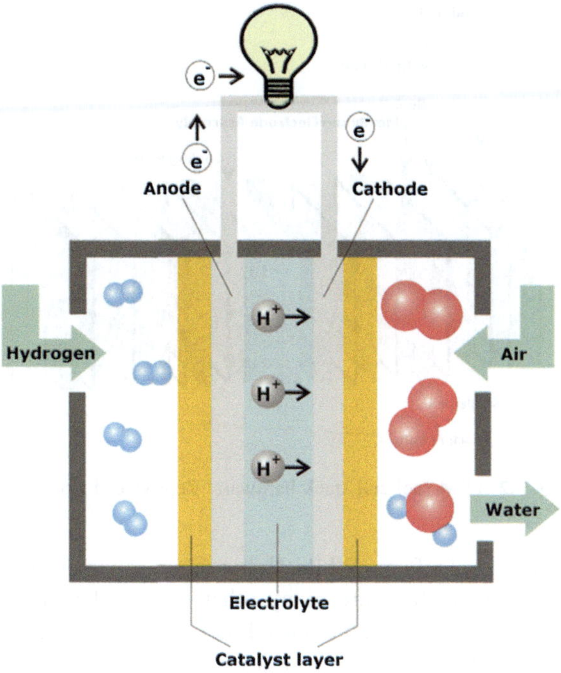

Fig. 3.1 Schematic of a hydrogen-fed Polymer Electrolyte Membrane Fuel Cell. H_2 is oxidized at the anode to protons releasing electrons which flow through an external electronic conductor to supply power to an electrical load. Protons flow through the electrolyte membrane to the cathode where they contribute together with the electrodes coming from the external circuit to provide *water* formation

3.2.1 Components of PEM Fuel Cell

The core of the fuel cell is the so-called membrane electrode assembly. Physically, an MEA is a foil where the solid electrolyte, the anode and cathode catalyst layers, and the gas diffusion layers are sandwiched together [3, 4] (Fig. 3.3). The MEA is the true "soul" of the fuel cell and the next subsections will be devoted to a somewhat detailed description of its main components.

3.2.1.1 Polymer Electrolyte Membrane (PEM)

The electrolyte is namely a polymer electrolyte membrane (PEM). Such a membrane is usually designed to conduct protons or hydroxyl ions while they should not be able to transport gaseous species such as hydrogen and oxygen or even liquid fuels such as methanol and other alcohols. This condition is essential to guarantee the separation of reactants and the conduction of ions. The phenomenon of transport of reactants through the membrane is known as "crossover" and must be avoided in order not to have efficiency losses. PEMs are constituted by pure polymer membranes or composite membranes. The latter are obtained by embedding other materials into the polymer matrix. The *state of the art* of the proton conducting membrane is the fluoropolymer (PFSA) Nafion^{TM} a trademark

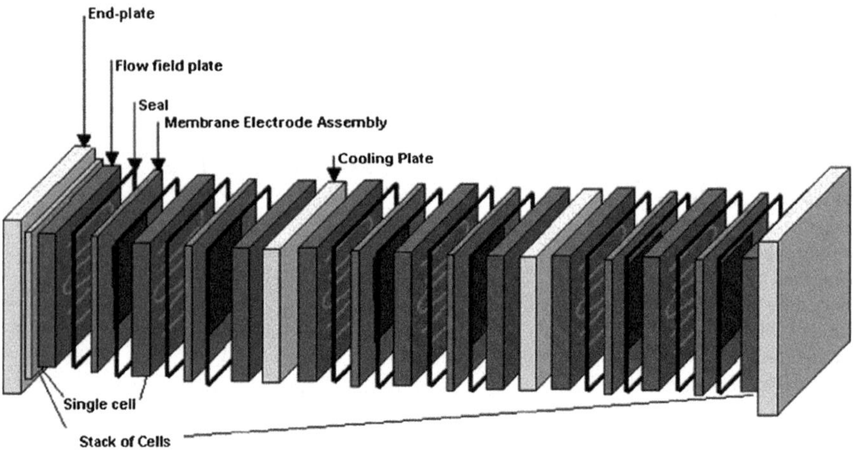

Fig. 3.2 PEM fuel cell stack hardware. Reproduced from Ref. [2] with permission of Elsevier

owned by DuPont. While Nafion is an ionomer with a perfluorinated backbone like Teflon, there are many other structural motifs used to make ionomers for proton exchange membranes. Lately, membranes capable of transporting hydroxyl ions have been developed allowing a better exploitation of fuel cell technology in alkaline environments. For a membrane to be a good candidate for application in fuel cells it has to fulfill the following requirements:

1. High conductivity
2. Low permeability for reactants
3. Good thermal stability (e.g., up to 100 °C and more)
4. Good mechanical resistance
5. Homogeneity of thickness.

A high conductivity guarantees low ohmic losses under cell operating conditions. The low permeability also guarantees high efficiency for two reasons: first there is no loss of fuel into the cathode gas flow and second there is no potential drop due the ability of the cathode electrocatalyst (usually Pt based) to operate both for the oxidation of the fuel and oxygen reduction. Thermal stability is required because cell temperatures may reach up to and above 80 °C and a durability of many thousands of hours is required. Mechanical stability is important for the manufacturability of the MEA. The homogeneity is also very important, especially for durability. Indeed if the thickness is not homogeneous paths with different resistance may occur through the MEA, leading to differences in the local current density and the formation of so called "hot spots" points where the temperature locally rises and that may lead to local membrane breakdown.

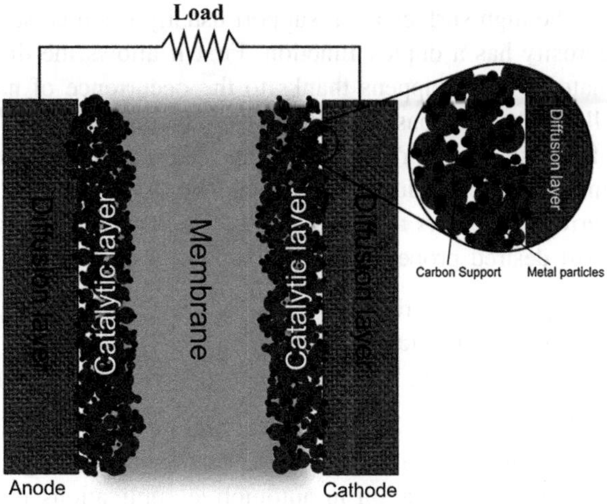

Fig. 3.3 Schematic of a membrane electrode assembly: from the *left* gas diffusion layer; anode catalytic layer; polymer electrolyte membrane; cathode catalyst layer; and cathode gas diffusion layer. In the up *left* exploded view of the catalyst layer an example of the schematic of carbon-supported metal nanoparticles electrocatalyst is reported

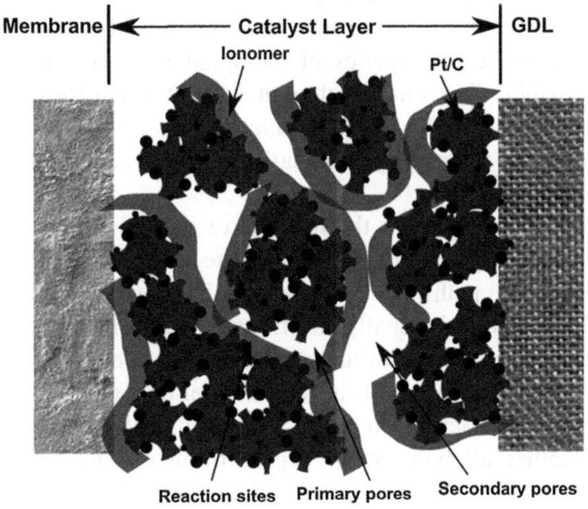

Fig. 3.4 Detailed schematic of the catalyst layer. The reaction sites correspond to the triple phase boundaries. Notice the presence of primary and secondary pores and the ionomer

3.2.1.2 Catalyst Layers

The catalyst layers are the actual fuel cell electrodes (Fig. 3.4) [5]. Both anode and cathode consist in two separated catalyst layers deposited onto the opposite sides of the PEM. Most of this book will be devoted to describing the technology of the catalyst layer as this is the place where the application of nanotechnology can help most in improving fuel cell performance. The catalyst layer is usually obtained by mixing the following elements:

1. High surface area support material, e.g., carbon black
2. Electroactive phase (the nanostructured or molecular electrocatalyst)
3. The ionomer
4. A binder.

The high surface area support usually has dual scale porosity. Such dual scale porosity has a duplex function. First it allows the dispersion of the electroactive material (this happens thanks to the occurrence of nanoscale porosity) second it allows the fast mass transport of products and reactants (thanks to the occurrence of porosity in the range of 100 nm and more). Furthermore, the support has to be an electronic conductor. Such conductivity is required to drive the electrons to the current collectors and in the end to the external electrical circuit. In the end, the most desired properties for the support material are:

1. High surface area
2. Good electronic conductivity
3. Stability (e.g., not prone to corrosion phenomena).

The problem of corrosion is, at present, one of the most challenging for fuel cell technology as one of the major factors hampering the fulfillment of durability criteria for application in automotive applications is the corrosion of the carbon support on the cathode side.

The electroactive phase is the electrocatalys (*e.g.,* Pt nanoparticles). Its function is to make the kinetics of the spontaneous electrochemical reactions as fast as possible. This is a crucial point as most of the energy efficiency losses are usually due to either the *oxygen reduction reaction* or the *liquid fuel oxidation*. On the contrary, the hydrogen oxidation reaction is rather fast and is not an issue for fuel cell efficiency.

The ionomer has the function of providing additional ionic conductivity to the catalyst layer. The presence of the ionomer in the catalyst layer provides low resistance pathways to the ions which have to be transferred to or from the PEM. The ionomer is usually constituted by a limited number of monomeric units of the polymer which constitute the MEA. The use of the ionomer may be avoided when operating with liquid fuels if the fuel contains an ionic conductor, e.g., KOH.

The binder has the role of holding all the components of the catalyst layer together allowing some degree of mechanical manufacturability and ensuring mechanical stability. Usually the preferred binder material is PTFE. While being sufficiently plastic to confer homogeneity to the catalyst layer it is also hydrophobic, allowing a control of the water affinity of the layer. The latter point is very important as water removal from the catalyst layer is essential to have fast kinetics for electrochemical reactions when gaseous reagents are used, *e.g.,* H_2 and O_2.

The catalyst layer is the location of the so called "Triple Phase Boundary" (TPB) [3]. The triple phase boundary is the point where the reactant gas meets the catalysts and ionomer. This is actually the "core" of the electrochemical activity of a fuel cell. This is because at the TPB the catalyst provides the reaction and both electrons and ions are collected from two separated conductivity pathways leading to the external circuit for the electrons and to or from the PEM for the ions. When fuel cells are fed with liquid fuel with high ionic conductivity, this may occur via a simpler two-phase boundary as the fuel solution itself can provide sufficient ionic conductivity.

3.2.1.3 Gas Diffusion Layer (GDL)

The gas diffusion layer distributes the reactant gases homogeneously from the flow field to the catalyst layer. With a homogeneous distribution of the gases it is possible to prevent local hotspots and catalyst flooding by removing heat and excess water from the electrode. GDLs are usually made of carbon-based fiber materials such as carbon paper or cloth. The GDL is also an electronic conductor which provides the necessary conductivity to collect the electrons to or from a metal collector located on the back of the gas diffusion layer.

A fuel cell is not complete without bipolar plates which are essential components in fuel cell stacks as they connect each single cell electrically, supply reactant gases, and remove reaction products from the MEA. Bipolar plates are usually fabricated from high-density graphite, even if in recent years experience based on the use of other cost-effective and feasible alternative materials have been reported. Particularly, metals and composites have received much attention mainly for manufacturability reasons. An MEA in a fuel cell stack is usually sandwiched between two bipolar plates.

3.2.2 Fuel Cell Key Performance Parameters

3.2.2.1 Polarization Characteristics

The theoretical optimum fuel cell voltage (e.g., 1.23 V for a hydrogen PEMFC) is never achieved under real fuel cell operation. The actual cell potential departs from its maximum thermodynamic value due to a variety of irreversible losses. The performance of a fuel cell depends on such losses. Hence their characterization is of fundamental importance in assessing the quality of fuel cell devices. Polarization curves are the key parameter used to measure cell performance. The polarization curve can be recorded connecting the fuel cell or the stack to a variable electrical load which can determine the potential output as a function of the applied electrical loading. The results are displayed in a chart reporting the potential against the current density (Fig. 3.5).

Figure 3.5 reports three distinct polarization regions, where the actual cell potential is dominated by three distinct losses. The first region is the so-called activation region , where the potential is almost entirely determined by electrochemical kinetics. The second region is the ohmic polarization region where fuel cell resistance plays a major role due to the increased current density. The third section of the polarization curve shows a sudden drop of the potential due to mass transport limitations. Next we give a short description of such polarizations.

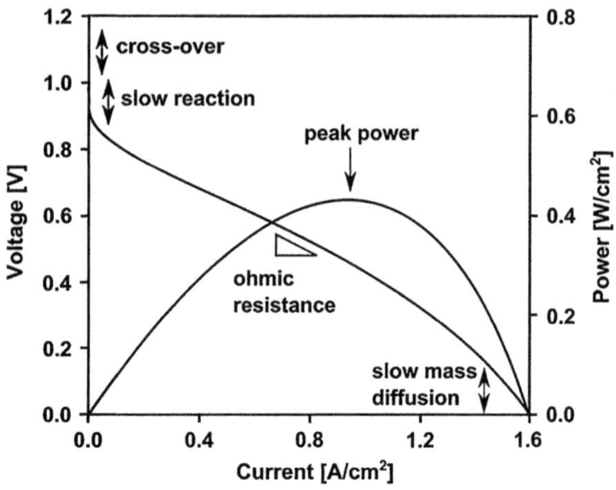

Fig. 3.5 Polarization *curve* for a generic hydrogen PEMFC

3.2.2.2 Activation Polarization

Activation polarization (η_{act}) is responsible for the potential losses occurring in the left portion of the polarization curve. It includes the kinetic overpotential for both the anode and cathode reactions. As described by the Butler-Volmer equations, the charge transfer resistance gets lower as the reaction overpotential gets larger. For this reason, activation overpotential dominates the polarization curve when the potential value is close to the open circuit potential.

3.2.2.3 Ohmic Polarization

Ohmic losses are the result of the electric resistance of both the electronic and ionic conductors as well as of the quality of the contacts. Because both the electrolyte and fuel cell electrodes obey Ohm's law, the Ohmic losses can be expressed by the Eq. (3.1):

$$\eta_{iR} = iR \tag{3.1}$$

3.2.2.4 Concentration Polarization

As reactants are consumed by the electrochemical reactions, there is a loss of potential (η_{act}) due to the depletion of concentration in proximity of the electrode. That is, a concentration gradient is formed. Due to the complex nature of the catalyst layer and the fuel cell architecture, several processes may contribute to concentration polarization: (i) slow diffusion of the fuel through the electrode pores, (ii) solution/dissolution of reactants/products into/out of the electrolyte, and

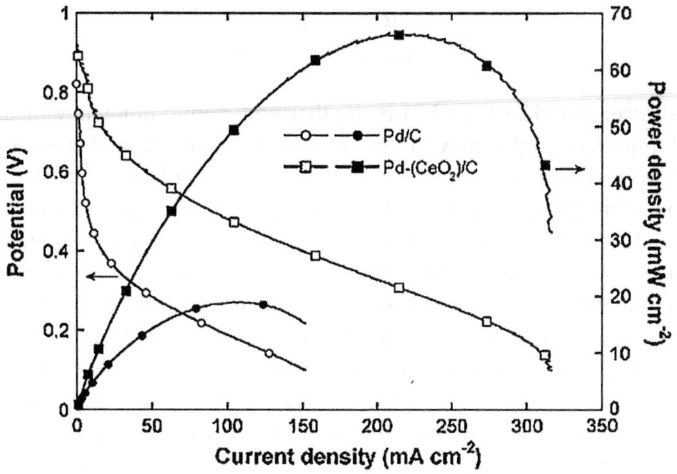

Fig. 3.6 Typical fuel cell polarization *curves* and the corresponding power density *curves*. Notice that potential values are reported on the *left* ordinate axis, while power density values are on the *right* ordinate axis. Reproduced from Ref. [6] with permission of John Wiley and Sons

(iii) the diffusion of reactants/products through the electrolyte to/from the electrochemical reaction site.

Formally, the curve reported in Fig. 3.4 can be described by the following relationship Eq. (3.2):

$$V = E_r - \eta_{\mathrm{act}} - \eta_{iR} - \eta_{\mathrm{conc}} \tag{3.2}$$

A different representation of the polarization curve is the power density curve. The power density curve is derived by multiplying the potential for the actual current value (this is the definition of electric power) and reporting such value on the y axis. The result is a curve which passes through a maximum (Fig. 3.6). This is used to define the maximum power output of a fuel cell. Very often polarization curves are sketched together with the power density curves in a same chart where the values for the polarization curve are on the left y axis, while the power density values are on the right y axis (Fig. 3.6).

3.2.2.5 Efficiency

As fuel cells have been proposed as power sources for a green and sustainable future, they are designed to exploit the energy content of the fuel in the most efficient way. Fuel cells may theoretically overcome the efficiency limitation of the thermomechanical cycles typical for internal combustion engines. Nevertheless, there are efficiency issues related to the fuel cell. The most comprehensive efficiency definition [3] is the following:

$$\varepsilon = \frac{\Delta G^o}{\Delta H_{HHV}} \times \frac{V}{E_r} \times \frac{M_f^c}{M_f^s} \qquad (3.3)$$

The first term found in Eq. (3.3) is the thermodynamic efficiency or the actual theoretical maximum efficiency. It is calculated by the ratio between the free energy of the reaction under standard conditions and the higher heating value for the reaction enthalpy (accounts for the condensation of vapor products such as water). Sometimes, the efficiency is expressed in terms of the lower heating value (LHV). The second term is the ratio between the actual operational voltage and the thermodynamic potential. Under operation fuel cells experience a voltage loss due to the activation overpotentials, the iR drop and concentration polarization according to Eq. (3.2). Such losses may dramatically reduce the efficiency. The third term is a fuel utilization factor which accounts for the fact that only a fraction of the fuel is usually oxidized. Such a partial fuel utilization may be also ascribed to losses through, for example, phenomena like crossover of the reactants through the PEM.

3.2.3 Main Operational Parameters

Fuel cell performance does not only depend on the nature of the materials and the fuel but they are also strongly affected by the operating conditions. What follows is a description of the main parameters affecting fuel cell performance:

3.2.3.1 Temperature

To improve sluggish kinetics an increase in the operation temperature may be applied (Fig. 3.7). Even if temperature increases are usually detrimental in terms of thermodynamic potential, actually the gains which can be obtained from increases in the temperature in terms of activation kinetics are much larger than thermodynamic losses. Furthermore, higher temperature increases the mass transport properties such as diffusion and even charge transport in ionic conductors leading to a reduction in the cell resistance. Increasing the temperature is by the way limited to about 100 °C. This is for a variety of reasons, e.g., PEMs may not be stable for long periods at high temperature. Furthermore, accumulation of water at the cathode may severely reduce the partial pressure of oxygen, reducing the cell performance by oxygen starvation. For practical purposes, the increase in operating temperature has a positive influence until the temperature approaches the boiling point of water. At present, for practical purposes it is widely recognized that safe operation of a PEM is usually around 80 °C at least for hydrogen-fueled PEMFCs.

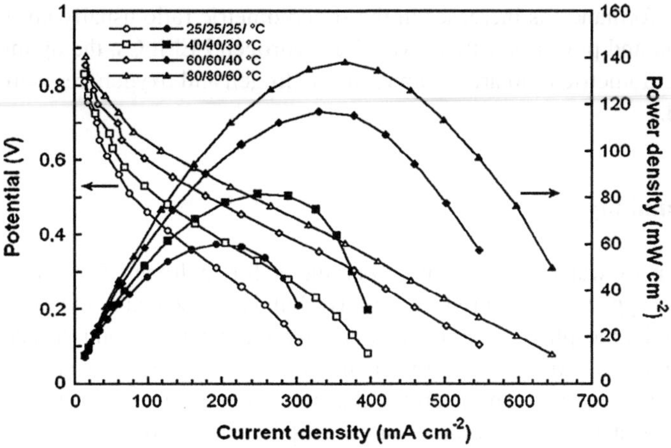

Fig. 3.7 Potentiodynamic (empty symbols) and power density (filled symbols) curves for active DEFCs fueled with a 2 M KOH and 10 wt % EtOH solution at different temperatures. Temperature of fuel (*left*), cell (*central*), and oxygen gas (*right*) were •: 25/25/25 °C, ◆: 40/40/30 °C, ▲: 60/60/40 °C, ▼: 80/80/60 °C. Reproduced from Ref. [6] with permission of John Wiley and Sons

3.2.3.2 Gas Pressure or Liquid Fuel Concentration

The rate of electrochemical reaction is strongly affected by partial pressure in the case of gas reactants and concentration for liquid reactants. Such dependence may be complex depending on the characteristics of chemical kinetics. Thus, the polarization curve shifts toward larger potentials by increasing the partial pressure or the concentration of the reactants (this may not be always true, e.g., for alkaline oxidation of DEFCs a balanced ratio between OH^- and ethanol is more relevant than the absolute ethanol concentration). It is worth noticing that on the side of the overall system efficiency, increasing the partial pressure may not be a good idea as higher pressure requires power to compress the reactant gases. Furthermore, operating at high pressure requires more sophisticated devices and may create problems like leakage. Often the gas pressure in fuel cell apparatus is in the range between 1 and 3 atm.

3.2.3.3 Stoichiometric Ratio

This is defined as the ratio of the amount of actual reactant to the amount of reactant needed to exactly complete a reaction. A stoichiometric ratio of 1.0 provides the number of molecules to theoretically complete the reaction. In hydrogen PEMFCs, a higher stoichiometric ratio increases the chance that sufficient numbers of hydrogen and oxygen react at the anode and the cathode, respectively. In the end, a stoichiometric ratio larger than 1 is usually required for optimal performance as increases in the stoichiometric ratio usually pushes up the polarization and power density curves. For hydrogen PEMFCs, the optimal values of the stoichiometric ratio are 1.4 and 2 for hydrogen and oxygen, respectively, at a given load.

3.2.3.4 Humidity

The content of water is an essential parameter for hydrogen PEMFC operation. Humidity is typically measured as "relative humidity" which is the ratio between the saturated atmosphere water content at a given temperature and the actual water content. Humidification allows the membrane to correctly transport protons as water molecules move with the proton during the ion exchange reaction. If there is not enough water, the membrane can get fragile leading to cracks and holes that lead to short circuits, hot spots, local gas mixing, and even the risk of explosion. On the other hand, excessive humidification leads to the condensation of liquid water and subsequent flooding within the flow field resulting in cell reversal, a phenomenon for which the affected cell might produce zero or even negative voltage. In practice, hydrogen fuel cells require gas supply with a precise control of the relative humidity. In the case of a cell fed with liquid fuel at the anode (e.g., alkaline DEFCs), there may be problems linked to cathode flooding and to fuel crossover from the anode compartment.

3.3 Major Low Temperature Fuel Cells

A variety of fuels have been proposed for use in direct fuel cells, each with their own advantages and drawbacks. This section will describe in some detail three classes of fuel cells employing polymer electrolyte membranes and operating with a variety of fuels. First a review of the hydrogen PEMFC is reported as, at least at present, the major candidate fuel cell technology in key sectors such as automotive. Next we review the various fuel cell technologies operating with liquid fuels. Liquid fuels have two major advantages as compared to gaseous H_2: (i) high energy density and (ii) simple storage and transportation. Nevertheless, the oxidation kinetics for methanol and other alcohols are much more sluggish than that of hydrogen. Inevitably, such systems cannot deliver the same power density of a hydrogen PEMFC. Furthermore, they require much more noble metal catalyst as compared to PEMFCs. Nevertheless they have been proposed for a variety of applications. Among the very many examples, the use of liquid fuel-fed cells for powering portable electronic devices where no electrical grid is present looks to be the most promising.

3.3.1 Hydrogen PEMFC

As described previously in PEMFCs, the hydrogen is oxidized to protons at the anode. The protons pass through the membrane while electrons produced by the oxidation flow through the external circuit and the electrical load to reach the cathode where they reduce oxygen to render water by combination with the protons (Fig. 3.1). PEMFCs are low temperature fuel cells with operating temperatures usually ranging between 60 and 100 °C. PEMFCs are compact systems with high weight power density. The output power of a state-of-the-art PEMFC stack may be up 250 kW and more. Among the very many strong points of the technology is the rapidity of the start-up and the expected long lifetime [7].

The above properties make PEMFCs suitable for powering automobiles in combination with an electric engine. Indeed this is the major explored application for such technology. At present, the total cost of a vehicle equipped with a PEMFC is in the range of a few hundreds of dollars (US) for each installed kW. Such a value is still much larger than that of internal combustion engine vehicles. Research on materials and manufacturing is still required to make PEMFCs competitive in the automotive sector. Most of the PEMFCs costs are found in the assembly process, bipolar plates, membrane and peripherals, and the platinum catalyst layers. With respect to the last point, nanotechnology offers a range of opportunities both for reducing the platinum loading and in some cases to completely remove noble metals from the fuel cell architecture.

The electrical efficiency of PEMFCs ranges between 40 and 50 %. It is worth mentioning that this is the efficiency of the electrical energy production only and does not account for the other energy losses related to the conversion of electrical energy into mechanical energy and the transmission of the mechanical energy from the engines to the wheel. When accounting for that one may find that the overall efficiency of a PEMFC vehicle can be as low as 30 % which is still much larger than a regular gasoline internal combustion engine [8]. In PEMFCs, there are no moving parts hence they are inherently low vibrating and silent. No lubrication oil is required eliminating a seriously polluting consumable which is required in internal combustion engines. It is worth mentioning that a conventional gasoline engine produces sulfur oxide (SO_2), nitrogen oxides (NO_x), carbon monoxide (CO), and carbon dioxide (CO_2) [8]. All of these compounds are the objects of severe regulations that are becoming stricter year after year. Furthermore, hydrogen is the fuel that can be produced with sustainable technologies which usually implies a renewable energy source (e.g. biomass, photovoltaic, wind etc…) coupled with an energy efficient process such as low and high temperature electrolytic water splitting. The main hurdle to the successful implementation of this approach is that hydrogen as an energy vector requires the design and realization of an entirely new and complex production and distribution infrastructure.

Most of the current research on electrocatalysts for PEMFC fuel cells is focused on the cathode. This is because the ORR is much more sluggish than the hydrogen oxidation reaction, requiring most of the loading of the noble metal catalyst.

Furthermore, most of the corrosion problems are located at the cathode and much work is required to increase the durability of the catalyst/support system, especially during transients and shutdown/startup cycles. At present, the platinum loading in PEMFCs accounting for both the anode and cathode catalyst is in the range of 0.3 mg cm^{-2} or in terms of amount of platinum per energy unit 0.3 g kW^{-1}. According to this, a 100 kW vehicle should use 30 g of platinum. Even if at first sight this number is not impressive it has to be considered that many millions of vehicles should be produced, leading to a net increase in the platinum request. From the point of view of scale of economy, this will naturally increase the price of the metal. This is not sustainable, in the sense that platinum as a mineral may experience a bell-shaped production curve analogous to the Hubbert peak for oil [9]. This could lead to the paradox that we are developing a technology to reduce our dependence on oil to fall down into an analogous one on platinum minerals. Hence platinum content reduction or elimination and stability improvements are the main material research focus.

3.3.2 *Direct Methanol Fuel Cells*

A direct methanol fuel cell (DMFC) (Fig. 3.8) is a particular PEMFC which operates with methanol or methanol solutions as the fuel and works at near room temperature. Methanol is oxidized at the anode with water, while oxygen is reduced at the cathode according to the scheme reported in Fig. 3.8. As for the hydrogen-fed PEMFC, oxidation of methanol produces protons which are transported by the proton conducting membrane to the cathode. The membrane also allows the electroosmotic drag of water, together with some methanol to the cathode, the so-called methanol crossover. Methanol crossover is one of the main issues when designing DMFCs as this leads to a large loss of efficiency for both the reduction of the fuel conversion and for the drop in the cell potential due to the activity of the cathode electrocatalyst toward the oxidation of methanol. In general, dilute methanol solutions are used as fuel in order to decrease methanol crossover in DMFCs. It has been reported that the concentration at which the maximum efficiency is obtained is in the 1 M range [10–15]. A number of reviews on DMFCs operating at various methanol concentrations have been reported [16–18]. The use of diluted solutions produces the serious drawback of lowering the energy density of the fuel. This counterbalances one of the main advantages of using methanol, i.e., the high volume energy density of the fuel as compared to hydrogen. The problem may be overcome by engineering the devices in such a way that methanol concentration is reduced in close proximity of the catalyst layer.

At present, the ideal application of DMFCs is in the powering of electrical portable devices. In this context, DMFCs can potentially compete with state-of-the-art technologies as methanol has many advantages: (i) easy storage; (ii) high energy density (6.1 kWh kg^{-1}); and (iii) ease of handling. On the other side, there are some drawbacks such as the high vapor pressure, flammability and toxicity.

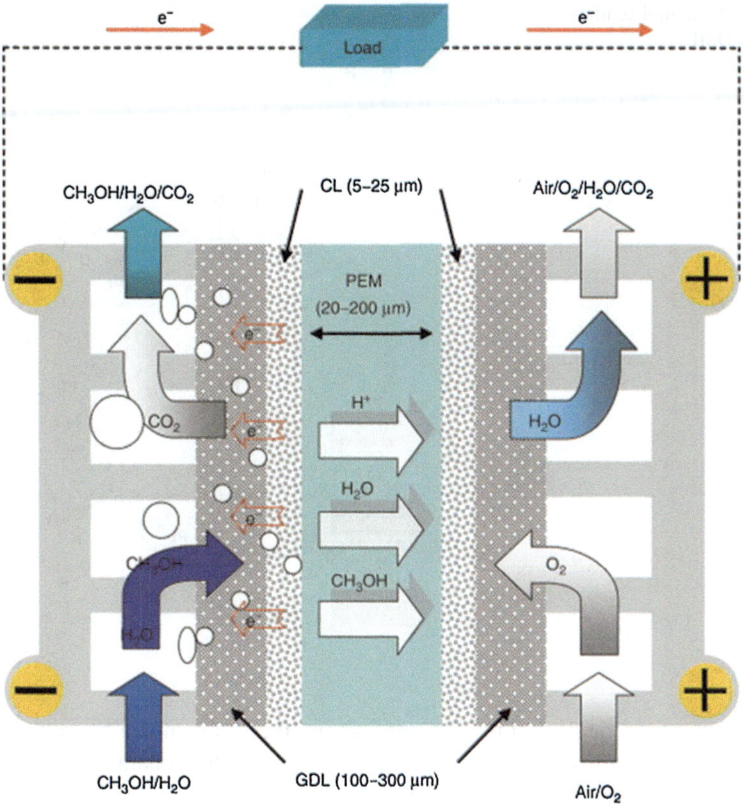

Fig. 3.8 DMFC fed with high concentration methanol solutions

DMFC systems are usually classified in terms of the oxidant supply mode. When methanol feeding occurs by pumps DMFCs are said to be active. When the methanol tank is directly connected with the anode side with no auxiliary devices, DMFCs are said to be passive. Active devices usually also allow the control of temperature and humidity and are equipped with a variety of sensors and controllers, allowing a complete control of all the operating parameters. On the other hand, passive devices are simpler, but the absence of auxiliary devices does not allow an accurate control of the operation. Sometimes hybrid devices where conditions are only controlled at the anode or the cathode may be encountered. Such devices are named semipassive. For very low power applications (10 W range), passive devices are usually preferred for the construction simplicity and the lower cost.

As extensively discussed in Chap. 2, the oxidation of methanol is usually performed with platinum-based electrocatalysts. Since methanol oxidation kinetics are even worse than ORR kinetics, a high content of Pt is usually required at the anode. Furthermore, Pt undergoes severe poisoning by CO adsorption, hampering

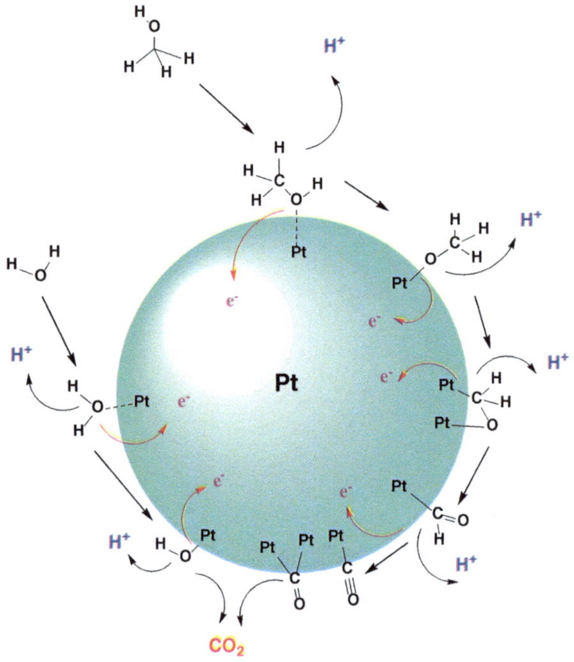

Fig. 3.9 Methanol oxidation on Pt in DMFC

the stability of the devices (Fig. 3.9). At present, the problem can be mitigated by the use of Pt–Ru alloys, that are the best performing electrocatalyst for the oxidation of methanol. Major research efforts are occupied with the development of new nanostructured materials capable of reducing the Pt content at the anode side and with enhanced tolerance to CO. A large research effort has also been devoted to the development of cathode catalysts' methanol tolerance in order to mitigate the effect of methanol crossover.

3.3.3 Direct Alcohol Fuel Cells

We have seen that methanol has several disadvantages as a fuel. The use of other alcohols with larger molecular weights as compared to methanol has been extensively investigated in the last decade. Among them ethanol has been the most considered because among other reasons, it shows a higher energy density (8.0 kWh kg^{-1}) than methanol (6.1 kWh kg^{-1}). Ethanol can be easily obtained from biomass through fermentation processes from renewable resources, such as sugar cane, wheat, corn, or even straw. Biomass production is a virtuous cycle in terms of greenhouse emissions. Indeed CO_2 emitted during the production process of ethanol and the CO_2 eventually released by ethanol oxidation in direct ethanol fuel cells (DEFC) can be reverted into biomass by the plants from which the fuel is produced. Hence ethanol is a sustainable energy source. On the other side, we should consider that life cycle analysis of bioethanol renders an energy payback ratio of the fuel in the range of 1 or lower if the production is from corn [19]. It is highly recommended not to use corn for producing ethanol. Sugar-rich plants such as rapeseeds and sugar cane can show a energy payback ratio for derived bioethanol ranging from 4 to 8. In the last few years, production of bioethanol from unused nonagricultural lands from spontaneous plants such as the "*arundo donax*" has also been reported [20]. In such cases it should be possible to get an energy payback ratio even larger than ten. At present, there is by the way still a lack of a shared opinion about such values and huge variations in the reported energy payback ratio exist between different investigations [19, 21]. Other alcohols also may be derived by a variety of processes from biomasses. Glycerol is now receiving much attention as it is a byproduct of biodiesel production. It can be oxidized in direct alcohol fuel cells to render energy and valuable chemicals.

Ethanol can be oxidized both in acidic and alkaline environments. The most relevant problem with ethanol and higher molecular weight alcohols is the difficulty in C–C bond cleavage [16]. In principle, ethanol oxidation could lead to the production of 12 e^- to produce carbon dioxide. However, the oxidation stops usually at the formation of acetaldehyde ($2e^-$) or acetic acid ($4e^-$). Recently, Adzic et al. reported a nanostructured catalyst able to selectively oxidize ethanol to CO_2. Nevertheless, no investigation on its application in a complete DEFC has been reported so far. Just as for methanol, alcohol oxidation in acidic environments requires the use of platinum-based catalysts and may suffer from the same CO poisoning problems already reported for DMFCs. Furthermore, kinetics are sluggish requiring noble metal loadings higher than 1 mg cm^{-2} to realize devices with power densities in the range of a few tenths of mW cm^{-2} and low energy efficiencies.

At present, the most performing DAFCs [22] are those operating in the alkaline environments [23] with alkaline ion-exchange polymeric membranes (Fig. 3.10). The kinetics of alcohol oxidation reactions and ORR are faster at high pH than in acidic media. It has been shown that when the acid electrolyte is changed to alkaline media, the fuel cell efficiency increases [24–26].

In alkaline environments, platinum can be replaced by nonnoble metal electrocatalyts at the cathode. This is an important advantage as these catalysts are not active toward alcohol oxidation, hence limiting the consequences of crossover. Furthermore, Pd-based materials have been proven to be the most effective electrocatalyst for the oxidation of alcohols in alkaline media being the most interesting candidates for practical applications.

The oxidation of ethanol on Pd in alkaline environments has been reported to produce power densities of over 100 mW cm^{-2} [6, 27]. Pd catalysts have been proven to be effective for the selective oxidation of ethanol to acetate in strongly alkaline electrolytes as they do not provoke C–C cleavage at all for pHs higher than 13 [28]. A reduction in pH does lead to an increase in the extent of the C–C cleavage in ethanol electrooxidation (Fig. 3.11).

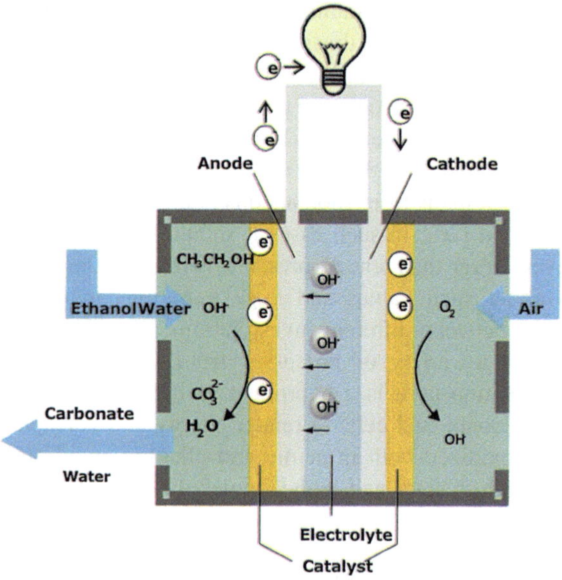

Fig. 3.10 Scheme of an alkaline DEFC

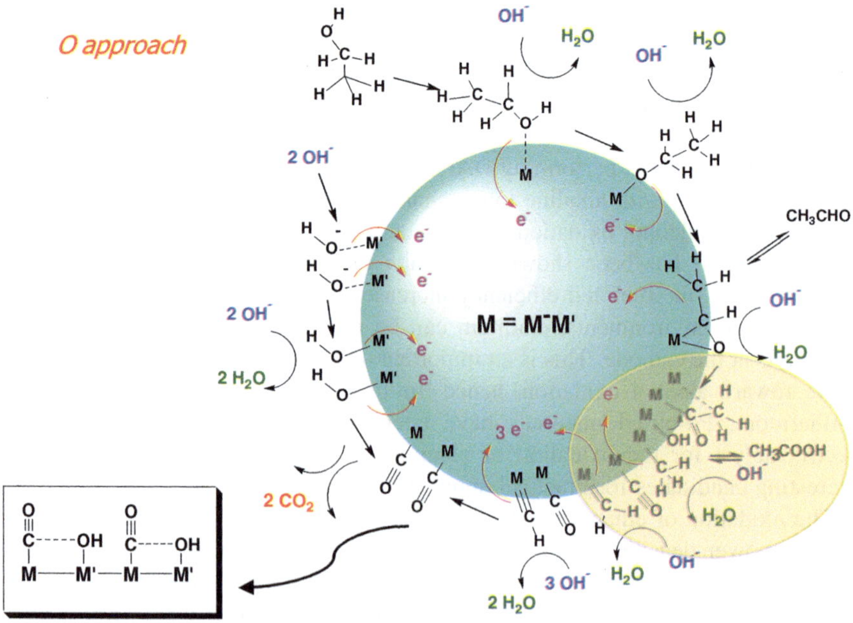

Fig. 3.11 Proposed mechanism reaction of ethanol oxidation at pHs lower than 13

Partial oxidation is a major drawback in the sense that only one third of the electrons can be extracted from the fuel. Another relevant issue of Pd-based DEFCs is the stability of performance. Pd may undergo oxidation at potentials close to the oxidation potential of alcohols. Indeed operating at high power densities may result in a dramatic DEFC performance degradation. It has been shown that the effect can be mitigated by the addition of a tiny amount of reducing agents to the fuel [29]. The oxidation of alcohols heavier that ethanol have also been extensively considered in the literature, leading to interesting approaches which can provide cogeneration of energy and high added value chemicals [30, 31].

Ethanol fuel cells still show a relatively low efficiency. At the moment, there are just a few reports that deal directly with the fuel efficiency of passive alkaline DEFCs. It has been found recently that ceria-promoted palladium catalysts can show fuel efficiencies as large as 6 % [6]. These values demonstrate clearly that DEFCs cannot at present be used as power sources for high power demanding applications. Nevertheless, they could be successfully applied in specialized niches such as micro fuel cells.

At the moment, most of the research efforts in direct alcohol fuel cells are in efficiency enhancement and in increasing the selectivity toward high value added products. For both targets the main issue is the anode electrocatalyst. Nanotechnology may provide a variety of approaches to accomplish the above-reported tasks such as increasing the specific activity of the noble metals, promoting catalytic activity through the addition of nanostructured oxides, and producing tailored nanoparticles with enhanced activity toward alcohol oxidation.

3.4 Electrolysis: General Background

A key issue in the realization of the so-called "hydrogen economy" will be the production of hydrogen from renewable energy sources. While hydrogen has been largely proposed as a clean and efficient energy vector it is not an energy source by itself. Electrolytic water splitting is the technology that enables hydrogen production from renewable energy sources. Hence the search for efficient electrolytic processes will play a major role in the future sustainability of our society and it is hard to underestimate its importance.

While fuel cells have not yet been fully exploited commercially, electrolysis is a mature technology used on a worldwide scale for the production of high purity hydrogen. Electrolyzers are available in a wide variety of sizes and it is easy to find them in research labs around the world for in situ hydrogen production. It is worth mentioning that at present most hydrogen is produced using fossil fuels such as natural gas. Hydrogen has been a key element for our society even in the past. Indeed it has been largely employed in atmospheric nitrogen fixation for the synthesis of chemical fertilizers, contributing to the demographic boom of the twentieth century. Electrolytic hydrogen production at present accounts for approx. 1 % of the overall world hydrogen production.

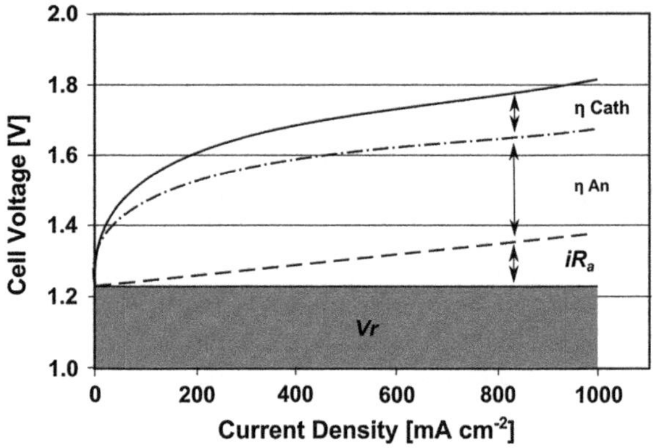

Fig. 3.12 A typical current density versus cell voltage diagram for an electrolyzer. The separate contributions of (i) ohmic drop (iR_A, including both electronic and ionic conductors contribution), (ii) OER activation (η_{An}) and HER activation (η_{Cath}) are also reported

What makes electrolysis particularly appealing in terms of sustainability is its ability to directly convert electric energy into molecular hydrogen. So any possible renewable sources of electricity may be used to drive electrolytic hydrogen production, leading to a valuable system for storing energy from intermittent sources.

The thermodynamic potential for water splitting is 1.23 V. This leads to a minimum energy consumption of 33.6 kWh per kg of hydrogen produced. Under real operating conditions, after contributions from iR drop and activation overpotentials for the anode and the cathode half reactions, the actual energy consumption usually is around 50 kWh kg^{-1}.

The performance of an electrolyzer is determined through the acquisition of a current density versus cell voltage curve (Fig. 3.12). The curve is obtained by polarizing the electrolyzer to a given potential and recording the current passing through the cell. At the same time, the amount of evolved hydrogen may be measured allowing the determination of the so called faradaic efficiency.

Table 3.1 lists the main characteristics of the most common water electrolyzer architectures and the actual cell architectures are shown in Fig. 3.13. Conventional electrolysis requires that the aqueous solution should be electrically conducting for the process to proceed. Hence conventional technology of water electrolysis makes use of alkaline solutions. In particular, an electrolyte of approx 30 wt % KOH is used and the cell temperature is usually about 80 °C. The use of KOH, although more expensive than NaOH, is dictated by two reasons: i) KOH is more conductive (about 1.3 times) than NaOH and ii) KOH is chemically less aggressive than NaOH. Alkaline electrolyzers typically operate with cell potentials around 2 V and current densities up to 300 mA cm^{-2}. Under these conditions voltage efficiencies in the range of 60 % [32] are achieved. The voltage efficiency is determined by

Table 3.1 Typical parameters of standard water electrolyzers

	Conventional alkaline	PEM	Advanced alkaline
E_{cell}	~2 V	1.8 V	<1.8 V
kWh (system)/N m^3 H_2	~5.0	4.3	<4.4
Current density (mA/cm^2)	>100	>1000	>500
Lifetime (years)	>10	~5	>10

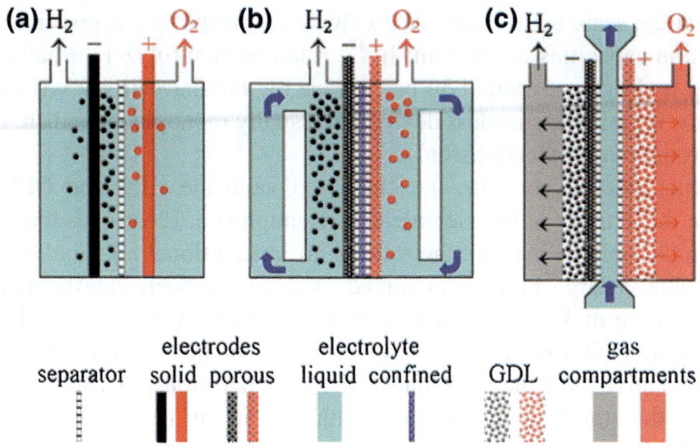

Fig. 3.13 Different electrolyzer cell architectures: **a** alkaline electrolyzer, **b** "zero-gap" or advanced alkaline configuration, **c** PEM electrolyzer. Reproduced from Ref. [33] with permission of Elsevier

dividing of the thermodynamic potential for water electrolysis by the actual cell voltage at a given current density [33].

The main advantage of alkaline electrolysis is its reliable set-up that has allowed lifetime of working longer than 10 years. The major limiting factor for distributed electrolytic generation of hydrogen is the capital cost rather than the electrical energy input. Indeed Marini et. al. in [33] state: "… capital cost, rather than efficiency and the price of electrical energy, is the major factor limiting a distributed electrolytic hydrogen production. In other words, we may be better off spending two times more energy per N m^{-3} H_2 if, at the same time, we can halve the capital costs by doubling the power density."

While noble metals are very effective in catalyzing the OER and the HER, they are not generally used in alkaline electrolysis. Most of large-scale electrolytic devices operating in alkaline conditions use Ni-based electrodes. The small losses due to larger activation overpotentials does not justify the increase in the cost which could come from the use of precious metal catalysts.

3.4.1 Alkaline Electrolysis

Figure 3.13a schematically shows the cell configuration for alkaline electrolyzers. In this configuration, anode and cathode are immerged together in a concentrated KOH solution. H_2 and O_2 gas bubbles evolve in separated electrolyte compartments. A gas impermeable membrane (the "separator") avoids mixing of hydrogen and oxygen. Bubble formation reduces the active surface of the electrodes and thus contributes to the electrolyte resistance. High current densities are not usually obtained but the scale up to large-scale volume H_2 production is possible (Module H_2 production capacities >100 N m^3 h^{-1}). Alkaline electrolyzers usually produce pressurized hydrogen at around 30 bar. Hence the bipolar cell stack is embedded in a pressure vessel. Pressure-less devices are usually monoplanar and are designed for low throughput applications only.

In alkaline electrolysis cells, the kinetics of both the HER and OER depend strongly on the activity of the electrocatalysts employed. Electrodes made of iron, nickel, or nickel-plated iron are the state of art in traditional alkaline systems. On the anode side, a large number of mixed oxides have been investigated. Mixed oxides containing nickel and cobalt, with a perovskite structure (e.g., lanthanum nickel oxide $LaNiO_3$) or a spinel structure (e.g., nickel cobaltite $NiCo_2O_4$) all display high activity for the OER under alkaline conditions. On the cathode side, Ni-Raney is the standard material used although it can be further improved by coating with different metals or alloys, e.g., Ni + Ru, Co + Mo, Pt +Mo and Ni + Fe [34, 35].

3.4.2 Zero Gap Electrolysis

The "zero-gap" configuration of an alkaline electrolyzer is reported in Fig. 3.13b. Such a design was developed in order to limit the effect of bubbling on the cell resistance. This was possible because in the zero gap system bubbles evolve in a region not relevant to ionic conduction. A thin spacer or separator constituted by a thin cellulose felt is introduced between the anode and cathode electrodes which are then tightly pressed together [33]. An anionic exchange membrane (AEM) [36] could also be used to replace the separators. In this way, the system would resemble a PEM electrolyzer . The anode and cathode structures are designed in such a way so they are porous and permeable to the electrolyte. As shown in Fig. 3.13b separate circulation of the electrolyte in the cathode and/or anode compartments is required. As the systems work with essentially the same electrolytes as the alkaline electrolyzers, the same electrocatalysts may also be used here.

3.4.3 The Proton Exchange Membrane Water Electrolyzer

This technology was first described in a patent by GE in 1970. The polymer electrolyte membrane employed here between the electrodes functions as both a gas separator and as a solid electrolyte. No liquid electrolyte is necessary and only deionized water is fed to the cell. PEM electrolyzers can operate with current densities exceeding 1 A cm^{-2} at cell potentials lower that 1.8 V (Table 3.1). This is possible thanks to the PEM which results in a much lower resistance as compared to the conventional alkaline electrolysis. A schematic of a PEM electrolyzer is illustrated in Fig. 3.13c. It may be noticed that the design is very similar to a hydrogen PEMFC, apart from the fact that liquid water is supplied to the anode and cathode compartments instead of gases. As reported by Millet et al. [37] "The inherent advantages of SPE technology over the alkaline one are clearly established: (i) greater safety and reliability are expected since no caustic electrolyte is circulated in the cell stack; (ii) tests made on bare membranes have revealed that some materials could sustain high pressure without damage and were efficient in preventing gas mixing; and (iii) the possibility of operating cells up to several amps per square centimeter with typical thicknesses of a few millimeters is afforded [38]."

The standard membrane material used in PEM water electrolysis units is Nafion™ 117 and is manufactured by DuPont. The cathode used in PEM electrolyzers usually consists of a porous graphite current collector with a Pt-based electrocatalyst which can be also a carbon-supported Pt/C material. Individual cells are stacked into bipolar modules with graphite-based separator plates providing the manifolds for water feed and gas evacuation.

In the case of PEM electrolyzers, the extreme anodic conditions (high potential and low pH) restrict the choice of electrocatalyst material to a few noble metal oxides or mixture of oxides such as: $IrO_2$with Ta_2O_5 or SnO_2. The most active and stable materials are mixed oxides of $IrO_2$and RuO_2*e.g.* $Ir_{0.6}Ru_{0.4}O_2$ [35, 39–41].

Figure 3.14 shows typical performances for each of the electrolyzers described in this section. It is clear that the PEM and zero gap electrolyzers are the best performers in terms of both current density and cell voltage. The conventional alkaline technology suffers from low current densities and high cell voltages.

3.4.4 Electrolysis with Anode Reactions Other than OER

The addition of a chemical species (or sacrificial agent) at the anode which is much more readily oxidized than water (e.g., ammonia or an alcohol) can make the thermodynamic contribution to hydrogen production very small thus dramatically lowering the energy demand for electrolytic hydrogen production. This is because of the high overpotential for O_2 evolution is substituted by that of a different anodic reaction [35]. The possibility of avoiding oxygen is also an advantage in terms of safety issues as hydrogen and oxygen mixtures are explosive in many concentration ranges. A variety of approaches have been recently attempted. Anodic O_2 evolution was replaced with oxidation of suspended carbonaceous materials [42] by Seerha et al. [43] who showed that such electrolysis can be performed at cell potentials as low as 0.5 V, although the current densities are very low.

Recently, the use of ammonia has been explored in some detail [44, 45]. The theoretical thermodynamic potential for electrolysis would in this case be 0.06 V with a net gain of 1.17 V as compared to that of water electrolysis (1.23 V). As a consequence, hydrogen can be produced with an amount of energy consumption which can be quantified at about 30 % of conventional alkaline electrolysis.

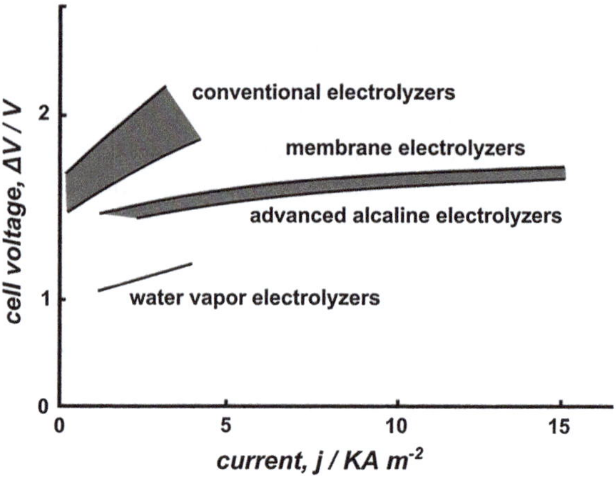

Fig. 3.14 Performance of the electrolyzer technologies discussed in the chapter. Notice the difference in the current densities between advanced (zero gap) and membrane electrolyzers as compared to the conventional alkaline

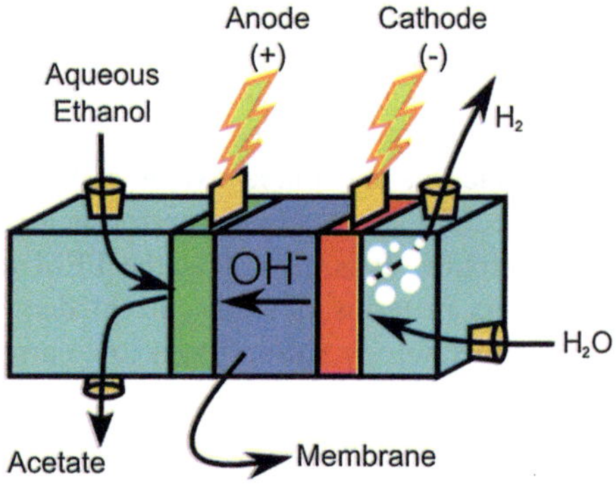

Fig. 3.15 Schematic illustration of a PME with an anion-exchange membrane, in which ethanol is converted into acetate at the anode and H_2 is produced at the cathode

The actual hydrogen production was indeed very small compared to conventional technologies [35] as the electrooxidation of NH_3 to nitrogen is a very demanding task for electrocatalysis. High loadings of noble metal Pt–Ir based anode catalysts were required to obtain significant current densities and the system suffered from severe poisoning phenomena associated with nitrogen-containing intermediates produced in the oxidation of ammonia.

Recently, a variety of attempts to reduce the consumption of electrolysis by using the electrooxidation of methanol have been performed, producing CO_2 at the anode [46, 47]. Methanol electrolysis suffers from problems analogous to those described for the oxidation of ammonia. The achievable current densities cannot compete with the best state-of-the-art technologies.

Recently, Vizza et al. [48] demonstrated the possibility of oxidizing ethanol at the anode of a electrolyzer equipped with an alkaline membrane (Fig. 3.15). According to these experiments, the electrolysis of a solution of ethanol (10 wt %) in 2 M KOH can proceed at effective potentials of 0.65–0.85 V with current densities ranging from 0.2 to 2 kAm^{-2} with an energy consumption of ca. 1.5–2.0 $kWhNm^{-3}$ H_2. The anode electrocatalyst was 1 mg cm^{-2} Pd on Vulcan

XC-72. The oxidation of ethanol resulted in the selective production of acetate. The same investigation also reported the possibility of using heavier alcohols (glycerol, ethylene glycol) for cogeneration of hydrogen and valuable chemicals. Such an approach offers the possibility of converting biomass-derived products such as these into both hydrogen and chemicals.

For each of the electrolytic processes described here for hydrogen production, there is ample possibility for the development of more effective electrocatalysts. They have to be designed accounting for a variety of constraints. Among them we may cite: (i) selectivity of the anode catalyst for the oxidation of a sacrificial agent; (ii) specific activity; (iii) precious metal loading; (iv) mass transport; and (v) manufacturability. It is expected that nanotechnology will strongly contribute to improve these technologies offering approaches for overcoming the current limitations of electrolytic hydrogen generation.

References

1. B. Dadda, S. Abboudi, A. Ghezal, Transient two-dimensional model of heat and mass transfer in a PEM fuel cell membrane. Int. J. Hydrogen Energy **38**, 7092 (2013)
2. V. Mehta, J.S. Cooper, Review and analysis of PEM fuel cell design and manufacturing. J. Power Sources **114**, 32–53 (2003)
3. R.P.O'Hayre, *Fuel Cell Fundamentals*. (John Wiley & Sons, Hoboken, 2006), pp. xxii, 409 p
4. M. Eikerling, A.A. Kornyshev, Modelling the performance of the cathode catalyst layer of polymer electrolyte fuel cells. J. Electroanal. Chem. **453**, 89 (1998)
5. S.M. Haile, Fuel cell materials and components. Acta Mater. **51**, 5981 (2003)
6. V. Bambagioni et al., Energy efficiency enhancement of ethanol electrooxidation on Pd–CeO2/C in passive and active polymer electrolyte-membrane fuel cells. ChemSusChem **5**, 1266 (2012)
7. A. Therdthianwong, P. Saenwiset, S. Therdthianwong, Cathode catalyst layer design for proton exchange membrane fuel cells. Fuel **91**, 192 (2012)
8. S. Mekhilef, R. Saidur, A. Safari, Comparative study of different fuel cell technologies. Renew. Sustain Energy Rev. **16**, 981 (2012)
9. U. Bardi, A. Lavacchi, A simple Interpretation of Hubbert's model of resource exploitation. Energies **2**, 646 (2009)
10. S.H. Seo, C.S. Lee, A study on the overall efficiency of direct methanol fuel cell by methanol crossover current. Appl. Energy **87**, 2597 (2010)
11. X. Li, A. Faghri, Review and advances of direct methanol fuel cells (DMFCs) part I: design, fabrication, and testing with high concentration methanol solutions. J. Power Sources **226**, 223 (2013)
12. D.H. Jung, C.H. Lee, C.S. Kim, D.R. Shin, Performance of a direct methanol polymer electrolyte fuel cell. J. Power Sources **71**, 169 (1998)
13. S. Hikita, K. Yamane, Y. Nakajima, Measurement of methanol crossover in direct methanol fuel cell. JSAE Rev. **22**, 151 (2001)
14. J. Liu, D. Liu, H.Y. Bai, P.S. Wu, X. Han, A new strategy for optimizing the parameters updating algorithm of fuzzy neural controller. Soft. Comput. **10**, 61 (2006)
15. G. Jewett, Z. Guo, A. Faghri, Water and air management systems for a passive direct methanol fuel cell. J. Power Sources **168**, 434 (2007)
16. C. Lamy et al., Recent advances in the development of direct alcohol fuel cells (DAFC). J. Power Sources **105**, 283 (2002)

17. R. Dillon, S. Srinivasan, A.S. Aricò, V. Antonucci, International activities in DMFC R&D: Status of technologies and potential applications. J. Power Sources **127**, 112 (2004)
18. T.S. Zhao, R. Chen, W.W. Yang, C. Xu, Small direct methanol fuel cells with passive supply of reactants. J. Power Sources **191**, 185 (2009)
19. D. Pimentel, T. Patzek, Ethanol production using corn, switchgrass, and wood; biodiesel production using soybean and sunflower. Nat. Resour. Res. **14**, 65 (2005)
20. D. Scordia, S.L. Cosentino, J.-W. Lee, T.W. Jeffries, Dilute oxalic acid pretreatment for biorefining giant reed (Arundo donax L.). Biomass Bioenergy **35**, 3018 (2011)
21. A.E. Farrell et al., Ethanol can contribute to energy and environmental goals. Science **311**, 506 (2006)
22. L. An, T.S. Zhao, Q.X. Wu, L. Zeng, Comparison of different types of membrane in alkaline direct ethanol fuel cells. Int. J. Hydrogen Energy **37**, 14536 (2012)
23. E. Antolini, E.R. Gonzalez, Alkaline direct alcohol fuel cells. J. Power Sources **195**, 3431 (2010)
24. L. An, T.S. Zhao, S.Y. Shen, Q.X. Wu, R. Chen, Performance of a direct ethylene glycol fuel cell with an anion-exchange membrane. Int. J. Hydrogen Energy **35**, 4329 (2010)
25. L. An, T.S. Zhao, R. Chen, Q.X. Wu, A novel direct ethanol fuel cell with high power density. J. Power Sources **196**, 6219 (2011)
26. Y.S. Li, T.S. Zhao, Z.X. Liang, Performance of alkaline electrolyte-membrane-based direct ethanol fuel cells. J. Power Sources **187**, 387 (2009)
27. A. Brouzgou, A. Podias, P. Tsiakaras, PEMFCs and AEMFCs directly fed with ethanol: a current status comparative review. J. Appl. Electrochem. **43**, 119 (2013)
28. C. Bianchini et al., Selective oxidation of ethanol to acetic acid in highly efficient polymer electrolyte membrane-direct ethanol fuel cells. Electrochem. Comm. **11**, 1077 (2009)
29. L. Wang et al., Sodium borohydride as an additive to enhance the performance of direct ethanol fuel cells. J. Power Sources **195**, 8036 (2010)
30. M. Simões, S. Baranton, C. Coutanceau, Electrochemical valorisation of glycerol. ChemSusChem **5**, 2106 (2012)
31. A. Marchionni et al., Electrooxidation of ethylene glycol and glycerol on Pd-(Ni-Zn)/C anodes in direct alcohol fuel cells. ChemSusChem **6**, 518 (2013)
32. J. Ivy, Summary of electrolytic hydrogen production–milestone completion report, NREL/MP-560–36734 (2004)
33. S. Marini et al., Advanced alkaline water electrolysis. Electrochim. Acta **82**, 384 (2012)
34. I.M. Kodintsev, S. Trasatti, Electrocatalysis of H_2 evolution on $RuO_2 + IrO_2$ mixed oxide electrodes. Electrochim. Acta **39**, 1803 (1994)
35. E. Guerrini, S. Trasatti, in *Catalysis for Sustainable Energy Production*. (Wiley-VCH Verlag GmbH & Co. KGaA, 2009), pp. 235–269
36. X. Li, F.C. Walsh, D. Pletcher, Nickel based electrocatalysts for oxygen evolution in high current density, alkaline water electrolysers. Phys. Chem. Chem. Phy. **13**, 1162 (2011)
37. P. Millet, F. Andolfatto, R. Durand, Design and performance of a solid polymer electrolyte water electrolyzer. Int. J Hydrogen Energy **21**, 87 (Feb, 1996)
38. P. Millet, R. Durand, M. Pineri, Preparation of new solid polymer electrolyte composites for water electrolysis. Int. J. Hydrogen Energy **15**, 245 (1990)
39. A. Marshall et al., Iridium oxide-based nanocrystalline particles as oxygen evolution electrocatalysts. Russ J Electrochem **42**, 1134 (2006)
40. A. Marshall, B. Børresen, G. Hagen, M. Tsypkin, R. Tunold, Electrochemical characterisation of $IrxSn_1—xO_2$ powders as oxygen evolution electrocatalysts. Electrochim. Acta **51**, 3161 (2006)
41. E. Slavcheva et al., Sputtered iridium oxide films as electrocatalysts for water splitting via PEM electrolysis. Electrochim. Acta **52**, 3889 (2007)
42. P. Patil, Y. De Abreu, G.G. Botte, Electrooxidation of coal slurries on different electrode materials. J. Power Sources **158**, 368 (2006)

43. M.S. Seehra, S. Ranganathan, A. Manivannan, Carbon-assisted water electrolysis: an energy-efficient process to produce pure H[sub 2] at room temperature. Appl. Phys. Lett. **90**, 044104 (2007)
44. M. Muthuvel, G. Botte, in *Modern Aspects of Electrochemistry*, vol. 45, ed. by R.E. White, (Springer, New York, 2009), pp. 207–245
45. F. Vitse, M. Cooper, G.G. Botte, On the use of ammonia electrolysis for hydrogen production. J. Power Sources **142**, 18 (2005)
46. T. Take, K. Tsurutani, M. Umeda, Hydrogen production by methanol–water solution electrolysis. J. Power Sources **164**, 9 (2007)
47. Z. Hu, M. Wu, Z. Wei, S. Song, P.K. Shen, Pt-WC/C as a cathode electrocatalyst for hydrogen production by methanol electrolysis. J. Power Sources **166**, 458 (2007)
48. V. Bambagioni et al., Self-sustainable production of hydrogen, chemicals, and energy from renewable alcohols by electrocatalysis. ChemSusChem **3**, 851 (2010)

Chapter 4
Factors Affecting Design

4.1 Key Concepts

The purpose of this chapter is to introduce the principles driving the design of new electrocatalytic materials.

The chapter starts with a review of the main targets defined by the U.S. Department of Energy (DOE) for both fuel cells and electrolyzers. The DOE is the most widely recognized authority in defining the goals for these technologies and periodically releases reports reviewing targets and definitions for fuel cells and hydrogen generation techniques.

After having stated the main objectives of energy conversion devices based on electrochemical processes, we will discuss the main parameters important in the definition of electrocatalyst activity. Sections 4.2 and 4.3 will review in some detail the properties desired from an electrocatalyst that can be potentially exploited in an electrochemical energy conversion and storage device. Particularly, a variety of aspects related to electrochemical kinetics, mass transport, and charge transport, which are connected to device efficiency will also be considered. These factors will be relevant when the reader tackles the chapters devoted to materials. Indeed, it will become clear that the state-of-the-art nanostructured electrode architectures are designed to best match the requirements laid down for fuel cell applications.

High surface area, high mass specific activity and capability to withstand high current densities without incurring mass transport limitations are some of the most relevant parameters which need to be accounted for when designing electrocatalytic materials.

And this is not the whole story. Electrode architectures that show outstanding performance may not be suitable candidates as other constraints must also be considered. Stability, possibility of being produced at reasonable costs, and manufacturability are further elements to be accounted for. Let us consider the following example. An electrode with a Pt loading of 1 mg cm^{-2} would certainly perform better than one with 0.1 mg cm^{-2}, provided that the support is the same as well as catalyst particle size, shape, and structure. Such a high Pt loading is not however feasible for a commercially viable system. At the same time it is clear

A. Lavacchi et al., *Nanotechnology in Electrocatalysis for Energy*, Nanostructure Science and Technology 170, DOI: 10.1007/978-1-4899-8059-5_4,

© Springer Science+Business Media New York 2013

that a small loss in performance is acceptable if the noble metal loading can be significantly reduced; a criteria important for producing a technology with acceptable costs and with some chance of commercial success.

At the very beginning of basic research it may not, of course, be possible to account for all these aspects. Nevertheless, we feel it is appropriate to consider these aspects in this book, to make the reader aware of the fact that a good candidate material, even if very promising on the lab scale, may not lead to successful technological or commercial exploitation. The proof of this is the fact that, till date, while very many nanostructured materials have been developed for hydrogen PEMFCs, together with their demonstration on the lab scale, the only ones with a reasonable technological development, and commercial viability are carbon black-supported Pt nanoparticles [1]. Furthermore, it may be worth mentioning that Pt/C after many decades is, in some sense, still under optimization as research and technology attempts to reduce the noble metal content are still in progress.

Other constraints will probably arise in the future as environmental and safety concerns about nanotechnology and nanoparticles become more evident. Recently, fundamental investigations into the toxicity of nanomaterials and in particular of nanoparticles have begun. This will probably lead to the application of regulations regarding the use and production of nanostructured materials, devoted both to hygiene and safety of the workplace and to the release into the environment of nanomaterials in any part of the life cycle of the device.

Even the energy used in the production process of nanomaterials plays a significant role. As illustrated in the introduction chapter (Chap. 1), the life-cycle assessment devoted to energy evaluation, accounts for the energy consumed in any stage of the life of the device, from the production of each single part through to decommissioning. Sometimes, approaches to the synthesis of nanostructured catalysts leading to excellent materials may be very energy intensive. This has a negative impact on the efficiency of the whole energy cycle.

This chapter ends with a short review on the nanotechnology approach in the search for a suitable match between technological constraints and performance targets for electrocatalytic materials.

4.2 Technology Targets

This section reviews in detail the targets imposed by the DOE that can be found in the following documents (see e.g. [2]). According to the DOE, tremendous research and technology efforts are required to produce commercially viable hydrogen-fueled PEMFCs. It has been pointed out that the most relevant issues to be tackled are *cost* and *durability*. Other relevant aspects are size and weight of the devices as well as the purity of feeding gases.

If we consider the main application for which PEMFCs are intended for, namely automotive, the situation is even more intricate. In the automotive sector size and weight constraints, as well as safety issues are more stringent than in any other application. Furthermore, operation occurs under intermittent conditions, with frequent start-up and shut-down cycles and continuous variation of delivered power. Such conditions put severe stress on fuel cell stack systems, posing limitations to durability.

In stationary power applications, operation may be performed at higher temperatures as compared to other PEMFC technologies improving heat and power cogeneration with beneficial effects to the overall energy efficiency. Nevertheless temperature may accelerate degradation of the MEA resulting, again, in durability issues.

The rise of portable electronic devices increases the demand for power supply capable of delivering power in the 2–250 W range. At present, such devices are powered by rechargeable batteries. Battery charge duration is limited in time, requiring frequent plugin to the electrical grid. This may be a problem if the grid is not present. Fuel cells are an excellent potential high-energy density alternative to the existing battery technologies. Nevertheless, research and technology development, with a special emphasis on the electrocatalysts is still necessary to fill the gap between energy demand and energy supply. Fuel cell systems for portable power applications (Portable Fuel cells, PFCs) fueled with methanol or hydrogen must have increased durability and reduced costs to compete with batteries.

Direct Methanol Fuel Cells (DMFCs) do not allow easy access to more than 100 W but PFCs fed with H_2 may release higher power. There is little doubt that a major issue hampering their commercial diffusion is the hydrogen source. Systems that make use of compressed hydrogen are not attractive for the need of carrying heavy and unsafe cylinders with limited autonomy (generally 1 or 2 normal cubic meters of hydrogen) and difficult recharging procedures.

On the side of hydrogen production, electrolysis is at present a commercial reality and electrolytic plants for water splitting are wide spread. Relative to electrolysis, the major research efforts needed are in the reduction of energy consumption and in the initial capital and operation and maintenance costs.

In the next section, we will focus on a synthetic and quantitative analysis of the main targets as defined by the DOE for the reported technologies. It is worth considering that many of the aspects presented can be related directly to the electrocatalyst and to the MEA that form the core components of an electrochemical device.

4.2.1 PEMFC

4.2.1.1 Durability

The current durability of hydrogen PEMFCs still does not meet DOE targets. Tolerance to air, fuel, and system-derived impurities needs to be improved. This is a very important condition, as in everyday operation, contamination of the fuel, oxygen, or air feed may occur. Some of the impurities are disastrous to the system (e.g., CO for PEMFCs) leading to quick degradation of the performance. Furthermore, operating under service conditions may include intermittent operation with frequent start-up and shutdown. This increases degradation phenomena associated with cathode corrosion. Materials for the mitigation of such phenomenon will be explicitly considered in Chaps. 5 and 6. Corrosion results from the presence of strongly oxidizing conditions due to the accumulation of oxidizing species at the cathode that ultimately lead to the oxidation of the carbon support of the platinum electrocatalysts.

Real systems show very large MEA geometric area, often in the range of a square meter or more. Local variations (usually on the mm scale) of the MEA ohmic resistance (often caused by aging or defects in the manufacturing process) may result in so-called hot spots. In the hot spots local current density may be significantly larger than the average current density, leading to localized quick deterioration of the MEA performance. This results in short circuits and mixing of products and reactants, phenomena that produce a reduction in the cell potential and ultimately the power output. In virtue of all these points, sufficient durability of fuel cell systems especially in automotive drive cycles, has not yet been demonstrated. As of 2011, no system was able to achieve operation longer than 2,500 h (Table 4.1), while the target for the application is 5,000 h. According to Table 4.2, by 2017, it will be considered acceptable a 40 % loss in the activity of MEA over the whole service life.

It is worth noting that not all the performance degradation mechanisms for PEMFC have been extensively investigated. Lack of information concerning degradation is an issue even for designing new materials. Degradation of PEMFC performance is still a huge and challenging research field [3] requiring the development of new investigation techniques capable of capturing phenomena at the nanoscale. Identical Location TEM (IL-TEM Fig. 4.1) is a technique very recently successfully applied to study the morphology evolution of cathode materials for PEMFCs [4, 5].

For stationary fuel cell systems durability has to exceed 60,000 h. This is the actual service life of state-of-the-art power generation systems. Here operating temperatures pose the major issues in terms of durability of the following fuel cells

Table 4.1 Status and targets for automotive hydrogen PEMFC according to the U.S. DOE

Characteristic	Units	2011 status	2017 targets	2020 targets
Energy efficiency 25 % of rated power	%	59	60	60
Power density	W/L	400	650	850
Specific power	W7 kg	400	650	650
Cost	$/kW	49		
Cold start-up time to 50 % of rated power				
@ −20 °C ambient temp	s	20	30	30
@ +20 °C ambient temp	s	<10	5	5
Start-up and shut-down energy	MJ	7.5	5	5
from −20 °C ambient temp	MJ		1	1
Durability in automotive drive cycle	h	2,500	5,000	5,000

Table 4.2 Status and targets for electrocatalytic material properties according to the U.S. DOE

Characteristic	Units	2011 Status	Targets	
			2017	2020
Platinum group metal total content (both electrodes)	g/kW (rated)	0.19	0.125	0.125
Platinum group metal (PGM) total loading	Mg PGM/cm electrode area	0.15	0.125	0.125
Loss in initial catalytic activity	% mass activity loss	48	40	40
Electrocatalyst support stability	% mass activity loss	10	10	10
Mass activity	A/mg Pt @ 900 mV iR-free	0.24	0.44	0.44
Non-Pt catalyst activity per volume of supported catalyst	A/cm^3 @ 800 mV iR-free	60 (measured at 0.8 V) 165 (extrapolated from 0.85 V)	300	300

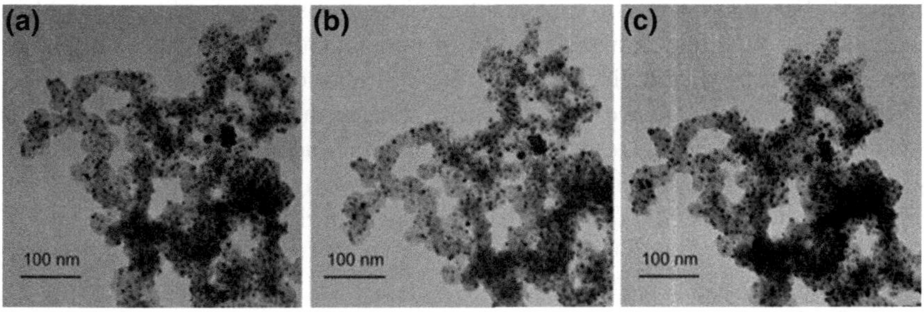

Fig. 4.1 IL-TEM micrographs of the carbon-supported Pt high surface area catalyst: **a** pristine sample before electrochemical measurements, **b** same layer after transfer to electrochemical cell and CO annealing for 60 min between 60 and 660 mV RHE, and **c** same layer after oxide annealing for 60 min to 1310 mV RHE with a respective scan speed of 20 mVs^{-1}. Reprinted from Ref. [5] with permission from Elsevier

materials: (i) electrolyte, (ii) electrolyte support and (iii) electrode. Improved durability under start-up and transient operation is also required. Research is also needed to understand failure mechanisms and to develop mitigation strategies. Suitable accelerated testing protocols have to be developed to enable projection of durability and to allow for timely iterations and improvements in the technology.

Challenges for fuel cells dedicated to portable power include (i) reducing cost (mainly by reducing precious metal catalyst loading), (ii) increasing efficiency (by reducing fuel crossover and increasing catalyst selectivity), and (iii) reducing the size of the system (Table 4.3). Total life cycle efficiency improvement would have a positive impact on emissions reduction during operation and disposal. The use of nonpolluting fuels such as ethanol, ethylene glycol and glycerol is appealing together their low inherent nontoxicity.

Table 4.3 Technical requirements for portable fuel cell devices according to the U.S. DOE

Characteristic	Units	2011 status	2013 targets	2015 targets
Specific power	W/kg	25	40	50
Power density	W/L	30	50	70
Specific energy	Wh/kg	250	440	640
Energy density	Wh/L	300	550	900
Cost	$/W	15	10	5
Durability	h	2,000	3,000	5,000
Mean time between failures	h	500	1,500	5,000

4.2.1.2 Cost

Both anode and cathode reactions in PEMFCs require precious metals to be effective. The minimum reported noble metal content as of 2011 is 0.15 mg cm^{-2} (Table 4.2). The target for 2017 is 0.125 mg cm^{-2}. There is little chance at the moment to replace noble metals in polymer electrolyte fuel cells. The fact that the technology operates in acidic conditions implies that just a few materials can withstand the corrosive environment. Operating in alkaline conditions could in principle be a solution, allowing the use of nonprecious metals especially at the cathode. Nevertheless, alkaline PEMFCs are still no more than a research subject and the technology is still very far from the state of the art in acidic conditions.

Nanotechnology may here provide solutions to reduce the amount of platinum by developing materials capable of withstanding long-lasting operations, even under intermittent operation, with enhanced mass specific activity.

4.2.1.3 Performance

Fuel cell systems for practical applications need to show efficiency as high as 60 % when operating at 25 % of the maximum rated power (Table 4.1). Poor cathode kinetics may be one of the major obstacles in reaching such efficiency. Overpotentials of 0.4 V or greater may occur in state-of-the-art PEM fuel cells operating under typical conditions. This overpotential represents a loss at the cathode of approximately one-third of the theoretically available energy from a fuel cell. This is the reason why a huge research effort is devoted worldwide to the development of novel nanostructured catalytic materials for the ORR.

4.2.2 Electrolysis

Issues in electrolysis are somewhat different than those regarding PEMFCs. No precious metals are usually employed in commercial alkaline electrolysis systems. This is because it is considered acceptable to have an electrode with lower activity but a much lower cost. Most of the energy losses in electrolytic devices are due to the ionic conductivity of the electrolyte and, especially in the case of the alkaline electrolysis, to the formation of a large amount of bubbles at the electrode surface and in the electrolyte. For these reasons, research in electrolysis is now directed to the development of polymer electrolyte electrolyzers and "zero gap" devices. Considering the state of the art of membrane technology PEM electrolyzers have to operate in acidic conditions, where, as in the case of PEMFC, corrosion is a problem for the vast majority of nonprecious metal catalytic materials. The most relevant issue in water electrolysis is the cost of produced hydrogen which, at present, cannot compete with reforming technologies. Following is a detailed description of the main issues actually limiting the diffusion of commercial electrolytic technologies.

4.2.2.1 Main Issues Hampering the Commercial Diffusion of Electrolysis

The cost of electrolytic devices is still very high and plant investment contributes significantly to the hydrogen cost. According to DOE (Table 4.4), capital costs have to be reduced by approximately 30 % by 2017.

In electrolytic hydrogen production the most relevant cost component is the electrical energy consumption. Improvements in efficiency are hence required to lower the energy consumption. As of 2011, the maximum efficiency of electrolyzers was approximately 65 %. Such a value needs to be increased to 75 % by 2017. On the side of electrical energy consumption for hydrogen production, this means that by 2017 hydrogen should be produced with an energy cost of 44 kWh kg^{-1}. To increase the volume energy density of hydrogen as a fuel, compression after production is required. The target for hydrogen storage in the automotive sector is 700 atm. This requires a compression stage after the electrolytic hydrogen production. Mechanical compression requires a lot of energy even if performed with modern multistage compressors. An increase of the operating pressure of electrolysis systems would reduce such a contribution, increasing the overall energy efficiency. The operations and maintenance costs for electrolysis are currently too high. Durability, maintenance, reliability, and demand management are

Table 4.4 Status and future technical requirements for electrolyzers according to the U.S. DOE

Characteristics	Units	2011 status	2015 targets	2020 targets
Hydrogen levelized cost (production only)	$/kg	4.20	3.90	2.30
Electrolyzer system capital cost	$/kg	0.70	0.50	0.50
	S/kW	430	300	300
System energy efficiency	% (LHV)	67	72	75
	kWh/kg	50	46	44
Stack energy efficiency	% (LHV)	74	76	77
	kWh/kg	45	44	43

similar to those of the distributed natural gas reforming systems. Operating efficiency, component durability, purification of water, and transients and changes in duty cycles need to be addressed. Barriers in control and safety include the efficiency of start-up and shut-down processes, turn-down capability, and the capability for rapid on–off cycling. Control and safety costs still remain high due to complex system designs and high-cost sensors. For commercialization of this technology, reliability and safety of these units is a key qualification target.

4.3 Main Electrocatalyst Aspects Affecting Design

The observed current density of an electrode at a given potential can be described by Eq. (4.1) [1].

$$J = f \cdot S \cdot i. \tag{4.1}$$

The current density J in a MEA is then the result of the product of the electrode's electrochemically active surface area S (in cm^2 of Pt per planar cm^2), the exchange current density i, and a [6] collision frequency scaling factor $f(\lambda, \rho_S)$. The collision frequency scaling factor, as stated in [1] "…depends on the mean free path length λ above the catalyst surface and on a spatial distribution function of surface area that to first order can be approximated by the catalyst surface area volumetric density ρ_S (in cm^2 Pt per cm^3)." The factor $f(\lambda, \rho_S)$, calculated from catalyst electrode physical properties, captures one way in which geometry has a differentiating role in comparing different electrocatalyst designs.

Kinetic activity is measured generally by half cell electrochemical measurements such as cyclic and linear voltammetry in stagnant electrolyte or with carefully controlled convection conditions (e.g., RDE, RRDE) [7–10]. Such approaches have been described in Chap. 2. Kinetic parameters may also be estimated by measuring properties "in situ", namely during fuel cell or electrolyzer operation. This offers the chance to monitor the evolution in the status of the catalysts under real operating conditions. "In situ" measurements are usually less reliable and objective but on the other side relating half-cell measurements to fuel cell performance is not always a straightforward task.

Changes in the surface structure of metals and alloys are known to produce variations in the catalytic and electrocatalytic activity of metals. Many classical electrochemical investigations in the past dealt with electrocatalysis at single crystal surfaces. Such single crystal investigations allowed us to understand that not all crystal faces behave the same in electrocatalysis. Electrocatalysts with selected terminations may exhibit faster kinetics with enhanced mass specific activity. This leads us to the conclusion that selecting the most active termination may lead to the chance of reducing the mass of noble metal in the electrocatalyst.

Transport issues in the PEMFCs catalysts layer may significantly condition the performance of the system. The three main mass transport issues are: (i) water flooding, i.e., liquid water entrapped inside the electrodes or flow channels which interrupts the flux of reactant/product gases; (ii) dilution of oxidant concentration due to the use of air instead of pure oxygen; (iii) depletion of reactants along the flow channel, which results in nonuniform current distribution over the whole electrode area, and is particularly severe in a large-scale fuel cell. Water flooding often acts as the main cause of serious performance drop at high current densities. The issue of diffusion, electronic and ionic conductivity are also of fundamental importance and it will be shortly discussed in Sect. 4.3.3.

4.3.1 Electrochemically Active Surface Area

If no mass transport limitation occurs the actual current density passing through an electrode at any given potential is roughly proportional to the electrode real surface area. In turn this means that at a given current density the activation overpotential for a given reaction will be lower meaning that a fuel cell operating with an electrode with a larger surface area will deliver larger power density. Large EASA is advantageous also in electrolysis, where with larger EASA a lower potential will be required to guarantee a given production of hydrogen.

The reactions occurring in fuel cells and electrolyzers take place at the topmost surface atomic layer of the electrocatalyst. Only the surface atoms determine the reactivity. Bulk atoms, at least in a first approximation, may be considered mere spectators. The consequence of this is that electrocatalysts have to show surface-to-volume ratios as large as possible. Such a condition allows the attainment of large mass specific activities. This makes also immediately clear why nanotechnology is so relevant to electrocatalysis.

The ability to control matter in such a way that at least one dimension is in the nanometer (10^{-9}) m range permits us to increase the surface-to-volume ratio. In the end with just a tiny amount of metal incredibly high surface areas may be obtained. The most widespread category of electrocatalyst used nowadays is the one based on Pt or Pt-alloy nanoparticles dispersed on standard or graphitized carbon black support particles. Pt nanoparticle diameters commonly range between 2 and 4 nm. The possibility of obtaining such small nanoparticles is strongly related to the morphology and the surface chemistry of the support material. In this sense high surface area carbon has been proved to be very effective. Platinum loading in such materials usually ranges between 20 and 60 % wt. State-of-the-art commercial Pt/C catalysts have surface areas ranging between 80 and 120 $m^2\ g^{-1}$ Pt with specific mass activities between 0.1 and 0.12 A mg^{-1} Pt as determined in MEAs from complete fuel cell tests. Homogeneous Pt-alloy nanoparticles on carbon have historically been observed to increase ORR activity over pure Pt/C by about a factor of 2 to 2.5. The commercially available, heat-treated 30 wt % PtCo/C system routinely provides MEA measurement values of specific and mass activity that are close to the DOE targets, for example, 1.2 mA cm^{-2} Pt and 0.39 A mg^{-1} Pt. For a wide range of carbon-based supports, initial Pt surface areas are from 20 to over 70 $m^2\ g^{-1}$ Pt. After stability testing these values converge to just 20–30 $m^2\ g^{-1}$ Pt [11].

The value of the EASA for Pt may be determined by electrochemical methods or physisorption and chemisorption. In electrochemistry, the charge due to the stripping of species known to specifically adsorb at the surface may be exploited. CO stripping is a common and reliable procedure for determining the platinum surface area. Analogous procedures have been developed for many other metals. All of them are based on the deposition or stripping of species which are known to have a limited growth and whose amount is a known function of the surface area. An analogous principle is applied to chemisorption experiments. Platinum specific surface area may be measured by determining the amount of CO adsorbed by a sample whose weight is precisely known.

Physisorption refers to interactions which are not specific to the nature of the surface. In materials such as platinum supported on high surface area carbon, physisorption provides the surface area of the whole sample, namely the area of carbon plus the area of platinum minus the contact surface which cannot be accessed by the probing gas (e.g., helium, nitrogen, or argon). Physisorption also allows the estimation of the pore size in a sample and for this reason is fundamental in assessing the morphological features of the multiscale porosity of electrodes, which as demonstrated by Bacon, are so important in fuel cells [12].

The knowledge of the real electrochemically active surface area is essential in determining the fundamental properties of electrocatalysts. Indeed normalization of current density against EASA is an important operation which allows us to discriminate between structural and promotional effects and simple surface extent effects when comparing different materials.

4.3.2 Surface Defects, Surface Structure and Particle Shape

The performance of nanocrystals used as catalysts depends strongly on the surface structure of facets enclosing the crystals. Thermodynamics usually ensures that crystal facets evolve to have the lowest surface energy during the crystal growth process. For fcc metals (such as Pt, Ag, Pd, Au, and most of transition metals), the most common crystal terminations are (111), (110), and (100). Such terminations show different atomic arrangement at the surface. Furthermore, they are not energetically equivalent. In terms of catalytic behavior they may show largely different behaviors. As an example, Pt (111) has been found to be the most active surface for ORR in alkaline environment [13].

The (100) surface is that obtained by cutting the *fcc* metal parallel to the front surface of the *fcc* cubic unit cell—this exposes a surface (lighted atoms) with an atomic arrangement of 4-fold symmetry. Figure 4.2 shows the conventional bird's-eye view of the (100) surface—this is obtained by rotating the preceding diagram through 45° to give a view which emphasizes the 4-fold (rotational) symmetry of the surface layer atoms. The tops of the second layer atoms are just visible through the holes in the first layer, but would not be accessible to molecules arriving from the gas phase. Each surface atom has four nearest neighbors in the first layer, and another four in the layer immediately below; a total of 8. This contrasts with the CN of metal atoms in the bulk of the solid which is 12 for a *fcc* metal. The (110) surface is obtained by cutting the *fcc* unit cell in a manner that intersects the *x* and *y* axes but not the *z*-axis—this exposes a surface with an atomic arrangement of 2-fold symmetry. The conventional bird's-eye view of the (110) surface—emphasizing the rectangular symmetry of the surface layer atoms is shown in Fig. 4.2. The diagram has been rotated such that the rows of atoms in the first atomic layer now run vertically, rather than horizontally as in the previous diagram. It is clear from this view that the atoms of the topmost layer are much less closely packed than on the (100) surface—in one direction (along the rows) the atoms are in contact, i.e., the distance between atoms is equal to twice the metallic (atomic) radius, but in the orthogonal direction there is a substantial gap between the rows. This means that the atoms in the underlying second layer are also, to some extent, exposed at the surface. The preceding diagram illustrates some of those second layer atoms, exposed at the bottom of the troughs. In this case, the determination of atomic coordination numbers requires a little more careful thought: one way to double-check your answer is to remember that the CN of atoms in the bulk of the *fcc* structure is 12, and then to subtract those which have been removed from above in forming the surface plane. If we compare this coordination number with that obtained for the (100) surface, it is worth noting that the surface atoms on a more open ("rougher") surface have a lower CN—this has important implications when it comes to the chemical reactivity of surfaces. Do the atoms in the second layer have the bulk coordination? No, indeed the fact that they are clearly exposed (visible) at the surface implies that they have a lower CN than they would in the bulk. Each surface atom has two nearest neighbors in the 1st layer, and another four in the layer immediately below, and one directly below it in the third layer; this gives a total of seven. To confirm this consider those that have been removed from the layers above—clearly there would have been four nearest neighbors in the layer immediately above the surface layer (equivalent to the four in the layer immediately below). In addition, there would have been one nearest neighbor directly above each surface atom (equivalent to the one directly below in the third layer). Hence, 7 (present) + 5 (removed) = 12—which is correct! The (111)

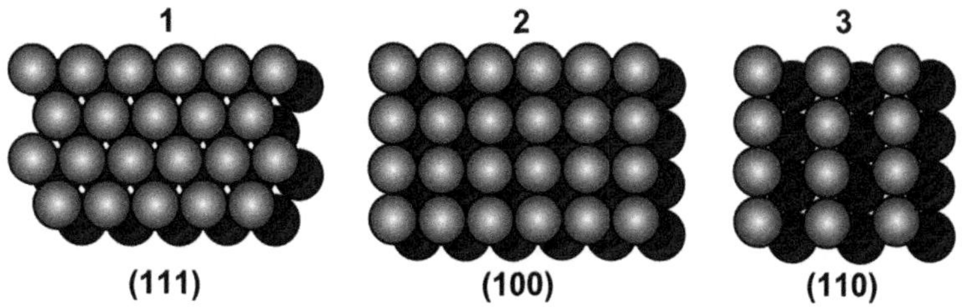

Fig. 4.2 Ball model of low index crystal termination

surface is obtained by cutting the *fcc* metal in such a way that the surface plane intersects the *x*-, *y*-, and *z*- axes at the same value—this exposes a surface with an atomic arrangement of 3-fold (apparently 6-fold, hexagonal) symmetry. This layer of surface atoms actually corresponds to one of the close-packed layers on which the *fcc* structure is based. The diagram below shows the conventional bird's-eye view of the (111) surface—emphasizing the hexagonal packing of the surface layer atoms. Since this is the most efficient way of packing atoms within a single layer, they are said to be "close-packed." Each surface atom has six nearest neighbors in the first layer, and another three in the layer immediately below; a total of nine.

The following features are worth noting;

1. All surface atoms are equivalent and have a relatively high CN.
2. The surface is almost smooth at the atomic scale.
3. The surface offers the following adsorption sites:
   - On-top sites.
   - Bridging sites, between two atoms.
   - Hollow sites, between three atoms.

The surface of crystals usually exhibits a variety of defects in the atomic arrangement that may play a significant role in catalysis. Usually, the atoms where the defect is located show a lower surface coordination leading to significant differences in the reactivity of the site. According to what is reported in Fig. 4.3 atoms at the crystal surfaces may be located in a variety of positions classified as: (i) terrace, (ii) step, or (iii) kink (adatom, atomic island located at terraces, may also occur but are not relevant to catalysis). Let us examine the case of the 111 termination on a *fcc* crystal (most metals of relevant catalytic interest show this structure). The coordination number of surface atoms on the terraces is nine (coordination number in bulk fcc is 12) (Fig. 4.4). Each step atom has four nearest neighbors in the surface layer of terrace atoms which terminates at the step, and another three in the layer immediately below which result in a total of seven. The lowest coordination number is exhibited by atoms "on the outside" of the kinks in the steps. Such atoms have only three nearest neighbors in the surface layer of terrace atoms which terminates at the step, and another three in the layer

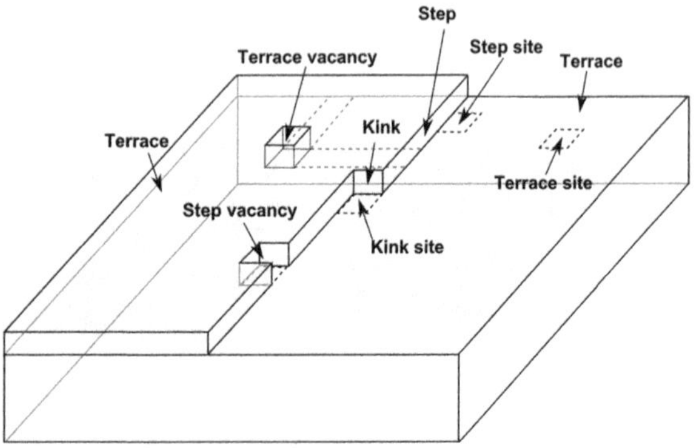

Fig. 4.3 Schematic view of a crystal surfaces with the main surface defects

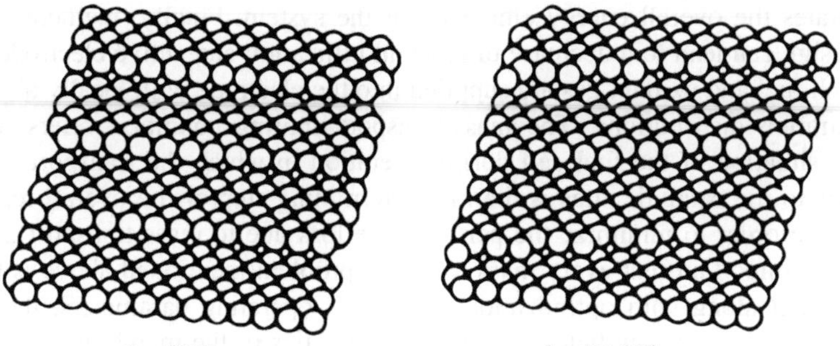

Fig. 4.4 High index termination surfaces examples

immediately below which results in a total of six. Hence atoms located at the surface are unsaturated as compared to the bulk and the saturation varies in dependence of the type if surface defect. Real vicinal surfaces do not, of course, exhibit the completely regular array of steps and kinks illustrated for the ideal surface structures, but they do exhibit this type of step and terrace morphology. The special adsorption sites available adjacent to the steps are widely recognized to have markedly different characteristics to those available on the terraces and may thus have an important role in certain types of catalytic reactions. This contrasts with the surface atoms on the terraces which have a coordination number of nine and the normal step atoms which have a coordination number of seven.

As particle dimensions reduce toward the nanoscale, the surface-to-volume ratio increases. Not all the surface sites have the same coordination. The catalytic performance of the nanostructured materials can also be adjusted by modification of the shape of nanoparticles, which determines surface atomic arrangement and coordination [14–16].

Understanding the nanoscale topography of surface sites, such as terraces, steps, kinks, adatoms, and vacancies, and their effects on catalytic and other physicochemical properties is the key to designing nanoscale functional materials by nanotechnology [15, 17–19].

4.3.3 Transport Issues

Once liquid water fills up the pores in the electrode, this liquid barrier produces a dead reaction zone and reduces the effective electrode area. Secondly, even though it is essential, from practical aspects, to develop air-operated PEMFCs, the power density produced in air-fed regimes is relatively low compared with that of pure oxygen operation. This is because the decreased partial pressure of oxygen slows down the oxygen reduction reaction and rapidly degrades cell performance. Furthermore, pressurization of the cathode side circumventing this problem deteriorates the overall energy efficiency of the system. Finally, inefficient flow field design can lead to a decrease of reactant utilization and active electrode area by which fuel and oxidant gas are supplied into the electrode with locally different concentration distributions. The mass-transport-limiting condition refers to the internal situation of the fuel cell during operation in which the transport rate of reactants to the catalytic sites is the rate-determining step of the overall reaction process. In order to minimize the performance loss due to mass transport limitations, it is important that every mass transport problem mentioned above is taken into account along with optimization of the cell operating parameters, the configuration of the gas flow field, and the characteristics of the membrane electrode assemblies (MEAs) [20]. The anode water removal technique developed by Voss et al. [21] and the interdigitated (dead-ended) flow field design devised by Nguyen and co-workers [22, 23] are two different approaches proposed for water management in PEMFCs. An equally important factor for efficient mass transport management is to tailor the electrode structure through the control of material properties, composition, preparation methods, etc. To design the electrode structure for PEMFCs, it should be recognized that the electrode is a field where multicomponent and multiphase flows are taking place. Vapor or liquid phase water is, along with fuel and oxidant gas, entering and exiting through the diffusion path in the electrode. Electrodes used for PEMFCs have, in general, two- or three-layered structures that can be divided into two parts, viz., catalytic and noncatalytic. The catalytic part, i.e., the catalyst layer, is formed by depositing a mixture of carbon-supported platinum catalyst and solubilized electrolyte on one side of the noncatalytic part; electrons, protons, and water molecules are produced in this region. By contrast, a hydrophobic carbon layer interposed between the catalyst layer and the backing substrate, i.e., the diffusion layer, though it has no catalytic reaction sites, is known to play a key role in providing good access of reactants toward catalytic sites and effective removal of product water from the electrode [24–32]. Among the cited investigations many have demonstrated the importance of morphology of diffusion layer. Nevertheless, there have been very different interpretations and experimental results regarding the relationship between the pore structure of the diffusion layer and the cell performance.

The need to minimize cell resistivity has a major impact on the selection and processing of the cell components. Cost-effective fabrication of porous electrode structures was achieved for the first time only about 40 years ago. The electrolyte, gaseous reactants, electrocatalyst, and current collector have to be brought into close contact within a confined spatial region termed the triple-phase-boundary interface. For the low-temperature systems, the introduction of hydrophobic polytetrafluoroethylene (PTFE or Teflon) greatly simplified the fabrication of porous, liquid-resistant gas-diffusion structures. Metal or carbon powders (or porous carbon papers) provided the electronic pathways, and to further reduce the Area Specific Resistance (ASR) of the electrode a metallic wire mesh or screen was usually incorporated into the structure. Further improvements in performance were obtained during the 1960s by depositing small crystallites (2–5 nm) of the electrocatalyst (usually platinum or Pt alloys) onto carbon powder or paper. In retrospect, this accomplishment was probably the first manifestation of an engineered nanostructure, and it is not surprising that its implementation more than 40 years ago was so difficult.

Although electronic conductivity, that is, the inverse of resistivity, should be measured in a direction perpendicular to the catalyst layer, this measurement is technically difficult because the probes for measuring resistance must be set accurately so as to sandwich the thin and soft catalyst layer. In addition, contact resistance between the probe and the catalyst layer cannot be eliminated when the two-probe method is used, and it is impossible to apply the four-probe method to such a thin layer. An alternative is to measure the in-plane resistance. This approach is valid if the resistance can be regarded as isotropic.

High ionic conductivities (1 S cm^{-1}) associated with the liquid KOH, phosphoric acid, and molten carbonate electrolytes ensured that, with appropriate design strategies, the ASR values of these components can be small. Although exhibiting lower specific ionic conductivity values, the Nafion membrane used in the PEMFC system can be fabricated relatively easily as a thick film (100 μm) to produce satisfactory ASR values, provided the water content of the film is controlled under the dynamic conditions of cell operation.

4.4 Constraints Affecting Design

4.4.1 Precious Metal Loading

The most effective electrocatalysts for fuel cell applications require precious metals, principally platinum and its alloys. A recent evaluation has shown that among PEMFC components around 56 % of the cost comes from the electrocatalyst [2]. At present, platinum is still irreplaceable due to both its high performance and excellent durability. This is one of the most relevant obstacle to the commercialization of fuel cells and in some sense may affect the sustainability of the technology. This has generated a strong push in the last few decades to the search for new and cheaper catalyst materials [33]. The quest for new catalysts is especially concentrated on the ORR as this is the half-cell reaction most important to fuel cell electrical energy production, limiting the performance due to its high overpotential due to sluggish reaction kinetics [34, 35]. As pointed out before, the cathodic catalyst is the one where most of the platinum is concentrated. Two main approaches have been pursued by researchers to mitigate this problem [36] (i) attempting to lower the precious metal content in the electrode without lowering performance and (ii) pursuing a radical approach consisting in the development of nonprecious metal catalysts (NPMCs) for the ORR. The former proposal aims at increasing the platinum utilization of the cathode catalyst such that the active electrochemical surface area of platinum can be maximized. Practically, this method has been carried out by utilizing several different techniques. One technique is to disperse platinum nanoparticles onto the carbon support such that the resulting platinum particle sizes are much smaller. This method has yielded only mediocre results [37]. Another method couples platinum with other materials and supports to increase platinum utilization and to assist in the removal of hydroxides from the platinum surface to increase the availability of platinum active sites. Materials such as ruthenium, tin, and tungsten, etc., have been used [38–42]. The work from these two methods has resulted in the development of ultra-low platinum loading catalysts for the ORR and show dramatic decreases in platinum loadings with minimal performance losses. However, due to the rapidly increasing cost of platinum, the work in this area has been negated overtime by the change of platinum price. Many research institutions have tried to develop NPMCs which are active for the ORR with performances similar to platinum. Of these two proposed alternatives, the latter choice is the optimal solution for long-term development of fuel cells catalysts. Still, nonprecious metal catalysts lack the performance and stability of platinum catalysts, an ongoing issue that is currently the subject of much research activity [43, 44].

4.4.2 Stability

The highly acidic environment (pH $\sim$ 0) combined with an electrochemical potential window of >1.23 V encountered at the oxygen electrode of both PEMFC and PEM water electrolyzers provides an extremely corrosive environment and most materials tend to be unstable under these highly stringent electrochemical conditions. It is desirable, therefore, that catalyst particles and supports exhibit high electronic conductivity combined with excellent chemical and electrochemical stability. Very few materials exhibit the desired electrical conductivity and electrochemical stability especially at 1.8–2.0 V. One of the main advantages of developing alkaline systems is that the range of stable materials is much wider.

4.4.3 Scale-up and Manufacturing

As fuel cell systems are now approaching true commercialization, especially in the automotive sector, issues such as *process scalability* and *quality control* of any new catalyst approach must be seriously considered from the outset. For many of the new catalyst synthesis methods detailed in this book, which have been developed at the "lab bench" stage, it is not clear at all whether these processes are at all feasible to be scaled to the levels required industrially. A recent review paper in nature magazine by Mark K. Debe entitled *Electrocatalyst approaches and challenges for automotive fuel cells,* very nicely defined the potential production scales necessary for MEAs in the automotive sector [1]. Debe outlines that an annual production of 15 million fuel cell vehicles each with a stack containing 300

MEAs (each around 300 cm^2) would require 4.50 billion MEAs a year. That means the required production rate is about 11,700 MEAs per minute. As far as Pt loadings go, the target loadings of about 0.1 mg Pt per cm^2 mean very thin electrode thicknesses (less than 2 mm) requiring precision coating methods with critical limits on debris and tolerances. Time-consuming process steps for electrode formation that require hot bonding, annealing, solvent evaporation, or drying steps will not be feasible. The quantity of Pt required in such processes is also certainly breath taking. At loading targets of even 0.125 mg cm^2 on MEAs with 300 cm^2 active areas, catalyst flow-through rates of 1.5 kg of Pt per hour are required—roughly US$2 million worth of Pt per day per manufacturing line, at a metal price of US$ 2,000 troy ounces. On-site recycling and recycling of Pt from fuel systems at end of life will be essential. Finally, issues associated with safety, environmental, and cost control will exclude certain coating processes such as those using flammable solvents. All of the above considerations regarding essential process requirements apply not only to catalyst synthesis methods but also to catalyst-membrane integration and the fabrication and integration of other stack components as well.

4.5 The Potential of Nanotechnology in Electrocatalyst Design

The basic designs for platinum catalysts both unsupported (Fig. 4.5) and supported (Fig. 4.6) can be categorized by overall geometry of the catalyst and its support and then further subdivided according to structural morphology and composition. It is important to emphasize that the nanoengineering approach differs from many traditional approaches to the preparation of supported catalysts in the enhanced abilities to control the particle size, shape, composition, phase, and surface properties. The fact that the bimetallic AuPt nanoparticle system displays a single-phase character different from the miscibility gap known for its bulk-scale counterpart and that phase-segregated or core–shell nanostructured bimetallic particles can be formed serve as important indications of the nanoscale manipulation of the phase properties of the catalysts. This finding and other related observations are being exploited for refining the design and preparation of various bimetallic or trimetallic catalysts. In particular, the insight gained from probing how the interactions between the nanoparticles and the planar substrate dictate the size evolution in the activation process serves as an important guiding principle for the design and control of different nanoparticle-support combinations. More importantly, the fact that some of the nanoengineered multimetallic nanoparticle catalysts exhibit electrocatalytic activities for fuel cell reactions which are 4–5 times higher than pure Pt catalysts constitutes the basis for exploration of a variety of multimetallic combinations.

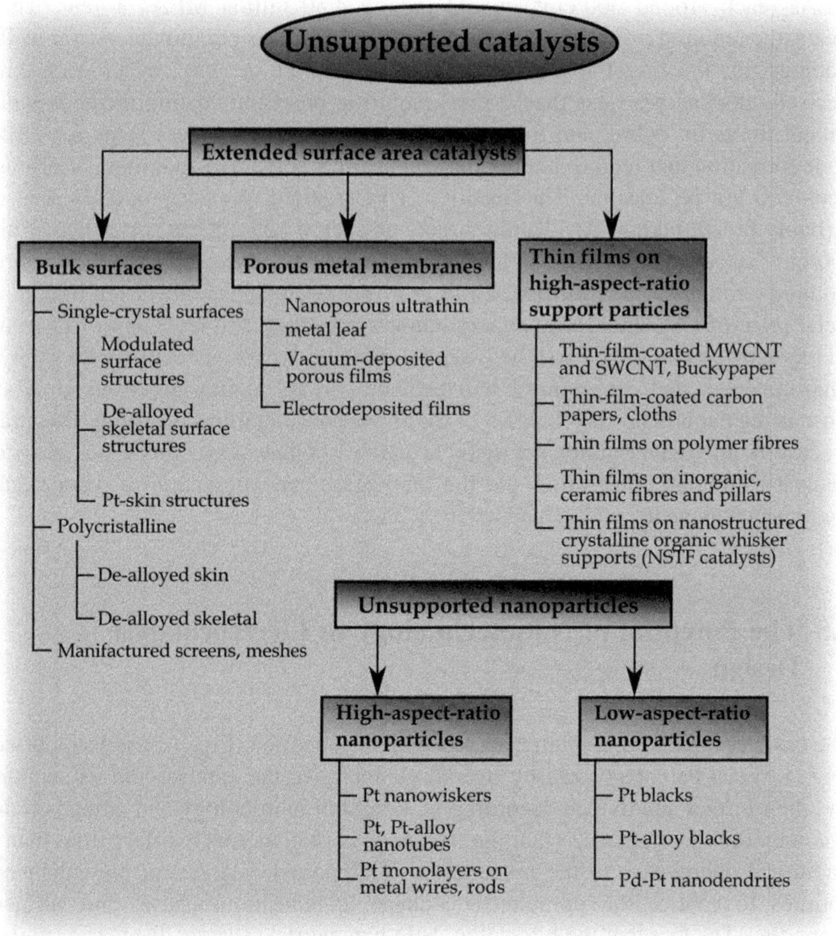

Fig. 4.5 A representative variety of nanostructured unsupported Pt electrocatalysts

An overview of the current status and future directions indicates that significant progress has been made in many areas of fundamental research into fuel cell engineering of nanostructured catalysts and the ultimate commercialization of fuel cells looks promising. However, the realization of this optimistic view has still to address challenges in several important areas. One of the most important areas is the need for a balanced and integrated approach to both fundamental and engineering aspects of the research and development of the catalysts. Fundamental investigations into the electronic and lattice structures of well-defined single crystal surfaces or selected bimetallic nanocrystals, and their characteristics such as lattice shrinking, skin effect, d-band vacancy, and so on, have generated

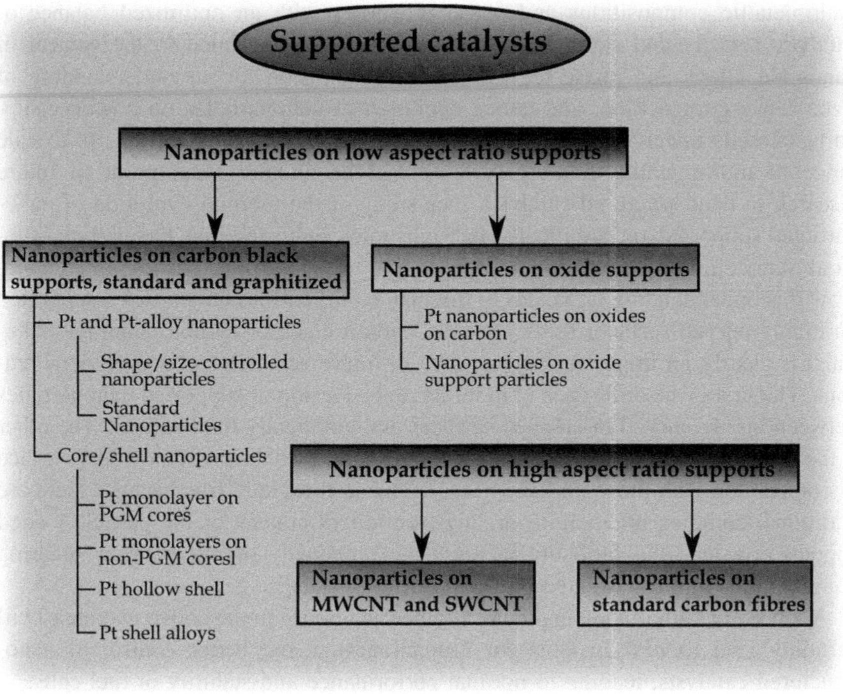

Fig. 4.6 A representative variety of nanostructured supported Pt electrocatalysts

valuable information for defining catalyst design parameters in order to achieve the highest electrocatalytic activity for fuel cell reactions.

However, insights from the well-defined single crystal systems are often subject to complications when dealing with practical evaluation of the catalysts where a balance of activity and stability is essential. On the other hand, research and development in fuel cell engineering has enriched the database for assessing the optimal performance of catalysts in terms of cell voltage, power density, durability, and cost. Such a database is very useful for implementing practical applications of the catalysts, but current efforts have provided limited feedback to help with the refinement of the catalyst design parameters. This situation is largely because of the lack of understanding of the detailed structures and degradation mechanisms of the catalysts in the membrane electrode assembly under fuel cell operation conditions. As already discussed in recent reports addressing some of the current challenges [45, 46], part of the future research on catalyst design will show an increased effort that focuses on optimizing the alloy compositions toward an optimal balance of catalytic activity and stability. In addition to combinatorial screening of bimetallic or trimetallic compositions from a more engineering perspective, the fundamental insights into the control of nanoscale alloy and phase-segregation or core–shell structures, as discussed for selected example systems in this book, will help scientists and engineers to identify combinations of multimetallic compositions and phase properties with an optimized balance of catalytic activity and stability. In another area closely related to the control of nanoscale alloys and phase-segregation, the expansion of our understanding of size, shape, composition, and lattice evolution of nanoparticles on powdery support materials under various thermal or chemical treatment conditions to a wide range of multimetallic catalyst systems will be another focal point of future research in nanostructured catalysts. The study of the thermal evolution of nanoparticles supported on atomically flat substrates indicates that the particle–substrate interaction is very important in addition to the particle–particle interaction. How this understanding translates to thermal evolution for nanoparticles supported on highly curved surfaces (e.g., powdery carbon black or carbon nanotube materials) is clearly an important subject area of future research for two obvious reasons. The first is the difference in mobility and adhesion of supported nanoparticles between highly curved or stepped surfaces and atomically flat surfaces. The other is the complication in probing the surface mobility or adhesion on powdery surface in comparison with those on flat surfaces. This is especially true because there are additional complex interactions arising from direct contact of the particle-loaded support powders that probably increase the propensity of interparticle sintering comparing to that for isolated particles on the surface.

Addressing both fundamental and engineering issues in the above two areas will certainly lead to new insights for rational design and better control of nanostructured catalysts, leading to optimal performance and stability of fuel cells.

In future studies, such insights will also be enriched by density functional theory computation [47, 48] fundamental electrochemical characterization of the electrocatalytic activity of the catalysts for fuel cell reactions under different conditions, and the evaluation of durability and degradation mechanisms of catalysts in fuel cells [49]. Such an integrated and balanced approach to the fundamental and engineering research into catalysts will serve as an effective vehicle for identifying low-cost, active, and robust catalysts for the ultimate commercialization of fuel cells. The same discussion can also be applied to catalyst development for electrolyzers.

References

1. M.K. Debe, Electrocatalyst approaches and challenges for automotive fuel cells. Nature **486**, 43 (2012)
2. M.K. Debe, *2009–2011 Annual Merit Reviews DOE Hydrogen and Fuel Cells and Vehicle Technologies Programs: Advanced Cathode Catalysts and Supports for PEM Fuel Cells*, (2011)
3. R. Borup et al., Scientific aspects of polymer electrolyte fuel cell durability and degradation. Chem. Rev. **107**, 3904 (2007)
4. K.J.J. Mayrhofer et al., Non-destructive transmission electron microscopy study of catalyst degradation under electrochemical treatment. J. Power Sources **185**, 734 (2008)

5. K.J.J. Mayrhofer, M. Hanzlik, M. Arenz, The influence of electrochemical annealing in CO saturated solution on the catalytic activity of Pt nanoparticles. Electrochim. Acta **54**, 5018 (2009)
6. M.K. Debe, Effect of electrode surface area distribution on high current density performance of PEM fuel cells. J. Electrochem. Soc. **159**, B54 (2012)
7. H.A. Gasteiger, S.S. Kocha, B. Sompalli, F.T. Wagner, Activity benchmarks and requirements for Pt, Pt-alloy, and non-Pt oxygen reduction catalysts for PEMFCs. Appl. Catal. B-Environ. **56**, 9 (2005)
8. K.J.J. Mayrhofer et al., Measurement of oxygen reduction activities via the rotating disc electrode method: from Pt model surfaces to carbon-supported high surface area catalysts. Electrochim. Acta **53**, 3181 (2008)
9. Y. Garsany, O.A. Baturina, K.E. Swider-Lyons, S.S. Kocha, Experimental methods for quantifying the activity of platinum electrocatalysts for the oxygen reduction reaction. Anal. Chem. **82**, 6321 (2010)
10. A.J. Bard, L.R. Faulkner, *Electrochemical methods : fundamentals and applications*, 2nd edn. (Wiley, New York, 2001), pp. xxi, 833 p
11. F.T. Wagner, Automotive challenges and opportunities for oxygen reduction catalysts. First CARISMA International Conference, La Grande Motte, France, 23 September 2008
12. F. T. Bacon, Fuel cells: Will they soon become a major source of electrical energy? Nature **186**, 589 La Grande Motte, France, 23 September 1960
13. R. Rizo, E. Herrero, J.M. Feliu, Oxygen reduction reaction on stepped platinum surfaces in alkaline media. Phys. Chem. Chem. Phys. **15**, 15416 (2013)
14. R. Narayanan, M.A. El-Sayed, Shape-dependent catalytic activity of platinum nanoparticles in colloidal solution. Nano Lett. **4**, 1343 (2004)
15. G.A. Somorjai, Surface science. Science **201**, 489 (1978)
16. F.J. Vidal-Iglesias et al., Shape-dependent electrocatalysis: ammonia oxidation on platinum nanoparticles with preferential (1 0 0) surfaces. Electrochem. Commun. **6**, 1080 (2004)
17. F. Tao, M. Salmeron, In situ studies of chemistry and structure of materials in reactive environments. Science **331**, 171 (2011)
18. D.L. Feldheim, The new face of catalysis. Science **316**, 699 (2007)
19. Y.-N. Wen, J.-M. Zhang, Surface energy calculation of the fcc metals by using the MAEAM. Solid State Commun. **144**, 163 (2007)
20. C.S. Kong, D.Y. Kim, H.K. Lee, Y.G. Shul, T.H. Lee, Influence of pore-size distribution of diffusion layer on mass-transport problems of proton exchange membrane fuel cells. J. Power Sources **108**, 185 (2002)
21. H.H. Voss, D.P. Wilkinson, P.G. Pickup, M.C. Johnson, V. Basura, Anode water removal: a water management and diagnostic technique for solid polymer fuel cells. Electrochim. Acta **40**, 321 (1995)
22. D.L. Wood Iii, J.S. Yi, T.V. Nguyen, Effect of direct liquid water injection and interdigitated flow field on the performance of proton exchange membrane fuel cells. Electrochim. Acta **43**, 3795 (1998)
23. T.V. Nguyen, A gas distributor design for proton-exchange-membrane fuel cells. J. Electrochem. Soc. **143**, L103 (1996)
24. D.M. Bernardi, M.W. Verbrugge, Mathematical model of the solid-polymer-electrolyte fuel cell. J. Electrochem. Soc. **139**, 2477 (1992)
25. Y.W. Rho, O.A. Velev, S. Srinivasan, Mass transport phenomena in proton exchange membrane fuel cells using O2/He, O2/Ar, and O2/N2 mixtures. I. Experimental analysis. J. Electrochem. Soc. **141**, 2084 (1994)
26. M.S. Wilson, J.A. Valerio, S. Gottesfeld, Low platinum loading electrodes for polymer electrolyte fuel cells fabricated using thermoplastic ionomers. Electrochim. Acta **40**, 355 (1995)
27. R. Mosdale, S. Srinivasan, Analysis of performance and of water and thermal management in proton exchange membrane fuel cells. Electrochim. Acta **40**, 413 (1995)

28. V.A. Paganin, E.A. Ticianelli, E.R. Gonzalez, Development and electrochemical studies of gas diffusion electrodes for polymer electrolyte fuel cells. J. Appl. Electrochem. **26**, 297 (1996)
29. D. Bevers, M. Wöhr, K. Yasuda, K. Oguro, Simulation of a polymer electrolyte fuel cell electrode. J. Appl. Electrochem. **27**, 1254 (1997)
30. L. Giorgi, E. Antolini, A. Pozio, E. Passalacqua, Influence of the PTFE content in the diffusion layer of low-Pt loading electrodes for polymer electrolyte fuel cells. Electrochim. Acta **43**, 3675 (1998)
31. L.R. Jordan et al., Diffusion layer parameters influencing optimal fuel cell performance. J. Power Sources **86**, 250 (2000)
32. E. Passalacqua, G. Squadrito, F. Lufrano, A. Patti, L. Giorgi, Effects of the diffusion layer characteristics on the performance of polymer electrolyte fuel cell electrodes. J. Appl. Electrochem. **31**, 449 (2001)
33. F.T. Wagner, B. Lakshmanan, M.F. Mathias, Electrochemistry and the future of the automobile. J. Phys. Chem. Lett. **1**, 2204 (2010)
34. H.A. Gasteiger, S.S. Kocha, B. Sompalli, F.T. Wagner, Activity benchmarks and requirements for Pt, Pt-alloy, and non-Pt oxygen reduction catalysts for PEMFCs. Appl. Catal. B **56**, 9 (2005)
35. N.M. Markovic, T.J. Schmidt, V. Stamenkovic, P.N. Ross, Fuel Cells **1**, 105 (2001)
36. J.K. Nørskov, T. Bligaard, J. Rossmeisl, C.H. Christensen, Towards the computational design of solid catalysts. Nature Chem. **1**, 37 (2009)
37. V. Mehta, J.S. Cooper, Review and analysis of PEM fuel cell design and manufacturing. J. Power Sources **114**, 32 (2003)
38. N.P. Brandon, S. Skinner, B.C.H. Steele, Recent advances in materials for fuel cells. Annu. Rev. Mater. Res. **33**, 183 (2003)
39. G.J.K. Acres et al., Electrocatalysts for fuel cells. Catal. Today **38**, 393 (1997)
40. G.Q. Lu, A. Wieckowski, Heterogeneous electrocatalysis: a core field of interfacial science. Curr Opin Colloid In **5**, 95 (2000)
41. J. W. Long, R. M. Stroud, K. E. Swider-Lyons, D. R. Rolison, How to make electrocatalysts more active for direct methanol oxidation—Avoid PtRu bimetallic alloys! J. Phys. Chem. B **104**, 9772 (2000)
42. S. A. Lee, K. W. Park, J. H. Choi, B. K. Kwon, Y. E. Sung, Nanoparticle synthesis and electrocatalytic activity of Pt alloys for direct methanol fuel cells. J. Electrochem. Soc.**149**, A1299 (2002)
43. R.P. O'Hayre, *Fuel cell fundamentals*, (John Wiley & Sons, Hoboken, NJ, 2006), pp. xxii, 409 p
44. A.K. Shukla, R.K. Raman, Methanol-rusistant oxygen-reduction catalysts for direct methanol fuel celms. Annu. Rev. Mater. Res. **33**, 155 (2003)
45. T. He, E. Kreidler, L. F. Xiong, E. R. Ding, Combinatorial screening and nano-synthesis of platinum binary alloys for oxygen electroreduction. J. Power Sources **165**, 87 (2007)
46. T. He, E. Kreidler, L. Xiong, J. Luo, C.J. Zhong, Alloy electrocatalysts—Combinatorial discovery and nanosynthesis. J. Electrochem. Soc. **153**, A1637 (2006)
47. J. Greeley et al., Alloys of platinum and early transition metals as oxygen reduction electrocatalysts. Nature Chemistry **1**, 552 (2009)
48. N. Dimakis, M. Cowan, G. Hanson, E. S. Smotkin, Attraction-repulsion mechanism for carbon monoxide adsorption on platinum and platinum-ruthenium Alloys. J. Phys. Chem. C **113**, 18730 (2009)
49. N. Ramaswamy, N. Hakim, S. Mukerjee, Degradation mechanism study of perfluorinated proton exchange membrane under fuel cell operating conditions. Electrochim. Acta **53**, 3279 (2008)

Part II
Support Materials

Chapter 5
Carbon-Based Nanomaterials

5.1 Key Concepts

In fuel cells both the anode and cathode electrocatalysts are generally composed of nanosized metal particles which are supported on high surface area materials. The electrochemically active surface area (EASA), of catalysts used in low temperature fuel cells, such as polymer electrolyte membrane fuel cells (PEMFC, fed with hydrogen), direct alcohol fuel cells (DAFCs, alcohols: ethanol, methanol, polyalcohols), and direct formic acid fuel cells (DFAFC), has been found to be greatly enhanced when high surface area carbon-based materials are used as the support. As described in detail in Chap. 3, at the anode of a PEMFC dihydrogen is oxidized yielding protons and electrons. The protons pass through the cation exchange membrane toward the cathode where they are used in the reduction of oxygen to water together with the electrons arriving from the anode. The use of an anion-exchange polymeric membrane as electrolyte, i.e., a membrane which allows only negative charges to pass, favors the production of negative ions, in this case OH^-, in the process of oxygen reduction at the cathode, while the overall electrochemical process is left unvaried, as well as the reversible voltage of the cell. In PEMFCs, the polymeric electrolyte is generally *Nafion*®, a proton-exchange fluorinated membrane, about 50–200 μm thick. This withholds negatively charged ions (usually sulfonate groups $-SO_3^-$) covalently bonded to the polymeric backbone and therefore allows the passage of protons. Electrons are therefore forced to flow through the outer circuit. *Nafion*®, like other proton-exchange polymeric membranes, is most efficient when it works between 70 and 100 °C, thus limiting the functionality of PEMFCs to low temperature operation.

In DAFCs at the anode alcohols are oxidized to yield protons, electrons, and CO_2 or carboxylates, depending on the nature of the alcohol used as a fuel, while the cathode process is wholly similar to the one that takes place in PEMFCs.

The most important component of a fuel cell is the membrane electrode assembly (MEA) that is composed of a *catalytic layer* where the electrochemical reactions occur, a *diffusion layer* providing access to the fuel and oxygen to the catalytic layer and a membrane where ions flow from one electrode to the other.

A. Lavacchi et al., *Nanotechnology in Electrocatalysis for Energy*, Nanostructure Science and Technology 170, DOI: 10.1007/978-1-4899-8059-5_5,

© Springer Science+Business Media New York 2013

The most important components of the MEA are the supported catalyst material, an ionomer (which favors the flow of ions (H^+ or OH^{-}) through the electrodes), and the ion exchange membrane.

Since the activity of an electrocatalyst increases as the reaction surface area increases, metal particles are generally reduced as much as possible in diameter in order to obtain higher activity, although in some cases the specific activity of metal nanoparticles has been found to decrease with decreasing the particles size [1–3]. The structure, morphology, and proper dispersal of these metal particles on a carbon support have made low loading catalysts feasible for fuel cell operation.

In addition to having a high surface area that can be obtained with a support having a high porosity, supports for fuel cell catalysts operating at low temperatures must also have a high electrical conductivity which allows the flow of electrons. In particular, carbon-based materials must meet the following requirements: (a) be chemically inert and pure; (b) provide high dispersion of metal nanoparticles and prevent agglomeration phenomena; (c) have high electron conductance to minimize ohmic losses associated with electron transport and have adequate porosity to ensure efficient mass transport of reactants and products; (d) be compatible with proton or anion conducting polymers so that the composites provide efficient proton or anion transport through the catalytic layer; (e) provide for efficient water management; (f) be corrosion resistant; (g) have high thermal resistance and thermal conductivity; (h) have elasticity, lubricity, and compressibility; and (i) allow gas and liquid adsorption.

5.2 Influence of the Carbon Support on the Catalytic Activity of Metal Nanoparticles

The correct choice of a carbon support is an aspect of primary importance in the realization of electrocatalysts for PEMFCs, DAFCs, and DFAFCs. In fact, there exists an interactive effect between the support material and the metal catalysts. The support may indeed modify the electronic character of the catalyst nanoparticles, thus affecting the reactions of oxidation and reduction on the catalytic surface [4]. The support material may also modify the shape of the catalyst particles and as a consequence the number of the active sites present on the catalytic surface [5].

Moreover the specific surface area of the support, pore size distribution, and surface properties strongly influence the metal particle size, the size distribution, the nanostructure, the alloying degree, and therefore the structure and morphology of the catalyst. It has been observed for the oxygen reduction reaction (ORR) [1, 6–8], in methanol [9, 10] and carbon monoxide oxidation [11, 12] that the specific catalytic activity per unit of metal surface area, decreases with decreasing particle size [1, 2, 10]. However, for the hydrogen and formic acid oxidation reactions, it has been found that the specific activity increases with reduction of the particle size [10, 13, 14]. Generally, the optimum size of metal particles for PEMFCs, DAFCs, and DFAFCs is around 2–4 nm.

High metal content favors metal coalescence (larger particle sizes), which has a negative consequence on catalytic activity, which is most significant for carbon supports with low surface areas, although it has been observed for some catalytic and electrocatalytic reactions that particle coalescence may bring about significant positive consequences [15–18].

A number of publications reported metal-support interactions to explain the differences in electrocatalytic activity on nanoparticles grown on various carbon materials. Unfortunately, the conclusions are ambiguous [19–21]. The only thing certain is that the strong interactions of some surface functional groups (e.g., species like OH^-, COO^-, or N) with the catalyst metal precursors can affect the dispersion of the resulting catalysts [22].

Some carbon materials are known to contain sulfur as an impurity. The poisoning of a Pt-based anode electrocatalyst by sulfur found in Vulcan XC-72 has been reported [23, 24] while on the other hand the presence of sulfur has also been observed to enhance the catalytic activity for ORR [25]. The difference in the behavior for anode and cathode electrodes with respect to the presence of sulfur can be explained by the different oxidation state of sulfur at the different potentials of the two electrodes.

The influence of metal loading and thus the catalytic layer thickness on cell performance is a very important aspect to consider. Generally, the current increases with the increased metal loading up to a point where either mass or proton transfer (for a PEMFCs) become limiting [25–27]. It is therefore necessary to maintain the catalytic layer as thin as possible thus using a high metal loading per geometric surface area of the MEA [28]. Another important influence on cell performance is due to the content of Nafion ionomer in the catalytic layer [29, 30]. It has been observed that an increased content of Nafion decreases the ohmic losses by improving proton conductivity but does not affect significantly the electronic resistance. Conversely, mass transport losses increase strongly by decreasing gas diffusion in the Nafion film of a PEMFC electrode [31].

Metal particles that are not in direct contact with the proton-exchange polymeric membrane of a MEA can also participate in the electrochemical reactions. Thus, proton conduction is believed to occur in a network of hydrogen-bonded water molecules absorbed on a metal surface [32]. This suggests that Nafion distribution in the catalytic layer has a significant influence on cell performance. In this regard, an important role is played by the carbon support. In fact, the carbon support should have a high percentage of mesopores in the 20–40 nm region to provide a high accessible surface area to catalyst and to the monomeric units of the Nafion ionomer to boost the diffusion of chemical species [33–35]. The influence of both porosity and texture of the carbon support have influence on mass transport losses. A decrease in current density for ORR with an increase in the specific surface area of carbon support has been observed for supported Pt nanoparticles having the same particle size on carbon supports with BET surface areas ranging from 254 (Vulcan) to 1,475 m^2/g (Black Pearls). The pore structure and the wetting properties of the carbon supports have a dramatic influence on the water balance in the catalytic layers. At the cathode side of a PEMFC a danger of flooding exists due to the production of water but also as a consequence of the crossover of water from the anode side accompanying proton transport (electro-osmotic drag) to the cathode side. At the same time on the anode side there is a problem of electrode over-drying due to water loss.

The stability of the catalyst support in fuel cell environments is also of great importance. In addition to high surface area porosity and electrochemical conductivity, corrosion resistance is an important factor. In this regard we have to consider that carbon is thermodynamically unstable at the operating conditions of a PEMFC or DAFC cathode because the equilibrium potential of its electrooxidation to carbon dioxide [36] is:

$$C + 2H_2O = CO_2 + 4H^+4e- \qquad E = 2.07\ V\ (vs\ RHE)$$

The corrosion current normalized to the unit mass of carbon material increases with increasing the electrode potential, temperature, and water concentration. The generation of CO_2 is believed to proceed through the intermediate formation of carbon surface oxides [37]. It has been reported that formation of surface oxides and evolution of CO_2 occur simultaneously during the initial corrosion of high surface area carbon blacks. Similar conclusions have also been reported recently, even though the mechanism of corrosion is not yet fully understood [38]. A very important role seems to be played by the hydrophobicity of the carbon which decreases after long-term performance of fuel cell electrodes.

The susceptibility of carbon materials to corrosion increases with increasing A_{BET} [39] but electrochemical degradation is slowed significantly by carbon graphitization. For example, the rate of corrosion of Vulcan XC-72 graphitized at 2,970 K, decreases compared with the untreated material. This can be explained by considering that electrochemical corrosion occurs at the edge plane of graphite while the basal planes are relatively not involved. For this reason a pronounced difference in corrosion resistance has been observed between Pt supported on carbon nanotubes which are rolled-up graphene sheets compared to Pt supported on carbon blacks [40–42].

Conditions that favor the corrosion of carbon in a PEMFC are manifold: the positive value of electrode potential, the acid environment, the presence of dioxygen, water, and temperatures up to 363 K, as well as the presence of metal nanoparticles. The chemical corrosion of carbon in PEMFCs occurs mostly due to the presence of strongly oxidizing radicals such as $OH^{\cdot}$ or $HO_2^{\cdot}$ derived from the decomposition of H_2O_2 at the anode as a consequence of O_2 crossover from the cathode and through the reaction of H_2O_2 with Fe^{2+} or Cu^{2+} ions present as impurities in the carbon black. H_2O_2 also reacts with ions such as Pt^{4+} or Pt^{2+} produced by the corrosion of Pt nanoparticles producing $OH^{\cdot}$ or $HO_2^{\cdot}$ species at the cathode side [36]. These radical species are also corrosive for both the catalyst and membrane of the MEA. The gradual loss of carbon induces agglomeration of metal nanoparticles decreasing the EASA and worsening the cell performance [43, 44]. The nonuniform distribution of hydrogen at the anode and the start-up and shut-down of a PEMFC are also responsible for the degradation of the carbon support and this affects the durability and lifetime of PEMFCs. Due to the high potential, during the start-up and shut-down, the carbon support for the cathode catalyst is subject to oxidation by reacting with oxygen or water [45]. If the catalyst layers cannot maintain their structure over the lifetime of the fuel cell, changes in the morphology of the catalysts from the initial state will result in a loss of electrochemical activity. For these catalysts more severe requirements have to be met to obtain a long-term stability. The U.S. Department of Energy (DOE) lifetime targets for 2017 are 5,000 h for transportation power systems and 60,000 h for stationary power systems. Current PEMFC technology yields only 1,700 h and 10,000 h, respectively [46]. Stability issues including carbon corrosion have thus limited the commercialization of PEMFCs and as a consequence improvements will be needed to reach acceptable levels of durability.

In the next sections the main types of carbon supports used as electrocatalyst supports will be discussed. Although not exhaustive, the objective is to help the reader in understanding the relationship between carbon structure and catalyst performance in the main types of low temperature fuel cell operation.

5.3 Carbon Blacks

Carbon blacks are the group of materials most widely used as carbon supports in low temperature fuel cells such as PEMFCs, DAFCs, and DFAFCs. Based on the production method and raw materials employed these materials are classified into furnace blacks, channel blacks, lamp blacks, thermal blacks, and acetylene blacks [47, 48].

For the production of this class of materials a common preparation procedure is used: liquid or gaseous hydrocarbons are decomposed at elevated temperature under a reduced presence of oxygen. Nowadays, the most important carbon black is the furnace black in which hydrocarbons are partially combusted and immediately quenched with water. Commercial carbon materials such as Vulcan XC-72R, Black Pearls 2000, Ketjenblack EC-300 J, and Ketjenblack EC-600 J all belong to this category. The primary particles of carbon blacks consist of several graphene-like layers, which combine to form aggregates of different sizes with particle diameters ranging from 5 to 100 nm. These aggregates are usually assembled into three-dimensional superstructures of sizes between 1 and 100 μm that may be defined in terms of increasing degree of aggregation: spherical, elliptic, linear, and ramified. Depending on the starting material and synthesis, BET surface areas range from 10 to 1,500 m^2/g.

Acetylene black is produced by partial oxidation of acetylene gas at high temperature. Due to the synthetic procedure, it shows a high aggregate structure and crystal orientation. The main characteristics of some carbon blacks are reported in Table 5.1.

Several studies have been carried out in order to determine how the characteristics of different carbon blacks may affect the dispersion of supported metals

Table 5.1 Carbon blacks: main characteristics

Carbon Black	Surface area ($m^2\ g^{-1}$)	Primary particle diameter (nm)
Vulcan XC-72	254	20–30
Black pearls 2000	1,500	10–15
Ketjenblack EC-600 JD	1,300	35–40
Ketjenblack EC-300 J	800	30–40
Denka black	65	40
Acetylene black	64	42

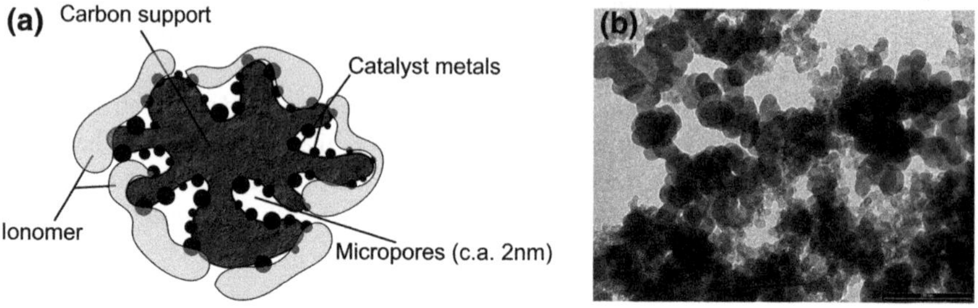

Fig. 5.1 **a** Schematic representation of catalyst—ionomer-Vulcan XC-72; **b** TEM image of Vulcan XC-72

and as a consequence their electrocatalytic activity [3, 49–52]. It has been observed for Pt/C catalysts, synthesized using the sulphite-metal complex method, that platinum particles size decreased with increasing the specific surface area of the carbon black [34]. Instead, when the metal deposition on the carbon support is performed by impregnation methods, the specific surface area of the carbon support seems to have only a little effect on Pt dispersion [53]. By supporting platinum particles on acetylene black, Watanable et al. [54] observed that the Pt catalyst had a larger mean Pt particle size with respect to Pt particles supported on furnace black supports. The catalyst supported on acetylene black exhibited a higher activity for methanol oxidation. The reason for this behavior has been attributed to the fact that acetylene black has a higher amount of pores of 3-8 nm than furnace black, which are useful for methanol diffusion, even if the platinum in these pores do not contribute to the electrooxidation reaction. Ionomer particles are much larger than the pore diameter of 3–8 nm and the Pt inside these pores cannot contact the ionomer. For supports with pore sizes smaller than 3 nm such as Vulcan XC-72, the activity of the catalyst may be limited for this reason (Fig. 5.1a). This result indicates that the pore size affects directly the catalytic performance of an anode in a DMFC. The contact between the metal nanoparticles and the Nafion® ionomer micelles in the catalytic layer of the gas-diffusion electrode is crucial and is also affected by the carbon pore size and its distribution. Uchida et al [55] reported that the Nafion® ionomer has micelles of about 40 nm in size. The nanoparticles, which sink into the carbon pores with diameters below 40 nm, have no access to the Nafion® ionomer and thus do not contribute to the total catalytic activity. Thus catalytic utilization of the metal nanoparticles is determined more by the electrochemical accessibility and not by the carbon surface area.

In addition to the size of the pores of the support, also the specific surface area is of fundamental importance for the specific mass activity of the catalysts. Rao et al. [51] reported novel catalysts for an anode of a DMFC made of PtRu nanoparticles supported on carbons with specific surface areas ranging from 6 to 415 $m^2\ g^{-1}$. Low surface area of carbon-supported catalysts having a surface area of 72 $m^2\ g^{-1}$ showed superior mass specific activities, exceeding that of a commercial 20 wt % PtRu catalyst by nearly a factor of 3. The catalyst utilization for low surface area carbons is close to 100 %, which may be explained by the compatibility between the size of the pores in carbon supports and Nafion® micelles. Superior mass activity of low surface area carbon supports was attributed to the high PtRu surface utilization, and to a facilitated diffusion in macropores. Hence, these authors concluded that an increase in the carbon surface area did not increase the catalytic activity. This work demonstrates the potential that exists for the optimization of carbon supports for use low temperature fuel cell electrocatalysts. More work is needed in order to determine the optimum pore structure and texture of carbon supports and to explore how surface area influences long-term stability of fuel cell catalysts. On the basis of what has been described above, three distinct interfaces are responsible for improving or deteriorating the catalytic performance of a fuel cell electrode: (1) metal–C interface, (2) metal–Nafion® interface, and (3) C–Nafion® interface. Consequently, the development of electrocatalysts has to take into account the search for new materials with optimized physical and chemical conditions at these important interfaces.

5.3.1 Activation and Functionalization of Carbon Blacks

To improve metal dispersion and catalytic activity, generally before their use, carbon supports are "activated" via oxidative treatments. The functional groups present on the carbon surface, for example carboxylic acid, phenolic, lactonic, and etheric groups, are responsible both for their acid/base and redox properties [56]. These oxidative treatments that can be performed using different oxidants such as HNO_3, H_2O_2, O_2, or O_3, lead to the destruction of basic sites and increase the surface acidic groups. Contradictory results regarding the effect of pretreatments of carbon on the resultant platinum dispersion have been reported [56–58] but, as expected, carbon treatments, which increase metal dispersion, generally also increase their electrocatalytic activity [59–62].

Improvements in electrode structures, which are composed of catalysts and a sulfonated polymer, can greatly improve performances by increasing the efficiency of electrocatalysts [63–68]. A porous carbon support, Ketjenblack EC 300 J (KB), used as catalyst for DMFCs has been functionalized to give sulfonated polystyrene (sPS) groups on the carbon surface [69]. The authors showed that electrodes with the Pt catalysts on sPS KB showed 25 % higher DMFC performance at 30 mW/ cm^2 (50 C) compared to conventional electrodes (Pt on KB with sulfonated polymers). This work highlights that support materials for fuel cell catalysts can be designed in such a way as to function both as a support as well as to promote proton conduction in the electrode. It was also suggested that by anchoring a phenyl-sulphonic group on the Vulcan XC-72R support of a platinum catalyst enhances the performance in PEMFCs. This was achieved by increasing the utilization of Pt by enhancing the three-phase boundary in the catalyst layer [70]. In both cases reported, the functionalized carbon plays dual roles of promoting mass transport and as a catalyst support. The improved performance of fuel cells with the electrodes containing these functionalized carbons was ascribed to a better mass transport which maximizes the catalytic activity.

Another type of functionalization that has been reported is the introduction of nitrogen into the Vulcan XC-72 support which has been shown to lead to better dispersion of Pt nanoparticles. A platinum electrocatalyst supported on N-doped Vulcan XC-72 has been shown to have a higher methanol oxidation activity than a standard Pt on Vulcan XC-72R catalyst in terms of mass-normalized activity [71].

5.4 Other Carbon Nanostructured Materials

5.4.1 Mesoporous Carbon

Nonconventional carbon supports have attracted much interest in electrocatalysis because of their versatility in pore size, pore distribution, and mechanical properties. To this class of materials belong the ordered mesoporous carbons (OMCs), which are constructed of regular arrays of uniform mesopores. These materials which present a different morphology than carbon blacks in terms of their pore texture and form, are expected to be advantageous for efficient diffusion and transport of reactants and by-products in the application of fuel cells.

To date, the nanocasting strategy, from preformed hard templates and the direct synthesis strategy induced by supramolecular assembly of the block copolymer soft templates are the two main routes to prepare these ordered mesoporous carbons [71–73].

According to the definition of IUPAC (International Union of Pure and Applied Chemistry), mesoporous carbon materials are defined as porous carbon materials whose pore size is in the range 2–50 nm. Compared to carbon blacks having pores <2 nm (micropores), mesoporous carbon present higher surface area and a lower amount or absence of micropores. In this support the metal catalyst particles are distributed on the surface or in pores of the mesoporous carbon. A large mesopore surface area, particularly with pore size >20 nm, gives rise to a high dispersion of Pt particles, which results in a large effective surface area of Pt with a high catalytic activity which is also facilitated by an improved mass transportation.

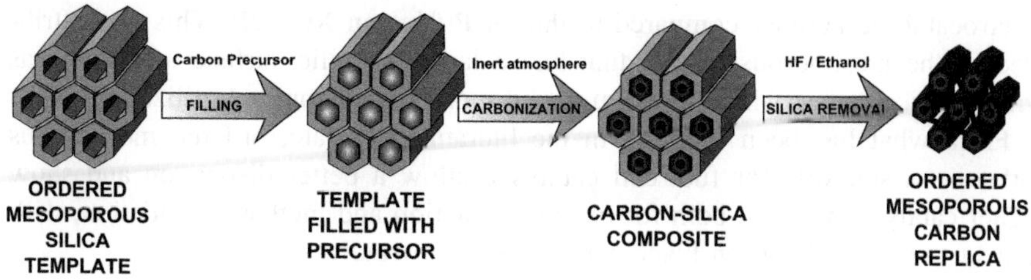

Fig. 5.2 Schematic representation of the synthesis procedure for OMC from ordered mesoprous silica template

Mesoporous carbon materials can be classified into two categories according to their preparation methods and final structures: ordered mesoporous carbon [72, 74–77] and mesoporous carbon with irregular pore structure [78]. Synthetic strategies for ordered mesoporous carbons have been reviewed by Chang et al. [79]. In brief, the synthesis follows this path: a carbon precursor such as sucrose, furfuryl alcohol, phenolic resin, acetylene gas, acenaphthene, acrylonitrile, pyrrole, or mesophase pitch is infiltrated into the pores of an ordered silica template. The carbon source is then polymerized and converted to a rigid carbon structure by pyrolyzing in a nonoxidizing atmosphere which results in the silica–carbon composite. In the final step, the mesoporous carbon replica is obtained by treating the composite in an ethanol–water solution of HF or NaOH to remove the silica template (Fig. 5.2). The advantages are a widely tunable pore size in the range of several nanometers to several micrometers achieved by choosing the size of the silica template, and a three-dimensional ordered uniform structure composed of interconnected voids with pore sizes in the range 10 to about 1,000 nm. The resulting ordered mesoporous carbons have BET surface areas ranging from 400 to 1,000 m^2/g and pore volume from 0.7 to 3.8 cm^3/g.

Nanoporous carbon with a large surface area of 1,300 m^2/g was also synthesized by Nam et al. [80] using a SBA-15 silica template. It was successfully used as the catalyst support for Pt in a DMFC cathode. The single cell test exhibited reasonably good DMFC performance with 61 mW/cm^2 as maximum power density, which was slightly better than that with platinum black under similar operation parameters.

Ordered uniform porous carbons having pore sizes in the range of 10–1,000 nm have been reported by Yu et al. [81] and Chai et al. [82] as supports for DMFCs that showed a much improved catalytic activity. The authors ascribed this to the high surface area, large pore volume, and three-dimensionally interconnected uniform pore structure, which allow a higher dispersion of the catalysts and efficient diffusion of the reagents. Carbons described by Chai et al. comprised a large fraction of micropores, whose contribution to the overall surface area amounted to about 25 %. These micropores were nearly fully blocked when Pt–Ru particles were supported on these carbon materials.

Mesoporous carbon with a surface area of 890 m^2/g and pore sizes of 10-50 nm as carbon-supported Pt catalyst in a direct alcohol fuel cell, showed three times the electrocatalytic activity compared to that of Pt/Vulcan XC-72R. This was attributed to the mesoporous carbon that had a larger specific surface area and was without micropores, which gave improved metal dispersion and utilization [83].

From what has been reported in the literature till date, ordered mesoporous carbons, as supports for fuel cell catalysts, allow a better dispersion and show higher catalytic activity, both for oxygen reduction and methanol oxidation, with respect to standard carbon black supported electrocatalysts.

5.4.2 Carbon Gels

Carbon used in industrial applications or in electrochemical devices, such as carbon blacks, most often display inappropriate pore textures with regard to mass transport. The texture of activated carbons, for instance, is generally microporous, with low macropore or mesopore volumes, which often induces diffusion limitations during catalytic and adsorption processes. Carbon blacks, which are composed of porous near-spherical particles (20–60 nm), coalesced together as aggregates (1–100 μm), remain the elected material for proton-exchange membrane fuel cell electrodes; however, the pore texture of the electrodes is quite dependent on the electrode processing because it is to a large extent defined by the packing of the carbon aggregates, which depends on the electrode manufacture technique. These drawbacks require the development of carbon materials with controllable and tunable pore texture.

Among the numerous new carbon materials developed, carbon gels, named also carbon xerogels, aerogels, and cryogels, have been successfully used at a laboratory scale as alternatives to commercial carbons in several processes. Carbon gels are a new form of carbon that posses fully interconnected mesopores. Their surface area, pore volume, and pore size distribution, can be controlled by the synthetic process [84]. Most studies report preparation by the carbonization of organic gels, which are prepared by a sol–gel procedure. Generally, a mixture of resorcinol and formaldehyde in water and sodium carbonate as basification agent are gelated and aged. The resulting gel which is composed of interconnected spherical nodules, delimited voids filled with the solvent. This is followed by drying and then pyrolysis. A large variation in the final pore volume is determined by the method used for drying which is performed by either evaporation, supercritical drying, or freeze drying. Pyrolysis induces the development of microporosity within the nodules. The final material obtained shows a bimodal texture consisting of mesopores and macropores delimited by spacing between the micropores.

Basically, the three variables that determine the pore texture of the material are the pH of the precursor solution, the dilution ratio of water and reactants, and the drying technique.

However, the relationship between the pore texture and the synthesis drying is not straightforward. Generally, the pore size and pore volume of carbon gels can range

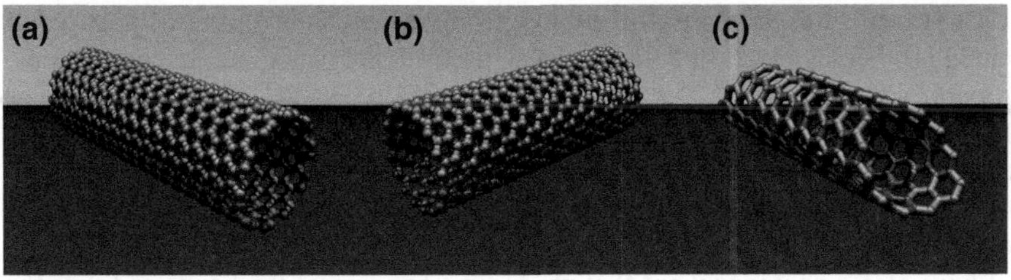

Fig. 5.3 **a** chair, **b** zigzag, and **c** chiral manotubes

from a few nm to a few μm, and from 0 to 5–6 cm^3/g, respectively. The surface area can vary from 40 to 1,290 m^2/g. This leaves many possibilities for adjustment as regards the final application. Recently, these materials were use to prepare mainly cathode electrodes for PEMFCs (H_2/O_2), [85–88] and DMFCs [89, 90].

The use of Pt–Ru catalysts supported on a carbon aerogel as an anode for DMFCs has been reported [91]. In this work the authors claimed that with much less metal loading on the carbon aerogel, the membrane electrode assemblies had the same power density as that of commercial catalysts. This was attributed to the mesopore texture of the carbon aerogel, which facilitated methanol transportation in the electrode.

To conclude, far too many variables influence the final performance of the carbon support materials used in fuel cells investigated up to now. Therefore, the optimization of both electrode architecture and carbon pore texture still requires further investigation.

5.4.3 Carbon Nanotubes

Carbon nanotubes are one of the most studied materials in nanoscience and nanotechnology because they possess special properties that make them suitable for many potential applications as materials for energy storage, electronics, catalysis, electrocatalysis as well as reinforcements for composites.

The discovery of carbon nanotubes (Carbon Nano Tubes, CNTs) is attributed to Sumio Iijima in 1991 [92, 93]. These structures, however, had already been observed previously, although the importance of this discovery was not fully realized [94].

CNTs can be considered as fullerenes elongated in one direction, synthesized from a graphene sheet folded back on itself to form a cylinder, as shown in Fig. 5.1b. It is possible to obtain individual cylinders (single-walled carbon nanotubes, SWCNTs) or more coaxial cylinders (multiwalled carbon nanotubes, MWCNTs). The surface is composed essentially of hexagons and three possible arrangements can be obtained, depending on the carrier network (Fig. 5.3): a chair (armchair), zigzag, or chiral.

These structures, however, are not free from defects. In fact, zones composed of pentagons, heptagons, or fullerenes emerging from the nanotube can change some properties of the material. The electronic conductivity of SWCNTs, for example, is closely linked to the surface structure passing from a conduction typical of a semiconductor p-type to a conduction typical of metals [95, 96]. Instead, MWCNTs, have a very high conductivity (4×10^9 Acm^{-2}, 1,000 times that of copper) [97]. The carbon atoms in CNTs are not only sp^2 hybrids, as in graphene, but have also a component sp^3 that produces a bending of the bonds and thus the cylindrical structure, which causes an increase in the electron density toward the outside of the cylinder depending on the curvature of the surface [98, 99].

This particular sp^2 orbital configuration allows the coordination of transition metals that may influence the formation of metal particles, by controlling their size and dispersion [100].

Also modifications of the surface which involve the introduction of oxygenated groups, due to reactions with strong oxidizing agents and under extreme conditions, increase the capacity of coordinating metal complexes and consequently exercise a control both of size and dispersion of metal particles [101]. These surface modifications (or functionalizations) also have a strong influence on the electronic bands and on Fermi levels of CNTs [102]. In particular, for SWCNT it can be stated that the effect of the functionalization depends on the type of conduction (semiconductor, quasi-metallic, or metallic) and can be positive or negative [103], while for MWCNTs functionalization produces always an increase in the electronic conductivity.

A quick comparison with Vulcan XC-72R (conductivity $= 4$ S cm^{-1}, specific surface area $= 240$ m^2 g^{-1}), shows that CNTs not only meet all requirements to be excellent support material, but also have a very high conductivity. Pores in MWNCT are mainly divided into inner hollow cavities of small diameter (3–6 nm) and mesopores (diameter of 10–40 nm but also from 5 to 200 nm) formed by interaction of isolated MWNTs, length from a few to hundreds of μm [104]. Instead, SWNCTs show a microporous nature: a diameter of 0.5–2 nm, length from few to 20 cm. Total surface area of SWCNTs range from 400 to 900 m^2 g^{-1}, whereas MWCNT values range from 150 to 450 m^2 g^{-1} [105]. The formation of polar groups on the surface also greatly influences the wet ability and the hydrophobicity of the material: untreated CNTs are hydrophobic, while those that have been modified, are more hydrophilic. The methods commonly used for this treatment are oxidations carried out with a variety of reagents such as HNO_3, aqua regia, H_2SO_4, OsO_4, HF-BF_3, or $KMnO_4$.

The interaction of carbon nanotubes with gases or liquids adsorbed either on their internal or external surfaces, is attracting increasing attention due to the possible influence of such adsorption on CNT electronic properties and to the possibility of using these materials for efficient hydrogen storage [106]. Moreover, they show particularly strong bonds in the curved graphene sheets that are stronger than in diamond. This makes CNTs particularly stable toward deformation with a Young's modulus on the order of terapascal [107] and a resistance to traction of 250 GPa [108].

The synthesis of CNTs is carried out mainly with arc discharge, [109] laser ablation [110], and chemical vapor deposition, [111] but there are also other techniques which include pyrolysis of plastics, [112] or combustion synthesis. In addition to the production process, particular importance is given to the purification processes which are carried out first with the partial oxidation of amorphous carbon and subsequently with the removal of the catalyst with acid minerals.

The arc discharge technique is based on the formation of an electric arc between two graphite electrodes at close range (about 1 mm) in an inert atmosphere of He or Ar and at reduced pressure (between 5 and 70 kPa). The discharge produces a plasma at high temperature (about 3,000 °C) that heats the graphite electrode inducing a rapid process of sublimation and condensation with the formation of CNTs and other carbon products.

The laser ablation technique is similar to arc discharge: a source laser heats to high temperature a graphite target in an inert atmosphere, thus repeating the same process of sublimation and condensation. The technique of chemical vapor deposition (CVD), is essentially a thermal dehydrogenation reaction of organic compounds (alcohols or hydrocarbons) in the presence of metal catalysts (Ni, Fe, Co). The procedure involves two basic steps: preparation of the catalyst and the reaction proper. The catalysts are generally constituted of nanosized metallic particles synthesized by sol–gel, impregnation, co-precipitation, or with organometallic precursors. The latter method, in particular, can utilize Fluidized Bed Chemical Vapour Deposition (FBCVD) because it allows for the in situ production of the catalyst. In fact, many organometallic compounds have a good volatility by sublimation or evaporation, and are easily degradable by thermal oxidation producing nanosized particles.

A significant development in the use of CNTs is undoubtedly in the field of fuel cells as supports for both anode and cathode electrocatalysis in DAFCs and PEMFCs. As described previously the catalyst support material must have a high electronic conductivity, a high level of mesoporosity, a suitable morphology to optimize and stabilize the three-phase boundary reactive sites, and a good hydrophobicity for water removal and mass transport improvement. In comparison with the more widely used Vulcan XC-72R carbon black support, CNTs have significantly higher electronic conductivities and present higher mesoporous volumes for comparable or higher surface areas. However, most of the CNTs are chemically inert making it difficult to anchor precursor metal ions or nanoparticles, which usually lead to poor metal dispersion. Thus, functionalization to introduce surface oxygen groups has been commonly performed in order to make the surface more hydrophilic and improve the catalyst support interaction.

The functionalization of the CNT surface can be performed together with metal deposition on the carbon. Pt nanoparticles supported on multiwalled carbon nanotubes grown directly on carbon have been prepared by a new method using glacial acetic acid as a reducing agent. In this case, the glacial acetic acid acts not only as reducing agent but has the capability of producing a high density of oxygen-containing functional groups on the surface of CNTs that leads to high density and monodispersion of Pt nanoparticles [113].

Nitrogen-containing CNTs (N-CNTs), which have about 87.2 wt % carbon and 6.6 wt % nitrogen have been prepared [114]. Platinum nanoclusters of around 3 nm were deposited onto these N-CNTs. According to the authors, the nitrogen-containing carbon nanotubes obtained in their study contained heterocyclic nitrogen which favored the deposition of Pt particles.

The ability of metal-free nitrogen doped CNTs (N-CNT) to facilitate ORR was reported in 2009 by Gong et al. [115] who described nitrogen-containing carbon nanotubes that can be used in metal-free electrodes with a much better electro-catalytic activity, long-term operation stability, and tolerance to the crossover effect than platinum for oxygen reduction in alkaline fuel cells. Since then several studies of CNTs for use in fuel cell cathodes have been published [116–118].

CNTs have now been widely employed as catalyst supports and studied for the major electrochemical reactions. They have been employed mostly in methanol electrooxidation (anode catalyst) and for oxygen reduction and to a lesser extent, hydrogen electrooxidation. Platinum alone is the most employed metal, but also a wide variety of binary (e.g., Pt–Ru, Pt–Co, Pt–Fe) as well as ternary catalyst (e.g., Pt–Ru–Pd, Pt–Ru–Ni, Pt–Ru–Os) systems using both noble and non-precious metals have been obtained [119–121].

An efficient fabrication method for carbon nanotube (CNT)-based electrodes with a nanosized Pt catalyst as cathode of a PEMFC has been reported by Tang et al. [122]. The integrated Pt/CNT layer was prepared by in situ growth of a CNT layer on carbon paper and subsequent direct sputter-deposition of the Pt catalyst. Both scanning electron microscopy (SEM) and transmission electron microscopy (TEM) demonstrated that the Pt/CNT layer consists of a highly porous CNT layer covered by well-dispersed Pt nanodots with a narrow size distribution. Compared with conventional gas-diffusion layer assisted electrodes, the CNT-based electrode with a Pt/CNT layer acting as a combined gas-diffusion layer and catalyst layer showed pronounced improvement in polarization tests. A very high maximum power density of 595 mW cm^{-2} was observed for a low Pt loading of 0.04 mg cm^{-2} that was significantly higher than the Pt/Vulcan XC-72R-based conventional electrode (435 m Wcm^{-2}) with equal Pt loadings.

Pt nanoparticles supported on carbon nanotubes (Pt/CNTs) have been synthesized from sulfur-modified CNTs impregnated with H_2PtCl_6 as Pt precursor [123]. The dispersion and size of Pt nanoparticles in the synthesized Pt/CNT nanocomposites are remarkably affected by the amount of sulfur modifier (S/CNT ratio). An S/CNT ratio of 0.3 gave well-dispersed Pt nanoparticles on CNTs with an average particle size of less than 3 nm and a narrow size distribution. The catalyst showed highest electrochemically active surface area (88.4 m^2g^{-1}) and highest catalytic activity for the methanol oxidation reaction. The mass-normalized methanol oxidation peak current observed for this catalyst (862.8 Ag^{-1}) was $\sim$6.5 folds higher than Pt deposited on pristine CNTs (133.2 A g^{-1}) and $\sim$2.3 higher than commercial Pt/C (381.2 A g^{-1}).

When compared to Pt supported on Vulcan XC-72, the CNT-based catalyst showed higher retention of electrochemical surface area, a smaller increment in interfacial charge transfer resistance and a slower degradation of the fuel cell performance. This study confirmed the higher corrosion resistance of MWCNTs and also a stronger interaction with the Pt nanoparticles. It was also observed that highly corrosion-resistant MWCNT prevented the cathode catalyst layer from severe water flooding by maintaining the electrode structure and hydrophobicity for a long period under continuous anodic potential stress [124].

Wu and Xu [125] reported a detailed comparison between multiwalled (MWNT) and single-walled carbon nanotubes (SWNT) in an effort to understand which can be the better candidate of a future supporting carbon material for electrocatalysts in direct methanol fuel cells. Pt particles were deposited via electrodeposition on MWNT/Nafion and SWNT/Nafion electrodes to investigate effects of the carbon materials on the physical and electrochemical properties of the Pt catalyst.

CO stripping voltammograms showed that the onset and peak potentials on Pt-SWNT/Nafion were significantly lower than those on the Pt-MWNT/Nafion catalyst, revealing a higher tolerance to CO poisoning of Pt in Pt-SWNT/Nafion. In the methanol electrooxidation reaction, the Pt-SWNT/Nafion catalyst was characterized by a significantly higher current density, lower onset potentials, and lower charge transfer resistances using CV and EIS analysis. Therefore, SWNTs present many advantages over MWNTs. The remarkable advantages of SWNT were explained by its higher electrochemically accessible area and easier charge transfer at the electrode/electrolyte interface, richness in oxygen-containing surface functional groups and highly mesoporous tridimensional structure.

In contrast, Carmo et al. [126] presented opposing results in the study of the activity of PtRu supported on SWCNT, MWCNT, and Vulcan XC-72R carbons as anode materials in DMFCs. The MOR activity observed was in the order PtRu/MWCNT > PtRu/C > PtRu/SWCNT.

Conjugated polymers such as polypyrrole (PPy) and polyaniline (PANI) can form covalent bonds between Pt atoms and N atoms enabling strong adhesion of Pt nanoparticles onto the polymer. Cyclic voltammetry studies and accelerated degradation tests revealed high electroactivity and excellent electrochemical stability displayed by Pt/polymer modified MWCNT electrodes when compared to nonfunctionalized MWCNT and commercial CB supports [127, 128].

In 2009, the first study involving the use of a Pd catalyst in a real DAFC was reported by P. Serp, F. Vizza et al. [129]. Up till then studies had been limited to investigations of alcohol oxidation in half cells. Palladium and platinum–ruthenium nanoparticles supported on MWCNT are prepared by the impregnation-reduction procedure. The materials obtained, Pd/MWCNT (Fig. 5.4) and Pt–Ru/MWCNT, were characterized by TEM, ICP-AES, and XRPD. Electrodes coated with Pd/MWCNT were scrutinized for the oxidation of methanol, ethanol, and glycerol in 2 M KOH solution in half cells. The catalyst was found to be very active for the oxidation of all alcohols, with glycerol providing the best performance in terms of specific current density and ethanol showing the lowest onset potential. Membrane electrode assemblies were fabricated using Pd/MWCNT anodes, proprietary Fe–Co/Ketjenblack cathodes [22] and an anion-exchange membrane and evaluated in both single passive and active direct alcohol fuel cells

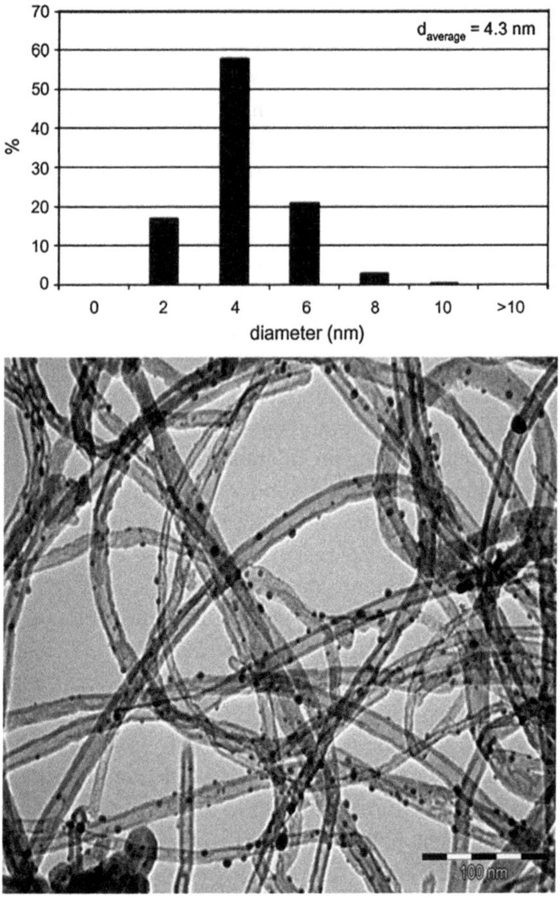

Fig. 5.4 TEM micrographs of Pd/MWCNT (*scale bar* 100 nm) and histogram of particle distribution versus diameter. Reprinted from reference 129 with permission from Elsevier

fed with aqueous solutions of 10 wt % methanol, 10 wt % ethanol, or 5 wt % glycerol. The results obtained have highlighted the excellent electrocatalytic activity of Pd/MWCNT in terms of both peak current density, as high as 2,800 A (g Pd)$^{-1}$ with glycerol, and onset oxidation potential, as low as −0.75 V versus Ag/AgCl/KCl $_{sat}$ with ethanol. Such a remarkable electrocatalytic activity of Pd/MWCNT has been associated to the high dispersion of the metal particles on MWCNTs. MEAs containing a Pd/MWCNT anode, a Fe–Co/C cathode, and a Tokuyama A-201 anion-exchange membrane have provided excellent results in monoplanar fuel cells. The MEA performance has been evaluated in both passive and active DAFCs fed with aqueous solutions of methanol, ethanol, and glycerol.

In view of the peak power densities obtained in the temperature range from 20 to 80 °C, at Pd loadings at the anode as low as 1 mg cm^{-2}, one can safely conclude that Pd/MWCNT exhibits unrivaled activity as an anode electrocatalyst for DAFCs (Fig. 5.5).

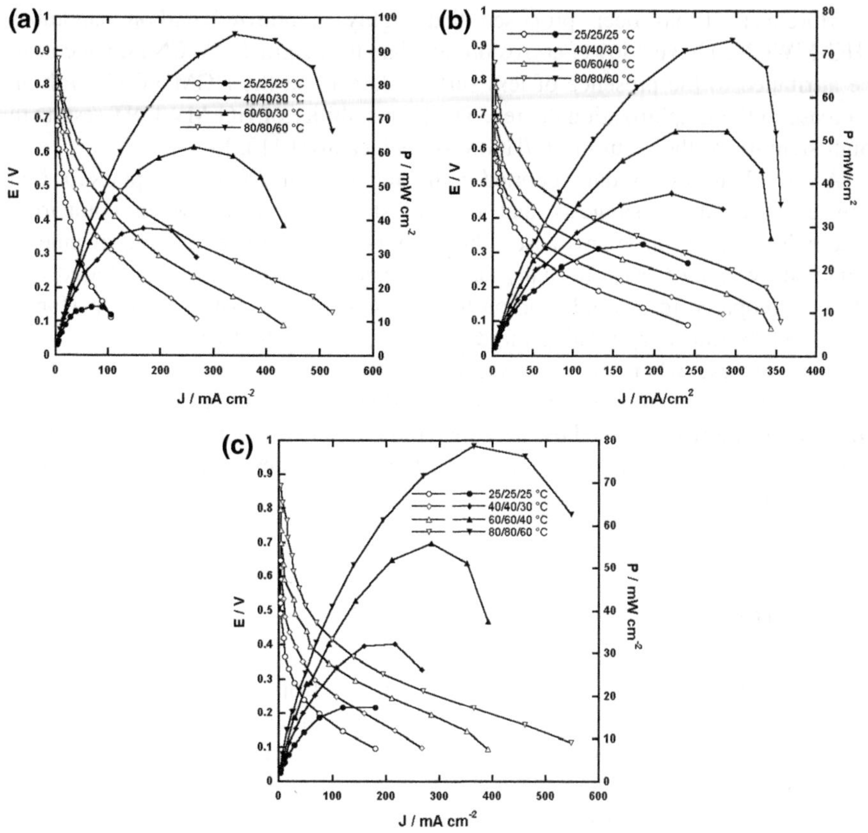

Fig. 5.5 Polarization and power density curves at different temperatures of active DAFC with a Pd/MWCNT anode (metal loading 1 mg cm^{-2}), fuelled with an aqueous 2 M KOH solution of **a** methanol (10 wt %); **b** ethanol (10 wt %); **c** glycerol (5 wt %). *Inset* report the temperatures of fuel (*left*), cell (*central*), oxygen gas (*right*). Reprinted from reference 129 with permission from Elsevier

The anode exhausts of galvanostatic experiments showed that ethanol was selectively oxidized on Pd/MWCNT to acetic acid, which was converted to the acetate ion in the alkaline media of the fuel cell, while methanol yields a mixture of carbonate and formate. A much wider product distribution, including glycolate, glycerate, tartronate, oxalate, formate, and carbonate was obtained from the oxidation of glycerol.

Numerous studies have focused on the electrochemical stability of electrodes for both PEMFC and DAFC revealing that catalysts prepared on CNTs are more corrosion resistant than those prepared in traditional carbon black support [130]. When attacked by oxidative acids, only the outside graphene layers of MWCNTs were found to be damaged, creating surface defects with edges on the CNT surface [131, 132].

Moreover, it has been proposed that highly graphitized carbon nanotubes (HG-MWCNT) have a lower corrosion rate than the original MWCNTs, which can be attributed to the presence of less surface defects on the HGMWCNT with an increase in the graphitization degree. The high stability of HG-MWCNT results in high stability of the resulting Pt/HG-MWCNT catalyst [133].

A careful analysis of data reported in the literature on CNTs as supports for fuel cell electrocatalysts suggests that caution should be taken when comparing materials because in many cases the authors overestimate the improvements obtained by using CNTs. In many cases the performance of the CNT-based catalyst is compared to a simple carbon black-based catalyst that in reality performs rather poorly thus exaggerating the difference between the two.

In summary, when an improvement really exists, they can be generally attributed to (1) the improved mass transport due to the peculiar three-dimensional mesoporous network formed by these materials, (2) the possibility of reaching high metal dispersion and high electroactive surface area values, and (3) their excellent conducting properties.

5.4.4 Graphene

Graphene belongs to the group of known carbon allotropes which consists of a flat monolayer of sp^2 bonded carbon atoms in a two-dimensional (2D) honeycomb lattice. It is the basic building block of all the “graphitic” materials such as fullerenes (0D), carbon nanotubes (1D), and graphite (3D) [134].

In 2004 Geim and co-workers [135] at Manchester University first isolated a single-layer, two-dimensional crystal from graphite. This discovery led to an explosion of interest, and much research has been conducted on the structure and property characterization of graphene [136–139]. The combination of the high surface area (2,630 $m^2\ g^{-1}$) and high conductivity makes graphene sheets a promising candidate as low temperature fuel cell catalyst support. In comparison with CNTs, graphene not only possesses similar stable physical properties but also larger surface area. Moreover, the production cost of graphene in large quantities is much lower than that of CNTs. Thus, fuel cell catalysts supported on graphene have been synthesized and characterized, their electrocatalytic activity and durability have been evaluated by half-cell measurements and tests in single fuel cells have also been performed.

The methods of preparation of graphene sheets are divided into two categories: “bottom-up” and “top-down” approaches. The bottom-up approach which is based on the growth of graphene from organic precursors such as methane, ethylene, and other hydrocarbon, [140] includes epitaxial growth [141] and chemical vapor deposition (CVD) [142]. The top-down approach includes mechanical or chemical exfoliation as well as electrochemical or liquid phase exfoliation of natural or synthetic graphite into the mixture of a single and a few layer graphene platelets [143–145]. However, the most commonly top-down method to obtain

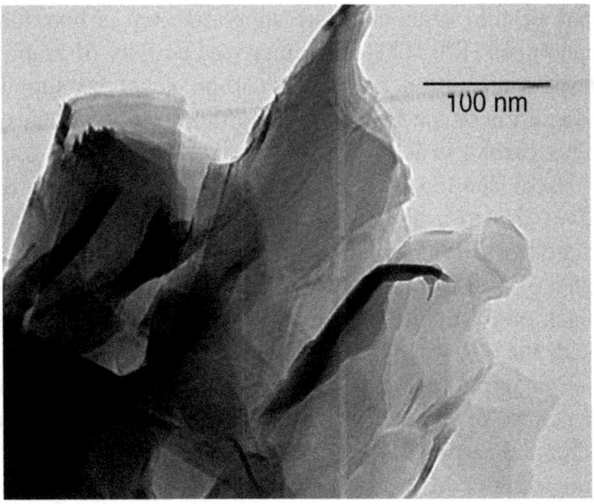

Fig. 5.6 TEM micrograph of few layer graphene

graphene is the oxidation of graphite to graphene oxide (GO), followed by exfoliation and reduction of GO to graphene nanosheets (GNS) [146–148].

Saner et al. [149] reported that the best method for the production of mostly exfoliated (minimum number of layers) graphene nanosheets is the oxidation of the sonicated graphite flake, ultrasonic treatment of GO, and chemical reduction of sonicated GO samples. The mild procedure applied was capable of reducing the average number of graphene sheets from 86 in the raw graphite to 9 in graphene-based nanosheets. The effective surface area of graphene sheets as catalyst support material relies on the layer number. When the layer number in the graphitic structure decreases, the effective surface area increases and thus increases the metal-support interaction.

Generally, the reduced graphene subjected to dying processes, tends to form irreversible agglomerates via van der Walls interactions that lead to a loss of surface area or even to reform graphite. To prevent this re-aggregation, and to increase catalyst dispersion, modified-graphene with epoxy, hydroxyl, and carboxyl groups, [149–154] poly(diallyldimethylammonium chloride) and nitrogen [155] have been largely investigated as catalyst supports.

The methods for preparing catalysts supported on graphene are essentially based on: (I) simultaneous reduction of mixtures consisting of GO and metal precursor; or (II) sequential reduction of GO and metal precursor. In the first case, metal precursors are deposited on GO and then are reduced simultaneously with different reducing agents such as ethylene glycol, [156–159] $NaBH_4$ [160], PVP/hydrazine, [161] sodium citrate, [162] microwave heating [163], and microwave polyol synthesis [164, 165]. Another synthesis method consists in the electrochemical reduction of GO, following by the electrochemical deposition of nanoparticles [137].

Pt and Pt-based anode electrocatalysts supported on graphene nanosheets (M-G) present higher catalytic activity in half cells for methanol oxidation than that of the same catalysts supported on carbon blacks [166–169] and carbon nanotubes [170, 171]. The improved activity of graphene supported catalysts with respect to carbon black and carbon nanotube supported catalysts was ascribed essentially to the lower metal particle size than that of the catalysts supported on other carbon structures, and to the presence of oxygenated groups on the graphene surface. The lower metal particle size, due to the large surface area of graphene, gives rise to a higher electrochemically active surface area of the catalyst. The catalytic activity of M-G (M = Pt, Pd, Pt-Ru) for the oxidation of hydrogen, ethanol, and formic acid [172] in half cells is higher compared with metal–carbon black and graphite. Conversely to the oxidation of ethanol, methanol, or formic acid, the ORR activity of M-G indicated contradictory results and in many cases the catalytic activity was lower [153, 173, 174]. It was believed that the sheet structure of graphene might slightly block oxygen diffusion compared with spherical carbon black particles. Instead, nitrogen doped graphene (N-G) has been shown to yield promising results for the sluggish cathodic ORR reaction [138, 175]. MEAs fabricated using Pt/N-G and Pt/G as the ORR catalyst showed a maximum power density of 440 and 390 $mWcm^{-2}$, respectively. The improved performance of Pt/N-G was attributed to the formation of pentagons and heptagons due to the incorporation of N in the C-backbone leading to an increase in the conductivity of the neighboring C atoms [176].

In the study of the methanol oxidation and hydrogen oxidation reactions research has also shown that Pt nanoparticles supported on graphene sheets exhibit high resistance to CO poisoning compared to Pt/C [139]. One possible explanation for the excellent CO tolerant HOR activity of the Pt-G is the presence of sub-nano-Pt clusters on G that may promote CO tolerance. Moreover, the chemical effect due to the modification of the Pt electronic structure by the G support may cause a difference in the catalytic activities. However, when the activity of Pt-G was compared with that of Pt/C, the performance of Pt-G was similar or even considerably lower than that of Pt/C. During the MEA fabrication, graphene sheets tend to horizontally stack, owing to their 2D structure, resulting in a decrease of active sites. One strategy to increase catalyst utilization is the addition of a spacing material such as carbon backs or CNTs to the catalyst layer. With this purpose, graphene sheets can be combined with various other components to form functional composites with enhanced properties. Within this area, graphene composites containing carbon nanotubes (CNTs) are now becoming an active research field and have already found applications in transparent conductors [177], energy storage (Li ion batteries and supercapacitors) [168], and energy conversion [178]. Their enhanced performances are often attributed to the fact that CNTs act as spacers between graphene layers, thus hindering the graphene restacking thus increasing the active surface areas, while at the same time providing good conductivity. Graphene composites with carbon blacks, carbon nanotubes carbon fibers, polypyrrole (PPy), and tin oxide (SnO_2) have been synthesized to prevent the face-to-face aggregation of the graphene nanosheets that may occur during the membrane electrode assembly (MEA) manufacturing [179].

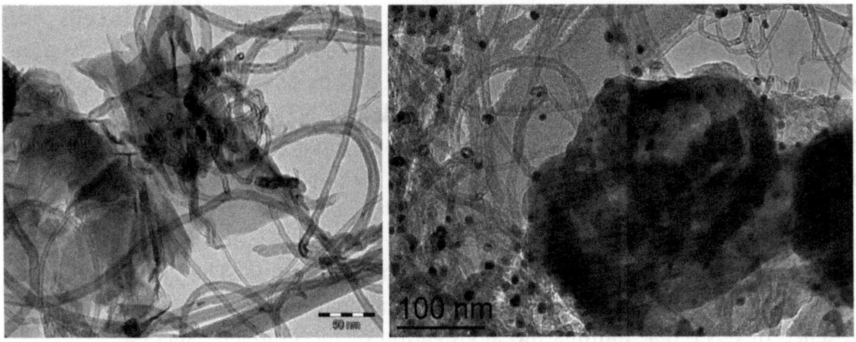

Fig. 5.7 TEM micrographs of (*left*) FLG-CNT composite, (*right*) Pd/FLG-CNT

In fact, the performance of a single cell PEMFC using graphene sheets-supported Pt catalyst mixed with carbon black as the cathode material considerably increased with respect to that of a PEMFC using Pt-G without carbon black, but was similar to that of commercial Pt/C. For methanol oxidation an improved single cell performance has been obtained by A. Cao, F. Kang et al. [180] with a N-doped carbon nanotube-graphene hybrid nanostructure. In this material, the graphene layers are distributed inside the CNT inner cavities that were designed to efficiently support noble metal (e.g., PtRu) nanoparticles. Compared to conventional CNTs and commercial catalysts, a much better catalytic performance was achieved by a synergistic effect of the hierarchical structure (graphene-CNT hybrid) and electronic modulation (N-doping) during the methanol electrooxidation reaction. Improved single cell performances with long-term stability were also demonstrated using CNT-G as catalyst support. Palladium nanoparticles (NPs) supported on the G-CNTs compared to Pd/Vulcan XC-72R carbon, Pd/G, or Pd/CNT catalysts, exhibit excellent electrocatalytic activity and stability for formic acid electrooxidation when the mass ratio of GO to CNTs is 5:1. The improved performance has been attributed to the inhibition of the agglomeration of graphene sheets and increased electrical conductivity brought about by CNTs. It is believed that the stability and electrical conductivity of G-CNT composite are increased by CNTs, on the other hand, the aggregation or restacking of graphene to form graphite platelets is effectively prevented by the presence of CNTs.

Recently P. Serp, F. Vizza et al. [181] have reported the electrooxidation in alkaline media of ethanol, ethylene glycol, and glycerol in half cell and passive direct ethanol fuel cells, on anode catalysts made of Pd nanoparticles supported on few layer graphene (FLG, Fig. 5.6) carbon nanotube, and a mixture of carbon nanotube-few layer graphene. Upon Pd deposition, a closer interaction was found to occur between metal and nanotube-graphene composite as the mean particle size was significantly smaller in this material (6.3 nm, Fig. 5.7), when compared with CNTs and FLG alone (8 and 8.4 nm, respectively), indicating a possible synergistic effect between CNT and FLG. CV experiments conducted with Pd/CNT, Pd/FLG, and Pd/CNT-FLG in 10 wt % ethanol in 2 M KOH solution, showed a high specific current. The oxidation of EG and G has also been investigated in half cells with Pd/CNT-FLG. The results obtained have highlighted the excellent electrocatalytic activity of the catalyst in terms of peak current density as high as 3.70 $mA{\cdot}\mu g^{-1}$ Pd for EG and 1.84 $mA{\cdot}\mu g^{-1}$ Pd for G. As a result Pd/CNT-FLG can be considered as among the best performing electrocatalysts ever reported for EG oxidation, especially considering the very low metal loading used in that work.

The higher catalytic activity observed in half cell with Pd/CNT-FLG electrocatalysts, toward the oxidation of ethanol and EG may be attributed to the specific proprieties of the FLG support such as its high conductivity. The synergistic effect of the mixture CNT-FLG as support for Pd nanoparticles can be attributed to the presence of CNT that inhibits the aggregation of the graphene sheets.

In conclusion, we can say that performances of DAFCs and DFAFCs, with anode catalysts supported on G alone or G-CNT composites are higher than catalysts supported on carbon blacks. On the other hand, their performance for hydrogen oxidation or the ORR reaction are similar or lower than conventional catalysts supported on carbon blacks materials. This may be due to a lower gas diffusion in these materials with respect to carbon blacks.

References

1. A. Kabbabi, F. Gloaguen, F. Andolfatto, R. Durand, Particle-size effect for oxygen reduction and methanol oxidation on Pt/C inside a proton-exchange membrane. J. Electroanal. Chem. **373**, 251 (1994)
2. K. Kinoshita, Particle-size effects for oxygen reduction on highly dispersed platinum in acid electrolytes. J. Electrochem. Soc. **137**, 845 (1990)
3. M. Watanabe, H. Sei, P. Stonehart, The influence of platinum crystallite Size on the electroreduction of oxygen. J. Electroanal. Chem. **261**, 375 (1989)
4. E. Antolini, Carbon supports for low-temperature fuel cell catalysts. Appl. Catal. B-Environ **88**, 1 (2009)
5. R.L. Augustine, *Heterogeneous Catalysis for the Synthetic Chemist* (Marcel Dekker Incorporated, New York, 1996)
6. F. Gloaguen, F. Andolfatto, R. Durand, P. Ozil, Kinetic-study of electrochemical reactions at catalyst-recast ionomer interfaces from thin active layer modeling. J. Appl. Electrochem. **24**, 863 (1994)
7. M.L. Sattler, P.N. Ross, The surface-structure of Pt crystallites supported on carbon-black. Ultramicroscopy **20**, 21 (1986)
8. K. Kinoshita, *Carbon: Electrochemical and Physicochemical Properties* (Wiley, New York, 1988)
9. T. Frelink, W. Visscher, J.A.R. Vanveen, Particle-size effect of carbon-supported platinum catalysts for the electrooxidation of methanol. J. Electroanal. Chem. **382**, 65 (1995)
10. K. Yahikozawa, Y. Fujii, Y. Matsuda, K. Nishimura, Y. Takasu, Electrocatalytic properties of ultrafine platinum particles for oxidation of methanol and formic-acid in aqueous-solutions. Electrochim. Acta **36**, 973 (1991)
11. M. Arenz et al., The effect of the particle size on the kinetics of CO electrooxidation on high surface area Pt catalysts. J. Am. Chem. Soc. **127**, 6819 (2005)
12. F. Maillard, E.R. Savinova, U. Stimming, CO monolayer oxidation on Pt nanoparticles: Further insights into the particle size effects. J. Electroanal. Chem. **599**, 221 (2007)

13. O. Antoine, Y. Bultel, R. Durand, P. Ozil, Electrocatalysis, diffusion and ohmic drop in PEMFC: Particle size and spatial discrete distribution effects. Electrochim. Acta **43**, 3681 (1998)
14. S. Park, Y. Xie, M.J. Weaver, Electrocatalytic pathways on carbon-supported platinum nanoparticles: Comparison of particle-size-dependent rates of methanol, formic acid, and formaldehyde electrooxidation. Langmuir **18**, 5792 (2002)
15. A.N. Gavrilov et al., On the influence of the metal loading on the structure of carbon-supported PtRu catalysts and their electrocatalytic activities in CO and methanol electrooxidation. Phys. Chem. Chem. Phys. **9**, 5476 (2007)
16. H. Gleiter, Nanostructured Materials. Adv. Mater. **4**, 474 (1992)
17. F. Maillard et al., Effect of the structure of Pt-Ru/C particles on COad monolayer vibrational properties and electrooxidation kinetics. Electrochimica Acta **53**, 811 (2007)
18. F. Maillard et al., Influence of particle agglomeration on the catalytic activity of carbon-supported Pt nanoparticles in CO monolayer oxidation. Phys. Chem. Chem. Phys. **7**, 385 (2005)
19. Y. Okamoto, Density-functional calculations of icosahedral M-13 (M = Pt and Au) clusters on graphene sheets and flakes. Chem. Phys. Lett. **420**, 382 (2006)
20. S.C. Roy, A.W. Harding, A.E. Russell, K.M. Thomas, Spectroelectrochemical study of the role played by carbon functionality in fuel cell electrodes. J. Electrochem. Soc. **144**, 2323 (1997)
21. A.K. Shukla et al., Electrooxidation of methanol in sulfuric-acid electrolyte on platinized-carbon electrodes with several functional-group characteristics. J. Electrochem. Soc. **141**, 1517 (1994)
22. V. Bambagioni et al., Single-site and nanosized Fe-Co electrocatalysts for oxygen reduction: Synthesis, characterization and catalytic performance. J. Power Sour. **196**, 2519 (2011)
23. K.E. Swider, D.R. Rolison, The chemical state of sulfur in carbon-supported fuel-cell electrodes. J. Electrochem. Soc. **143**, 813 (1996)
24. K.E. Swider, D.R. Rolison, Reduced poisoning of platinum fuel-cell electrocatalysts supported on desulfurized carbon. Electrochem. Solid State Lett. **3**, 4 (2000)
25. S.C. Roy, P.A. Christensen, A. Hamnett, K.M. Thomas, V. Trapp, Direct methanol fuel cell cathodes with sulfur and nitrogen-based carbon functionality. J. Electrochem. Soc. **143**, 3073 (1996)
26. C.C. Boyer, R.G. Anthony, A.J. Appleby, Design equations for optimized PEM fuel cell electrodes. J. Appl. Electrochem. **30**, 777 (2000)
27. M. Eikerling, A.A. Kornyshev, Electrochemical impedance of the cathode catalyst layer in polymer electrolyte fuel cells. J. Electroanal. Chem. **475**, 107 (1999)
28. Q.P. Wang, D.T. Song, T. Navessin, S. Holdcroft, Z.S. Liu, A mathematical model and optimization of the cathode catalyst layer structure in PEM fuel cells. Electrochim. Acta **50**, 725 (2004)
29. E. Antolini, L. Giorgi, A. Pozio, E. Passalacqua, Influence of Nafion loading in the catalyst layer of gas-diffusion electrodes for PEFC. J. Power Sour. **77**, 136 (1999)
30. J.M. Song, S.Y. Cha, W. M. Lee, Optimal composition of polymer electrolyte fuel cell electrodes determined by the AC impedance method. J. Power Sour. **94**, 78 (2001)
31. G.C. Li, P.G. Pickup, Ionic conductivity of PEMFC electrodes—effect of Nafion loading. J. Electrochem. Soc. **150**, C745 (2003)
32. U.A. Paulus et al., Fundamental investigation of catalyst utilization at the electrode/solid polymer electrolyte interface—part I. Development of a model system. J. Electroanal. Chem. **541**, 77 (2003)
33. R. Fernandez, P. Ferreira-Aparicio, L. Daza, PEMFC electrode preparation: Influence of the solvent composition and evaporation rate on the catalytic layer microstructure. J. Power Sour. **151**, 18 (2005)
34. M. Uchida et al., Influences of both carbon supports and heat-treatment of supported catalyst on electrochemical oxidation of methanol. J. Electrochem. Soc. **142**, 2572 (1995)

35. M. Watanabe, M. Tomikawa, S. Motoo, Preparation of a high-performance gas-diffusion electrode. J. Electroanal. Chem. **182**, 193 (1985)
36. O. Antoine, R. Durand, RRDE study of oxygen reduction on Pt nanoparticles inside Nafion (R): H2O2 production in PEMFC cathode conditions. J. Appl. Electrochem. **30**, 839 (2000)
37. K. Kinoshita, J. Bett, Electrochemical oxidation of carbon black in concentrated phosphoric acid at 135 °C. Carbon **11**, 237 (1973)
38. K.H. Kangasniemi, D.A. Condit, T.D. Jarvi, Characterization of vulcan electrochemically oxidized under simulated PEM fuel cell conditions. J. Electrochem. Soc. **151**, E125 (2004)
39. H. Binder, A. Köhling, K. Richter, G. Sandstede, Über die anodische oxydation von aktivkohlen in wässrigen elektrolyten. *Electrochimica Acta* **9**, 255 (1964)
40. L. Li, Y. Xing, Electrochemical durability of carbon nanotubes in noncatalyzed and catalyzed oxidations. J. Electrochem. Soc. **153**, A1823 (2006)
41. Y.Y. Shao, G.P. Yin, Y.Z. Gao, P.F. Shi, Durability study of Pt/C and Pt/CNTs catalysts under simulated PEM fuel cell conditions. J. Electrochem. Soc. **153**, A1093 (2006)
42. Y.Y. Shao, G.P. Yin, J. Zhang, Y.Z. Gao, Comparative investigation of the resistance to electrochemical oxidation of carbon black and carbon nanotubes in aqueous sulfuric acid solution. Electrochim. Acta **51**, 5853 (2006)
43. E. Guilminot et al., Membrane and active layer degradation upon PEMFC steady-state operation—I. Platinum dissolution and redistribution within the MEA. J. Electrochem. Soc. **154**, B1106 (2007)
44. J. Xie et al., Durability of PEFCs at high humidity conditions. J. Electrochem. Soc. **152**, A104 (2005)
45. Y. Yu et al., A review on performance degradation of proton exchange membrane fuel cells during startup and shutdown processes: Causes, consequences, and mitigation strategies. J. Power Sour. **205**, 10 (2012)
46. T. Payne, *Fuel Cells Durability and Performance* (The Knowledge Press Inc., US Brookline, 2009)
47. J.B.B. Donnet, R.C.; Wang, M.J., *Carbon Black* (Marcel Dekker, New York, 1993)
48. J.B.V. Donnet, A., *Carbon Black: Physics, Chemistry and Elastomer Reinforcement* (Marcel Dekker, Inc., New York, 1976)
49. E. Antolini, F. Cardellini, E. Giacometti, G. Squadrito, Study on the formation of Pt/C catalysts by non-oxidized active carbon support and a sulfur-based reducing agent. J. Mater. Sci. **37**, 133 (2002)
50. H. Gharibi, R.A. Mirzaie, E. Shams, M. Zhiani, M. Khairmand, Preparation of platinum electrocatalysts using carbon supports for oxygen reduction at a gas-diffusion electrode. J. Power Sour. **139**, 61 (2005)
51. V. Rao et al., The influence of carbon support porosity on the activity of PtRu/Sibunit anode catalysts for methanol oxidation. J. Power Sour. **145**, 178 (2005)
52. Y. Takasu, T. Kawaguchi, W. Sugimoto, Y. Murakami, Effects of the surface area of carbon support on the characteristics of highly-dispersed Pt-Ru particles as catalysts for methanol oxidation. Electrochimica Acta **48**, 3861 (2003)
53. M.A. Fraga et al., Properties of carbon-supported platinum catalysts: Role of carbon surface sites. .J. Catal. **209**, 355 (2002)
54. M. Watanabe, S. Saegusa, P. Stonehart, Electro-catalytic activity on supported platinum crystallites for oxygen reduction in sulfuric-acid. Chem. Lett., 1487 (1988)
55. M. Uchida, Y. Fukuoka, Y. Sugawara, H. Ohara, A. Ohta, Improved preparation process of very-low-platinum-loading electrodes for polymer electrolyte fuel cells. J. Electrochem. Soc. **145**, 3708 (1998)
56. D.S. Cameron, S.J. Cooper, I.L. Dodgson, B. Harrison, J.W. Jenkins, Carbons as supports for precious metal catalysts. Catalysis Today **7**, 113 (1990)
57. S. Dong Jin, P. Tae-Jin, I. Son-Ki, Effect of surface oxygen groups of carbon supports on the characteristics of Pd/C catalysts. Carbon **31**, 427 (1993)

58. C. Prado-Burguete, A. Linares-Solano, F. Rodríguez-Reinoso, C.S.-M. de Lecea, The effect of oxygen surface groups of the support on platinum dispersion in Pt/carbon catalysts. J. Catal. **115**, 98 (1989)
59. S. Kim, S.J. Park, Preparation and electrochemical behaviors of polymeric composite electrolytes containing mesoporous silicate fillers. Electrochim. Acta **52**, 3477 (2007)
60. C.K. Poh, S.H. Lim, H. Pan, J.Y. Lin, J.Y. Lee, Citric acid functionalized carbon materials for fuel cell applications. J. Power Sour. **176**, 70 (2008)
61. H. Shioyama et al., C2F6 plasma treatment of a carbon support for a PEM fuel cell electrocatalyst. J. Power Sour. **161**, 836 (2006)
62. Z.B. Wang, G.P. Yin, P.F. Shi, Effects of ozone treatment of carbon support on Pt-Ru/C catalysts performance for direct methanol fuel cell. Carbon **44**, 133 (2006)
63. C.Y. Chen, C.S. Tsao, Characterization of electrode structures and the related performance of direct methanol fuel cells. Int. J. Hydrogen Energy **31**, 391 (Mar, 2006)
64. J.H. Kim, H.Y. Ha, I.H. Oh, S.A. Hong, H.I. Lee, Influence of the solvent in anode catalyst ink on the performance of a direct methanol fuel cell. J. Power Sour. **135**, 29 (2004)
65. J. Prabhuram, T.S. Zhao, Z.X. Liang, H. Yang, C.W. Wong, Pd and Pd–Cu alloy deposited nafion membranes for reduction of methanol crossover in direct methanol fuel cells. J. Electrochem. Soc. **152**, A1390 (2005)
66. J. Prabhuram, T.S. Zhao, Z.K. Tang, R. Chen, Z.X. Liang, Multiwalled carbon nanotube supported PtRu for the anode of direct methanol fuel cells. J. Phys. Chem. B **110**, 5245 (2006)
67. S.Q. Song et al., Direct methanol fuel cells: The effect of electrode fabrication procedure on MEAs structural properties and cell performance. J. Power Sour. **145**, 495 (2005)
68. Z.B. Wei et al., Influence of electrode structure on the performance of a direct methanol fuel cell. J. Power Sour. **106**, 364 (2002)
69. H. Kim, W. Lee, D. Yoo, Functionalized carbon support with sulfonated polymer for direct methanol fuel cells. Electrochim. Acta **52**, 2620 (2007)
70. G. Selvarani et al., A phenyl-sulfonic acid anchored carbon-supported platinum catalyst for polymer electrolyte fuel cell electrodes. Electrochim. Acta **52**, 4871 (2007)
71. B. Choi, H. Yoon, I.S. Park, J. Jang, Y.E. Sung, Highly dispersed pt nanoparticles on nitrogen-doped magnetic carbon nanoparticles and their enhanced activity for methanol oxidation. Carbon **45**, 2496 (2007)
72. J. Lee, J. Kim, T. Hyeon, Recent progress in the synthesis of porous carbon materials. Adv. Mater. **18**, 2073 (2006)
73. T.Y. Ma, L. Liu, Z.Y. Yuan, Direct synthesis of ordered mesoporous carbons. Chem. Soc. Rev. **42**, 3977 (2013)
74. T. Kyotani, Control of pore structure in carbon. Carbon **38**, 269 (2000)
75. J. Lee, S. Han, T. Hyeon, Synthesis of new nanoporous carbon materials using nanostructured silica materials as templates. J. Mater. Chem. **14**, 478 (2004)
76. A.H. Lu, F. Schuth, Nanocasting: A versatile strategy for creating nanostructured porous materials. Adv. Mater. **18**, 1793 (2006)
77. H.F. Yang, D.Y. Zhao, Synthesis of replica mesostructures by the nanocasting strategy. J. Mater. Chem. **15**, 1217 (2005)
78. R. Ryoo, S.H. Joo, Nanostructured carbon materials synthesized from mesoporous silica crystals by replication. Mesop. Cryst. Relat. Nano-Struct. Mater. **148**, 241 (2004)
79. H. Chang, S.H. Joo, C. Pak, Synthesis and characterization of mesoporous carbon for fuel cell applications. J. Mater. Chem. **17**, 3078 (2007)
80. J.H. Nam, Y.Y. Jang, Y.U. Kwon, J.D. Nam, Direct methanol fuel cell Pt-carbon catalysts by using SBA-15 nanoporous templates. Electrochem. Commun. **6**, 737 (2004)
81. J.S. Yu, S. Kang, S.B. Yoon, G. Chai, Fabrication of ordered uniform porous carbon networks and their application to a catalyst supporter. J. Am. Chem. Soc. **124**, 9382 (2002)
82. G.S. Chai, S.B. Yoon, J.S. Yu, J.H. Choi, Y.E. Sung, Ordered porous carbons with tunable pore sizes as catalyst supports in direct methanol fuel cell. J. Phys. Chem. B **108**, 7074 (2004)

83. V. Raghuveer, A. Manthiram, Mesoporous carbon with larger pore diameter as an electrocatalyst support for methanol oxidation. Electrochem. Solid State Lett. **7**, A336 (2004)
84. A.M. ElKhatat, S.A. Al-Muhtaseb, Advances in tailoring resorcinol-formaldehyde organic and carbon gels. Adv. Mater. **23**, 2887 (2011)
85. H.D. Du et al., Influences of mesopore size on oxygen reduction reaction catalysis of Pt/carbon aerogels. J. Phys. Chem. C **111**, 2040 (2007)
86. N. Job et al., Electrochemical characterization of Pt/carbon xerogel and Pt/carbon aerogel catalysts: first insights into the influence of the carbon texture on the Pt nanoparticle morphology and catalytic activity. J. Mater. Sci. **44**, 6591 (2009)
87. J. Marie et al., Highly porous PEM fuel cell cathodes based on low density carbon aerogels as Pt-support: Experimental study of the mass-transport losses. J. Power Sour. **190**, 423 (2009)
88. A. Smirnova, X. Dong, H. Hara, A. Vasiliev, N. Sammes, Novel carbon aerogel-supported catalysts for PEM fuel cell application. Int. J. Hydrogen Energy **30**, 149 (2005)
89. C. Arbizzani, S. Beninati, E. Manferrari, F. Soavi, M. Mastragostino, Cryo- and xerogel carbon supported PtRu for DMFC anodes. J. Power Sour. **172**, 578 (2007)
90. J.L. Figueiredo et al., Development of carbon nanotube and carbon xerogel supported catalysts for the electro-oxidation of methanol in fuel cells. Carbon **44**, 2516 (2006)
91. H.D. Du, B.H. Li, F. Y. Kang, R.W. Fu, Y.Q. Zeng, Carbon aerogel supported Pt–Ru catalysts for using as the anode of direct methanol fuel cells. Carbon **45**, 429 (2007)
92. S. Iijima, Helical Microtubules of Graphitic Carbon. Nature **354**, 56 (1991)
93. S. Iijima, T. Ichihashi, Single-shell carbon nanotubes of 1-Nm diameter. Nature **363**, 603 (1993)
94. M. Monthioux, V.L. Kuznetsov, Who should be given the credit for the discovery of carbon nanotubes? Carbon **44**, 1621 (2006)
95. J.W. Mintmire, B.I. Dunlap, C.T. White, are fullerene tubules metallic. Phys. Rev. Lett. **68**, 631 (1992)
96. R. Saito, M. Fujita, G. Dresselhaus, M.S. Dresselhaus, Electronic-structure of chiral graphene tubules. Appl. Phys. Lett. **60**, 2204 (1992)
97. H.J. Dai, Carbon nanotubes: Synthesis, integration, and properties. Acc. Chem. Res. **35**, 1035 (2002)
98. J.C. Charlier, Defects in carbon nanotubes. Acc. Chem. Res. **35**, 1063 (2002)
99. X. Lu, Z.F. Chen, Curved Pi-conjugation, aromaticity, and the related chemistry of small fullerenes (<C-60) and single-walled carbon nanotubes. Chem. Rev. **105**, 3643 (2005)
100. M. Menon, A.N. Andriotis, G.E. Froudakis, Curvature dependence of the metal catalyst atom interaction with carbon nanotubes walls. Chem. Phys. Lett. **320**, 425 (2000)
101. Y.C. Xing, Synthesis and electrochemical characterization of uniformly-dispersed high loading Pt nanoparticles on sonochemically-treated carbon nanotubes. J. Phys. Chem. B **108**, 19255 (2004)
102. C.H. Lau et al., The effect of functionalization on structure and electrical conductivity of multi-walled carbon nanotubes. J. Nanopart. Res. **10**, 77 (2008)
103. A.R. Govindarai, C.N.R., in *The Chemistry of Nanomaterials,* ed. by C. N. R. M. Rao, A. Müller, A.K. Cheetham (Wiley-VCH Verlag GmbH & Co., Weinheim, 2004)
104. P. Serp, M. Corrias, P. Kalck, Carbon nanotubes and nanofibers in catalysis. Appl.Catal.-Gen. **253**, 337 (2003)
105. P. Serp, J. L. Gigueiredo (Eds.) *Carbon Materials for Catalysis* (Wiley, Oboken, New Jersey, 2009)
106. M.S. Monthioux, P.; Flahaut, E. Laurent, C. Peigney, M. Razafinimanana, W. Bacsa, J.M. Broto, *Handbook of Nanotechnology* 2nd edn. (Springer, Heidelberg, 2007)
107. R. Dagani, Much ado about nanotubes. Chem. Eng. News **77**, 31 (1999)
108. M.M.J. Treacy, T.W. Ebbesen, J.M. Gibson, Exceptionally high Young's modulus observed for individual carbon nanotubes. Nature **381**, 678 (1996)

109. T.W. Ebbesen, P.M. Ajayan, Large-scale synthesis of carbon nanotubes. Nature **358**, 220 (1992)
110. A. Thess et al., Crystalline ropes of metallic carbon nanotubes. Science **273**, 483 (1996)
111. A. Jorio, G. Dresselhaus, M. S. Dresselhaus, (Eds.) Advanced Topics in the Synthesis, Structure, Properties and Applications Series: Topics in Applied Physics, Vol. 111 (Springer, New York, 2008)
112. Y.H. Chung, S. Jou, Carbon nanotubes from catalytic pyrolysis of polypropylene. Mater. Chem. Phys. **92**, 256 (2005)
113. M.S. Saha, R.Y. Li, X.H. Sun, High loading and monodispersed Pt nanoparticles on multiwalled carbon nanotubes for high performance proton exchange membrane fuel cells. J. Power Sources **177**, 314 (2008)
114. T. Maiyalagan, B. Viswanathan, U. Varadaraju, Nitrogen containing carbon nanotubes as supports for Pt—alternate anodes for fuel cell applications. Electrochem. Commun. **7**, 905 (2005)
115. K.P. Gong, F. Du, Z.H. Xia, M. Durstock, L.M. Dai, Nitrogen-doped carbon nanotube arrays with high electrocatalytic activity for oxygen reduction. Science **323**, 760 (2009)
116. Z. Chen, D. Higgins, H.S. Tao, R.S. Hsu, Z.W. Chen, Highly active nitrogen-doped carbon nanotubes for oxygen reduction reaction in fuel cell applications. J. Phys. Chem. C **113**, 21008 (2009)
117. Y.F. Tang, B.L. Allen, D.R. Kauffman, A. Star, Electrocatalytic activity of nitrogen-doped carbon nanotube cups. J. Am. Chem. Soc. **131**, 13200 (2009)
118. D.S. Yu, Q. Zhang, L.M. Dai, Highly efficient metal-free growth of nitrogen-doped single-walled carbon nanotubes on plasma-etched substrates for oxygen reduction. J. Am. Chem. Soc. **132**, 15127 (2010)
119. L. Gan, R.T. Lv, H.D. Du, B.H. Li, F.Y. Kang, High loading of Pt–Ru nanocatalysts by pentagon defects introduced in a bamboo-shaped carbon nanotube support for high performance anode of direct methanol fuel cells. Electrochem. Commun. **11**, 355 (2009)
120. Z. L. Liu et al., Preparation and characterization of platinum-based electrocatalysts on multiwalled carbon nanotubes for proton exchange membrane fuel cells. Langmuir **18**, 4054 (2002)
121. C.H. Wang et al., High performance of low electrocatalysts loading on CNT directly grown on carbon cloth for DMFC. J. Power Sour. **171**, 55 (2007)
122. Z. Tang et al., Enhanced catalytic properties from platinum nanodots covered carbon nanotubes for proton-exchange membrane fuel cells. J. Power Sour. **195**, 155 (2010)
123. R. Ahmadi, M.K. Amini, Synthesis and characterization of Pt nanoparticles on sulfur-modified carbon nanotubes for methanol oxidation. Int. J. Hydrogen Energy **36**, 7275 (2011)
124. S. Park et al., Polarization losses under accelerated stress test using multiwalled carbon nanotube supported Pt catalyst in PEM fuel cells. J. Electrochem. Soc. **158**, B297 (2011)
125. G. Wu, B.Q. Xu, Carbon nanotube supported Pt electrodes for methanol oxidation: A comparison between multi- and single-walled carbon nanotubes. J. Power Sour. **174**, 148 (2007)
126. M. Carmo, V.A. Paganin, J.M. Rosolen, E.R. Gonzalez, Alternative supports for the preparation of catalysts for low-temperature fuel cells: the use of carbon nanotubes. J. Power Sour. **142**, 169 (2005)
127. D.P. He et al., Polyaniline-functionalized carbon nanotube supported platinum catalysts. Langmuir **27**, 5582 (2011)
128. Y.C. Zhao, X.L. Yang, J.N. Tian, F.Y. Wang, L. Zhan, A facile and novel approach toward synthetic polypyrrole oligomers functionalization of multi-walled carbon nanotubes as PtRu catalyst support for methanol electro-oxidation. J. Power Sour. **195**, 4634 (2010)
129. V. Bambagioni et al., Pd and Pt–Ru anode electrocatalysts supported on multi-walled carbon nanotubes and their use in passive and active direct alcohol fuel cells with an anion-exchange membrane (alcohol = methanol, ethanol, glycerol). J. Power Sour. **190**, 241 (2009)

130. L. Li, Y.C. Xing, Electrochemical durability of carbon nanotubes in noncatalyzed and catalyzed oxidations. J. Electrochem. Soc. **153**, A1823 (2006)
131. A. Santasalo-Aarnio et al., Durability of different carbon nanomaterial supports with PtRu catalyst in a direct methanol fuel cell. Int. J. Hydrogen Energy **37**, 3415 (2012)
132. B. Viswanathan, Architecture of carbon support for Pt anodes in direct methanol fuel cells. Catal. Today **141**, 52 (2009)
133. J.J. Wang, G.P. Yin, Y.Y. Shao, Z.B. Wang, Y.Z. Gao, Investigation of further improvement of platinum catalyst durability with highly graphitized carbon nanotubes support. J. Phys. Chem. C **112**, 5784 (2008)
134. A.K.N. Geim, K.S., The rise of graphene. Nat. Mater. **6**, 9 (2007)
135. K.S. Novoselov et al., Electric field effect in atomically thin carbon films. Science **306**, 666 (2004)
136. S.J. Guo, S.J. Dong, E.K. Wang, Three-dimensional Pt-on-Pd bimetallic nanodendrites supported on graphene nanosheet: Facile synthesis and used as an advanced nanoelectrocatalyst for methanol oxidation. Acs Nano **4**, 547 (2010)
137. S. Liu et al., "Green" electrochemical synthesis of Pt/graphene sheet nanocomposite film and its electrocatalytic property. J. Power Sour. **195**, 4628 (2010)
138. L.T. Qu, Y. Liu, J.B. Baek, L.M. Dai, Nitrogen-doped graphene as efficient metal-free electrocatalyst for oxygen reduction in fuel cells. Acs Nano **4**, 1321 (2010)
139. N. Soin, S.S. Roy, T.H. Lim, J.A.D. McLaughlin, Microstructural and electrochemical properties of vertically aligned few layered graphene (FLG) nanoflakes and their application in methanol oxidation. Mater. Chem. Phys. **129**, 1051 (2011)
140. M.J. Allen, V.C. Tung, R.B. Kaner, Honeycomb carbon: A review of graphene. Chem. Rev. **110**, 132 (2010)
141. C. Berger et al., Electronic confinement and coherence in patterned epitaxial graphene. Science **312**, 1191 (2006)
142. K.S. Kim et al., Large-scale pattern growth of graphene films for stretchable transparent electrodes. Nature **457**, 706 (2009)
143. D.A. Dikin et al., Preparation and characterization of graphene oxide paper. Nature **448**, 457 (2007)
144. Y. Hernandez et al., High-yield production of graphene by liquid-phase exfoliation of graphite. Nat. Nanotechnol. **3**, 563 (2008)
145. S. Park, R.S. Ruoff, Chemical methods for the production of graphenes. Nat. Nanotechnol. **4**, 217 (Apr, 2009)
146. H.Y. He, T. Riedl, A. Lerf, J. Klinowski, Solid-state NMR studies of the structure of graphite oxide. J. Phys. Chem. **100**, 19954 (1996)
147. C. Hontoria-Lucas, A.J. López-Peinado, J. de. D. López-González, M.L. Rojas-Cervantes, R.M. Martín-Aranda, Study of oxygen-containing groups in a series of graphite oxides: Physical and chemical characterization. Carbon **33**, 1585 (1995)
148. A. Lerf, H.Y. He, T. Riedl, M. Forster, J. Klinowski, C-13 and H-1 MAS NMR studies of graphite oxide and its chemically modified derivatives. Solid State Ion **101**, 857 (1997)
149. B. Saner, F. Okyay, Y. Yürüm, Utilization of multiple graphene layers in fuel cells. 1. An improved technique for the exfoliation of graphene-based nanosheets from graphite. Fuel **89**, 1903 (2010)
150. S.M. Choi, M.H. Seo, H.J. Kim, W.B. Kim, Synthesis and characterization of graphene-supported metal nanoparticles by impregnation method with heat treatment in H_2 atmosphere. Synth. Metals **161**, 2405 (2011)
151. H.J. Kim et al., Efficient electrooxidation of biomass-derived glycerol over a graphene-supported PtRu electrocatalyst. Electrochem. Commun. **13**, 890 (2011)
152. R. Kou et al., Enhanced activity and stability of Pt catalysts on functionalized graphene sheets for electrocatalytic oxygen reduction. Electrochem. Commun. **11**, 954 (2009)
153. S. Park et al., Design of graphene sheets-supported Pt catalyst layer in PEM fuel cells. Electrochem. Commun. **13**, 258 (2011)

154. M.H. Seo, S.M. Choi, H.J. Kim, W.B. Kim, The graphene-supported Pd and Pt catalysts for highly active oxygen reduction reaction in an alkaline condition. Electrochem. Commun. **13**, 182 (2011)
155. Y. Shao et al., Highly durable graphene nanoplatelets supported Pt nanocatalysts for oxygen reduction. J. Power Sour. **195**, 4600 (2010)
156. H.W. Ha, I.Y. Kim, S.J. Hwang, R.S. Ruoff, One-pot synthesis of platinum nanoparticles embedded on reduced graphene oxide for oxygen reduction in methanol fuel cells. Electrochem. Solid State Lett. **14**, B70 (2011)
157. W. He et al., An efficient reduction route for the production of Pd–Pt nanoparticles anchored on graphene nanosheets for use as durable oxygen reduction electrocatalysts. Carbon **50**, 265 (2012)
158. C.V. Rao, A.L.M. Reddy, Y. Ishikawa, P.M. Ajayan, Synthesis and electrocatalytic oxygen reduction activity of graphene-supported Pt3Co and Pt3Cr alloy nanoparticles. Carbon **49**, 931 (2011)
159. C. Xu, X. Wang, J.W. Zhu, Graphene-metal particle nanocomposites. J. Phys. Chem. C **112**, 19841 (2008)
160. Y.C. Xin et al., Preparation and characterization of Pt supported on graphene with enhanced electrocatalytic activity in fuel cell. J. Power Sour. **196**, 1012 (2011)
161. Y.J. Hu, P. Wu, Y.J. Yin, H. Zhang, C.X. Cai, Effects of structure, composition, and carbon support properties on the electrocatalytic activity of Pt–Ni-graphene nanocatalysts for the methanol oxidation. Appl. Catal. B-Environ. **111**, 208 (2012)
162. Y.J. Wang, J.C. Liu, L. Liu, D.D. Sun, High-quality reduced graphene oxide-nanocrystalline platinum hybrid materials prepared by simultaneous co-reduction of graphene oxide and chloroplatinic acid. Nanoscale Res. Lett. 1–8 **6** (2011)
163. H. Zhang et al., Microwave-assisted synthesis of graphene-supported Pd-1 Pt-3 nanostructures and their electrocatalytic activity for methanol oxidation. Electrochim. Acta **56**, 7064 (2011)
164. P. Kundu et al., Ultrafast microwave-assisted route to surfactant-free ultrafine Pt nanoparticles on graphene: Synergistic co-reduction mechanism and high catalytic activity. Chem. Mater. **23**, 2772 (2011)
165. S. Sharma et al., Rapid microwave synthesis of CO tolerant reduced graphene oxide-supported platinum electrocatalysts for oxidation of methanol. J. Phys. Chem. C **114**, 19459 (2010)
166. L.F. Dong, R.R.S. Gari, Z. Li, M.M. Craig, S.F. Hou, Graphene-supported platinum and platinum-ruthenium nanoparticles with high electrocatalytic activity for methanol and ethanol oxidation. Carbon **48**, 781 (2010)
167. Y.M. Li, L.H. Tang, J.H. Li, Preparation and electrochemical performance for methanol oxidation of pt/graphene nanocomposites. Electrochem. Commun. **11**, 846 (2009)
168. E. Yoo et al., Enhanced electrocatalytic activity of Pt subnanoclusters on graphene nanosheet surface. Nano Lett. **9**, 2255 (2009)
169. Y.-G. Zhou, J–.J. Chen, F.-B. Wang, Z.-H. Sheng, X.-H. Xia, A facile approach to the synthesis of highly electroactive Pt nanoparticles on graphene as an anode catalyst for direct methanol fuel cells. Chem. Commun. **46**, 5951 (2010)
170. S.H. Lee, N. Kakati, S.H. Jee, J. Maiti, Y.-S. Yoon, Hydrothermal synthesis of PtRu nanoparticles supported on graphene sheets for methanol oxidation in direct methanol fuel cell. Mater. Lett. **65**, 3281 (2011)
171. R.N. Singh, R. Awasthi, Graphene support for enhanced electrocatalytic activity of Pd for alcohol oxidation. Catal. Sci. Technol. **1**, 778 (2011)
172. S. Bong, S. Uhm, Y.-R. Kim, J. Lee, H. Kim, Graphene supported Pd electrocatalysts for formic acid oxidation. Electrocatalysis **1**, 139 (2010)
173. H. Wu, D. Wexler, H. Liu, Durability investigation of graphene-supported Pt nanocatalysts for PEM fuel cells. J. Solid State Electrochem. **15**, 1057 (2011)
174. Y.S. Yun, D. Kim, Y. Tak, H.-J. Jin, Porous graphene/carbon nanotube composite cathode for proton exchange membrane fuel cell. Synth. Metals **161**, 2460 (2011)

175. S. Scire et al., Preparation of ceria and titania supported Pt catalysts through liquid phase photo-deposition. J. Mol. Catal. a-Chem. **333**, 100 (2010)
176. R.I. Jafri, N. Rajalakshmi, S. Ramaprabhu, Nitrogen doped graphene nanoplatelets as catalyst support for oxygen reduction reaction in proton exchange membrane fuel cell. J. Mater. Chem. **20**, 7114 (2010)
177. V.C. Tung et al., Low-Temperature solution processing of graphene-carbon nanotube hybrid materials for high-performance transparent conductors. Nano Lett. **9**, 1949 (2009)
178. S. Yang et al., Preparation and electrochemistry of graphene nanosheets–multiwalled carbon nanotubes hybrid nanomaterials as Pd electrocatalyst support for formic acid oxidation. Electrochim. Acta **62**, 242 (2012)
179. E. Antolini, Graphene as a new carbon support for low-temperature fuel cell catalysts. Appl. Catal. B: Environ. **123–124**, 52 (2012)
180. R. Lv et al., Open-ended, N-doped carbon nanotube–graphene hybrid nanostructures as high-performance catalyst support. Adv. Funct. Mater. **21**, 999 (2011)
181. F. Machado, A. Marchionni, R.R. Bacsa, M. Bellini, J. Beausoleil, W. Oberhauser, F. Vizza, P. Serp, Philippe, Synergistic effect between few layer graphene and carbon nanotube supports for palladium catalyzing electrochemical oxidation of alcohols. J. Energy Chem. **22**, 296 (2013)

Chapter 6
Other Support Nanomaterials

6.1 Key Concepts

Corrosion of carbon supports has been identified as one of the major factors hampering the durability of fuel cell electrocatalysts. In Chap. 5 we have seen that carbon support corrosion mainly occurs at the fuel cell cathode and is accelerated by the presence, even in traces, of hydrogen peroxide. While it has been undoubtedly demonstrated that mesoporous carbon, graphene, and carbon nanotubes [1–3] may improve carbon oxidation resistance as compared to carbon blacks, they do not completely resolve the problem. Oxidation of elemental carbon at the expense of molecular oxygen or hydrogen peroxide is indeed a thermodynamically spontaneous process. Carbon corrosion may be complete or partial. Oxidation may lead to the formation of CO_2 or CO, causing detachment of metal nanoparticles from the substrate and in the end lowering the device performance. On the other side, partial oxidation with the formation of oxygenated functional groups at the carbon surface has been demonstrated to increase the mobility of the metal catalyst nanoparticles. This favors particle contact ultimately leading to a progressive decrease of the catalyst electrochemically active surface area through agglomeration.

The quest for materials exceeding the stability properties of carbon blacks, nanostructured and mesoporous carbons must also seek materials with the following prerequisites:

(1) High surface area
(2) Controlled Porosity
(3) High electrical conductivity.

High surface area is required to guarantee an optimal metal nanoparticle dispersion. Porosity has to be tailored in order to allow an optimal flow of products and reactants without incurring mass transport limitations. It is worth mentioning that the first two properties may be obtained with materials showing multiscale porosity [4] with small pores in the nanometer scale range and large pores in the micrometer scale range. As a rule of thumb, at least for the case of cathode

A. Lavacchi et al., *Nanotechnology in Electrocatalysis for Energy*,
Nanostructure Science and Technology 170, DOI: 10.1007/978-1-4899-8059-5_6,

© Springer Science+Business Media New York 2013

catalysts for PEMFC, the support should allow dispersion of the Pt catalyst phase with surface area in the range of 100 $m^2\ g^{-1}$. Electrical conductivity is essential to correctly close the circuit on the load without substantial ohmic and hence efficiency losses.

These properties have of course to be combined with stability under operating conditions which implies both good adhesion of metal nanoparticles and resistance to strongly oxidizing, reducing, acidic, or alkaline conditions. Among the very many materials proposed as candidates for nanostructured electrocatalyst supports, in the present chapter we will review those we believe have been the most extensively investigated. These include conductive metal oxides, metal carbides other ceramic materials, and hybrid and composite materials. A brief review of the application of conductive polymers will also be considered. The literature in this field is quickly expanding and a variety of new materials come onto the scene each year. Necessarily, the present review cannot account for all the findings in the field. Rather it presents a variety of approaches that have been proved to be effective and potentially exploitable. In all the examined cases, nanotechnology plays a fundamental role for addressing application targets. Nanostructures are indeed required to control both porosity and high surface area. The use of support materials different from carbon can also lead to an actual improvement in the catalytic activity of metal nanoparticles through synergistic effects. Such synergistic effects usually involve strong metal to support interactions and spillover phenomena, which have been exploited in DAFCs [5] for enhancing the energy efficiency and in the enhancement of the ORR [6, 7].

Last but not least production processes using these alternative support materials should be suitable for a full technological exploitation, guaranteeing the manufacturability of suitable electrode architectures at reasonable cost and with good process reliability.

6.2 Inorganic Oxides

Most of the effort in the application of inorganic oxides as electrocatalyst support materials has been concerned with the development of suitable architectures of conducting metal oxides. For example mesoporous silica has been recently considered as a possible candidate material, nevertheless its application in fuel cells is still limited. Furthermore, an incredibly wild variety of methods for preparing nanoarchitectures of conducting metal oxides have been recently reported. According to Wang et al. [8] metal oxides show reasonable surface areas, mechanical strength, and both thermal and hydrothermal stability. In the discussion that follows, we show that some types of metal oxides are indeed good candidates as support materials as they can provide; homogeneous dispersion of metal nanoparticles, favorable metal to support interactions, and high catalytic activity, thus fulfilling the requirements for fuel cells applications. Our discussion

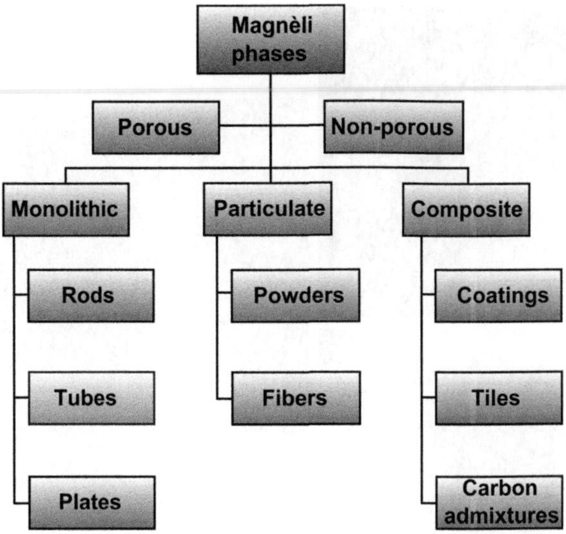

Scheme 6.1 The main physical forms in which magnèli phase are available

will mainly focuses on the application of two materials titanium and wolframium oxides, respectively. Other oxides are briefly considered in a separate section.

6.2.1 Sub-stoichiometric Titanium Oxides

Titanium is known to form a variety of oxides with tunable properties. On the side of the application as electrocatalyst support, certainly the most relevant ones are the so-called "magneli phases." Magneli phases have been produced in a wide variety of sizes and shapes (Scheme 6.1) and are also available commercially. Amongst all the possible titanium oxides, those with the stoichiometry Ti_4O_7 and Ti_5O_9 have been proved to show the highest stability toward corrosion under both acidic and alkaline conditions. Furthermore, they have high electronic conductivity ($>10^3$ S cm^{-1} for Ti_4O_7) [9] and $>10^2$ S cm^{-1} for Ti_5O_9) making them suitable for application in both fuel cells and electrolyzers [10].

Pure Ti_4O_7 has been prepared by reduction of ultrafine rutile TiO_2 under hydrogen at 1,050 °C for 50 min [11, 12]. Noble metal electrocatalysts, such as Pt, Ru, and Ir have been added to the thus obtained powders, resulting in electrode architectures with high catalytic activity and stability for fuel cell applications. The corrosion of such materials has been investigated in detail. It has been shown by Ioroi et al. [13] that the recorded anodic current at potentials larger than 1.0 V RHE is much smaller for Ti_4O_7 as compared to the conventional Vulcan XC-72 supports. Other accelerated corrosion tests performed by Ioroi et al. [14] demonstrated a significant improvement when Pt/Ti_4O_7 was used as a catalyst as

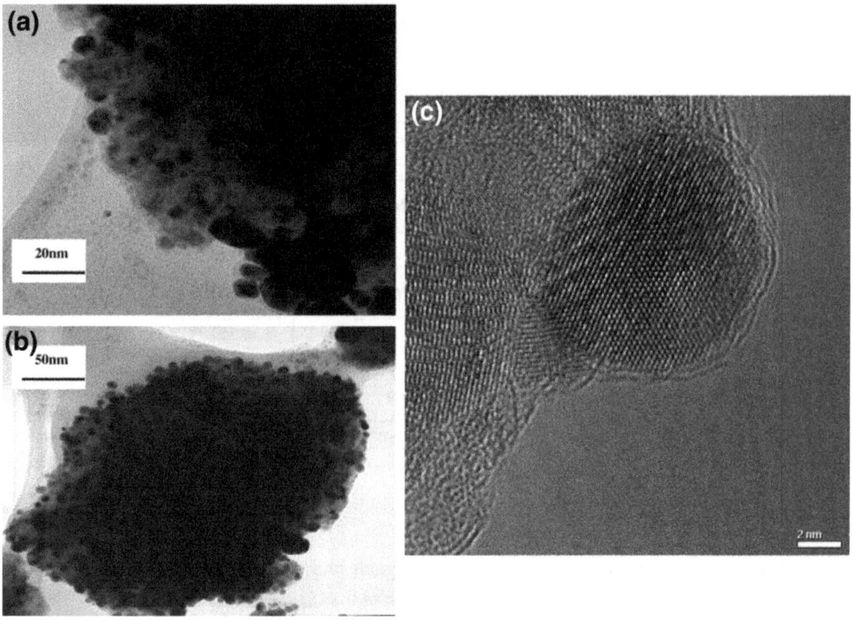

Fig. 6.1 TEM (**a**, **b**) and HRTEM (**c**) images of Pt nanoparticles on Ebonex^{TM} substrate. Low magnification overview showing the Pt particle distribution on Ebonex support. Reproduced from Ref. [18] with permission of Elsevier

compared to Pt/XC-72. Tests consisted of measuring the Pt EASA as a function of the electrode potential in the interval 1.0–1.5 V RHE in 0.1 V steps. The potential was kept constant for 1 h for each single step. At such potentials it has been found that carbon-based catalysts start to show significant reduction of the electrochemically active surface area, while titania-based materials retain the starting EASA values. The titania-based materials also demonstrated a superior stability in tests in a complete PEMFC, with a duration of 350 h at 80 °C feeding the cell with H_2 and O_2.

At present electrically conductive mixtures of Ti_4O_7 and Ti_5O_9 are commercially available. They are commercialized under the name of Ebonex^{TM}, produced by Ebonex Atraverda Ltd (Sheffield, UK). In agreement with what has been found for Ti_4O_7 Ebonex^{TM} has been reported to show good conductivity, and sufficient mechanical and chemical stability for application as an electrocatalyst support. Furthermore, Ebonex^{TM} is a microporous material which allows the deposition of metal nanoparticles with good dispersion (Fig. 6.1). A variety of researchers have reported aspecific metal nanoparticle interaction with the Ebonex^{TM} catalyst support [14, 15]. This has been confirmed by the investigations performed by Slavcheva et al. [16] through the synthesis of PtCo nanoparticles on Ebonex^{TM}. Because of the interaction with the support it has been demonstrated that the OER starts at an overpotential lower than that of bare PtCo. Even the ORR on such catalysts has been demonstrated to be very effective. PtCo/Ebonex^{TM} delivered better ORR performance than bare PtCo. Furthermore, Vracar et al. [17] found the ORR to be faster on Pt/Ebonex than on polycrystalline Pt. The electrochemical stability of the Pt/Ebonex^{TM} system in alkaline and acidic environment has been investigated [9, 12]. It has been found that at a Pt loading of 2 mg cm^{-2} stability has been achieved for 500 h in a flow electrolysis cell fed with a 1 M NaOH electrolyte. Under oxygen evolution conditions in a H_2SO_4, electrolyte stability has been found to be rather poor as compared to under alkaline conditions. It is probable that the use of Ebonex supported materials in acidic environments is likely to be unpractical.

6.2.2 Stoichiometric Titanium Oxides

The use of stoichiometric titanium dioxide, very often referred to as titania, has been recently extensively investigated in terms of its potential stability for fuel cell devices. Three main crystallographic varieties of titanium dioxide are known: rutile, anatase and brookite. Anatase and brookite are metastable phases that can be converted into rutile by heating. Furthermore, amorphous titanium oxides are often obtained by a variety of synthetic methods, requiring annealing to convert it into crystalline phases. Titania has been widely exploited in heterogeneous catalysis as a support for platinum and other noble metals with which specific interactions such as SMSI and spillover phenomena are known. Furthermore, titania-based materials are commercially available also being rather cheap. For example nanodispersed titania powder are to date commercialized by Degussa under the name P25. Such materials have also been exploited for their photocatalytic properties. Deposition of metal nanoparticles has been widely investigated on titania supports mainly for enhancing the photocatalytic activity. Among others, Huang et al. [19] mixed porous titanium dioxide, sodium borohydride, hexachloroplatinic acid and sodium dodecyl sulfate to obtain titania-supported platinum nanoparticles, and assessed the stability as electrode material for the ORR. After 50 h at 1.2 V versus RHE, the carbon-based materials started to lose activity, while the titania-supported electrocatalyst only showed a slight decrease in the activity after more than 200 h. Table 6.1 reports the evolution of the EASA and the particle diameter as a function of testing time. It was found that the average Pt particle size on the carbon support experienced a 4-fold increase while the particles on titania showed an increase of just 25 %. The increase in particle size resulted in a corresponding decrease of the electrochemically active surface area.

The increased stability of the TiO_2 materials deserves some more comments. The interaction between TiO_2 and metals is usually stronger than that occurring between carbon-based materials and metals. Furthermore the oxidation of carbon may reduce the surface interaction between the support and the particle. Under such conditions nanoparticles can move along the support, reaching other nanoparticles and leading to agglomeration. The strong interaction between metals and

Table 6.1 Evolution of average platinum particle size and electrochemically active surface area for Pt/TiO_2 and Pt/C during an accelerated degradation test

Catalyst	d_{pt} (nm)		EASA (m^2/g)	
	0 h	80 h	0 h	80 h
Pt/TiO_2	6.2	7.8	13.8	11.0
Pt/C	2.5	11.5	53.1	3.8

The test consisted in keeping the potential fixed at 1.2 V versus RHE for up to 80 h

titania, as well as a variety of other metal oxides prevents such migration and agglomeration phenomena and can be considered responsible for the larger stability of such classes of supports as compared to carbon blacks [20].

As all the crystallographic forms of TiO_2 are semiconductors with band gaps exceeding 3.0 eV, there is the need to improve the electronic conductivity [21–23], in order for them to be suitable as electrocatalyst supports. To improve titania support conductivity, Kim et al. [23] investigated the possibility of a post-treatment for commercial rutile powders. In their synthesis rutile was first dispersed in deionized water under sonication. The resulting suspension was then reacted with three different additives for 24 h at 200 °C; urea, thiourea, and hydrofluoric acid. A fourth sample had the same treatment but with no additive. After the reaction platinum nanoparticles were deposited using the sodium borohydride method. The resulting morphologies are shown in Fig. 6.2. TEM investigation showed that the sample treated with HF shows the best dispersion of the Pt nanoparticles. This is confirmed by the data reported in Table 6.2. The average Pt particle size is the lowest for the sample treated with HF and the highest for the sample without treatment. The same trend is followed by the EASA.

The electrocatalytic activity of these materials for the ORR was investigated by performing RDE measurements in oxygen-saturated 0.1 M $HClO_4$. The most active sample was the Pt/TiO_2 (HF) treated material, while the other samples showed poor performance. The increase in the activity cannot be ascribed to the largest surface area only. Band gap measurements also showed that the sample treated in HF showed the lowest band gap amongst all the samples suggesting that conductivity plays a fundamental role in determining the activity of the catalysts.

Nanostructured TiO_2 materials show at least one dimension of less than 100 nm. Many TiO_2 shapes have been obtained including spheroidal nanocrystallites and nanoparticles, nanosheets, and nanofibers [24].Very recently the possibility of obtaining TiO_2 nanotubes via a facile anodization process in fluoride containing media of a polycrystalline titanium surface has strongly reinforced interest in titania materials [25].

Due to their excellent stability, high surface area, and moderate electrical conductivity, titanium dioxide nanotubes (TONTs) have also been extensively considered for application as electrocatalyst supports primarily for application in fuel cells. TONT shows many advantages as a support for electrocatalysts, mainly associated with the specific geometry and the chemical nature of titanium dioxide. The structure of the arrays positively affects the dispersion of catalysts allowing

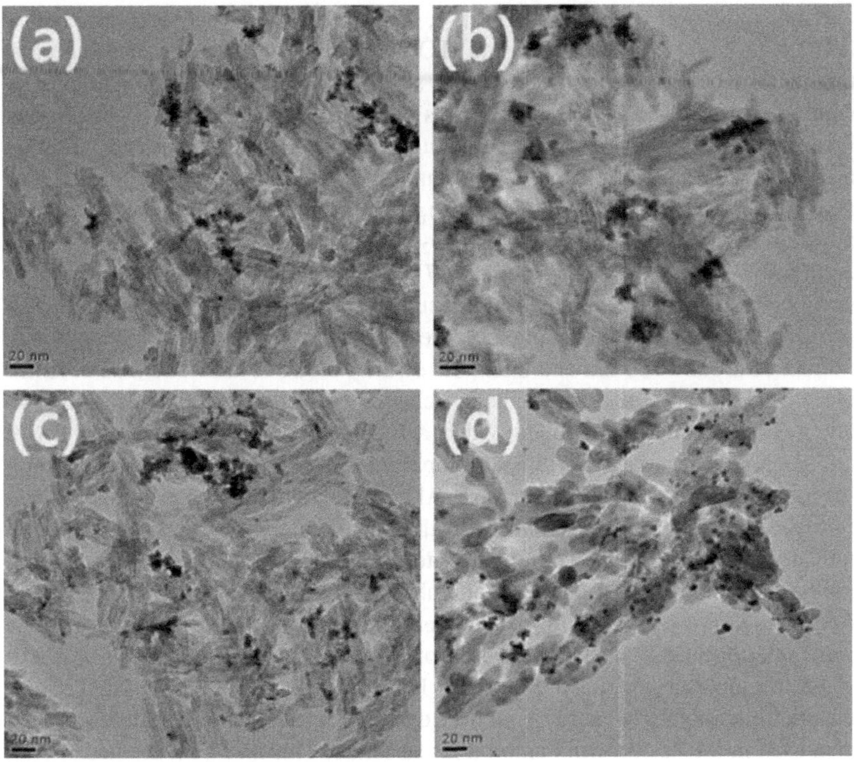

Fig. 6.2 Representative TEM images of Pt nanoparticles deposited on variously treated rutile samples: **a** TiO_2 (without additives), **b** TiO_2 (treated with urea), **c** TiO_2 (treated with thiourea), and **d** TiO_2 (treated with HF). Reprinted from Ref. [23] with permission of Elsevier

Table 6.2 Average particle size of catalysts for the platinum nanoparticles dispersion reported in Fig. 6.2

Catalyst	Average particle size (nm)	Surface area ($m^2\ g^{-1}$)
Pt/TiO_2 (w/o additives)	6.05	46.3
Pt/TiO_2 (treated with urea)	4.52	62.0
Pt/TiO_2 (treated with thiourea)	4.17	67.2
Pt/TiO_2 (treated with HF)	3.64	77.0

Particle size derived from X-ray diffraction

the high metal loadings while keeping small average particle sizes. The permeation of reactants through the porous structure of the materials is also in principle good [24] even if limitations in mass transport may occur, when the length of the tube largely exceeds the diameter of its mouth. TONT have been largely considered especially in relation to methanol oxidation and oxygen reduction processes for application in DMFCs. Pd,PtRu, PtNi, and PtCo nanoparticles have all been deposited on TONTs and characterized in terms of activity toward both the methanol oxidation reaction (MOR) [26, 27] and the oxygen reduction reaction (ORR) [28] and [29]. Free standing TONT synthesized by the hydrothermal method [30] have been considered as a support for palladium nanoparticles by Wang et al. [26]. The Pd/TONT catalysts were synthesized at 120 °C employing glycol as reducing agents for the palladium metal salts. Well-dispersed Pd nanoparticles were obtained on the titania nanotubes, leading to a high surface area value for the Pd. Pd/TONT showed a larger catalytic activity for the MOR than that of Pd black and Pd supported on TiO_2 nanoparticles. Ordered arrays of titania nanotubes were tested by Macak et al. [27]. The ordered arrays of TiO_2 nanotubes were obtained by anodization in an electrolyte containing sulphuric and hydrofluoric acids. The obtained nanotubes were approximatively 500 nm in length with a mouth of around 100 nm and thickness of the wall of 15 nm. The thus obtained layers were doped with Pt/Ru nanoparticles and tested for methanol electrooxidation. The obtained material provided a much larger electrochemically active surface area as compared to Pt/Ru nanoparticles at the same loading but immobilized on a conventional TiO_2 support. Also the catalytic activity resulted greatly improved and it significantly enhanced the electrocatalytic activity of Pt/Ru for methanol oxidation again as compared to compact TiO_2 supports. A crystallographic analysis revealed that the nanotube often shows an amorphous structure after anodization. An annealing at temperatures up to 400 °C produces a phase transformation leading to the formation of anatase. This transformation has been shown to enhance the catalytic activity. Indeed annealed TiO_2 nanotubular supports exhibit a higher enhancement effect during electrooxidation of methanol than when used in the "as-growth" amorphous structure. This may be attributed to both the better electronic conductivity of anatase and the strong interaction of the Pt/Ru nanoparticles with the TiO_2 surface. This interaction is promoted by the annealing, especially if performed under reducing atmosphere (e.g., H_2 stream).

Another approach used to increase the conductivity of the nanotube arrays consists in the addition of heteroatoms such as C [31–33], S [34] and N [31, 35–38]. Furthermore, it has been recently demonstrated that treating the TNTs in hydrogen atmosphere results in an increase in the oxygen vacancy density. Such vacancies are known to be highly beneficial for enhancing the conductivity of the nanotubes [39–41]. The excellent performance of PEMFCs equipped with nanotubes annealed under hydrogen has been ascribed to such an increased density of oxygen vacancies [42]. It has been observed that the conductivity of the TNTs increased by an order of magnitude after hydrogen treatment at 350 °C for 1 h as compared with samples treated under the same conditions but in air. The hydrogen-treated TiO_2 nanotubes (H–TNT) have been tested as nanostructured electrode supports with the purpose of significantly improving the electrochemical performance and durability of fuel cells. It has also been shown that the increase in the density of oxygen vacancies and hydroxyl groups after the treatment under hydrogen also greatly increases the anchoring of Pt atoms in the course of Pt electrodeposition. Deposition of Pt onto the H–TNTs has been performed using the SIAR method (Successive Ion Adsorption and Reaction). Such a method has been shown to enhance dispersion of Pt catalysts in electrodeposition processes. The SIAR method consists in adsorbing an activator (e.g., Pd) that provides an homogeneous distribution of nucleation sites for the Pt deposition. With the SIAR, it is possible to get high metal loadings while keeping the size of the nanoparticles small. In [42] it was shown that the Pt nanoparticles had a distribution centered on 3.4 nm and that the particles were uniformly distributed across the support. In terms of fuel cell performances the as-prepared electrodes performed excellently in terms of stability during accelerated durability tests, particularly for H–TNT-loaded Pt catalysts that have been annealed in ultrahigh purity H_2 for a second time. There was minimal decrease in the electrochemical surface area of the as-prepared electrode after 1,000 cycles compared to a 68 % decrease for a commercial JM 20 % Pt/C electrode after 800 cycles. X-ray photoelectron spectroscopy showed that after the H–TNT-loaded Pt catalysts are annealed in H_2 for the second time, the strong metal-support interaction between the H–TNTs and the Pt catalysts enhances the electrochemical stability of the electrodes. Fuel cell testing showed that the power density reached a maximum of 500 $mWcm^{-2}$ when this highly ordered electrode was used as the anode. When used as the cathode in a fuel cell with extra-low Pt loading, the new electrode generated a specific power density of 2.68 kW/gPt.

Titanium nanotube arrays can also serve as a support for the deposition of nanoparticles from vapor phase processes. Kang et al. [28, 29] used a conventional anodization process to produce TiO_2 nanotube arrays on a Ti substrate. They obtained nanotubes with a length of approximately 1 μm and a diameter of 120 nm. In order to deposit metal alloy nanoparticles they used dual-gun sputtering. Using Pt, Ni, and Co targets they were able to obtain PtNi (PtNi size 5–10 nm) and PtCo (PtCo size 3–4 nm) nanoparticles. Among the prepared catalysts PtNi showed poor ORR activity, in particular at high potentials. This has been related to the strong OH adsorption. Annealing of the same catalyst at 400 °C under hydrogen atmosphere resulted in a net improvement of the ORR activity. It is worth mentioning that the ORR activity was not largely affected by the presence of methanol. A positive shift of the onset potential for the ORR activity was achieved with $Pt_{70}Co_{30}$ supported on the nanotubes. The enhancement in activity has been attributed to the peculiar geometry of the nanotubes allowing diffusion of oxygen and the high dispersion of the PtCo catalyst both at the mouth and in the inner wall of the nanotubes. Nevertheless, no comparison with the same catalyst supported onto carbon blacks has been reported.

In the case of palladium nanoparticle deposited on titania nanotube arrays, the possibility of tailoring size and surface structure of the nanoparticle through electrochemical post-treatment has also been observed (Fig. 6.3) [43]. The post-processing that has been named Electrochemical Milling and Faceting (ECMF) consisted in the application of a pulsed potential which resulted in the production in fragmentation of the original nanoparticles. Furthermore particles have been observed to expose, preferentially, high index termination. An increase in the activity toward ethanol oxidation by almost an order of magnitude as a result of ECMF was found. The enhancement has been ascribed to both increase in the surface area and presence of high index terminations.

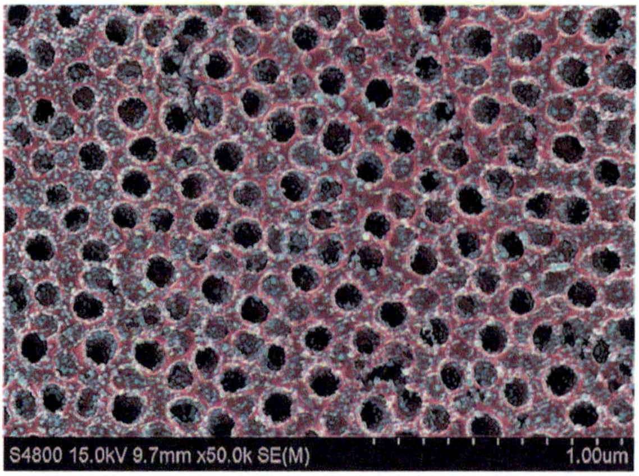

Fig. 6.3 Palladium nanoparticle dispersion on a nonindependent titania nanotube array. The material has been shown to be very effective for the electrooxidation of ethanol and the production of raw chemicals from the use of biomasses in electrolytic processes. Reprinted with permission from Ref. Reprinted from Ref. [43] with permission of Elsevier

6.2.3 Metal Doped Titanium Oxide

Modification of the physical properties of titanium oxides has been investigated by the addition of other metals. Sintering mixtures of NbO_2 and TiO_2 is a method that has been shown to be effective for producing electrically conducting materials [44, 45]. Nevertheless, the high temperatures required by the sintering process frequently lead to the formation of materials with low surface areas. Furthermore, TiO_2 undergoes a phase transition from anatase to rutile, a less catalytically active phase near 700 °C [46]. In order to retain large surface areas and to avoid phase transitions low-temperature synthesis route have been developed. $Nb_{0.1}Ti_{0.9}O_2$ has been synthesized via the surfactant templating method by Garcia et al. [47]. The synthesis consists in mixing octadecylamine as a template, niobium(V)ethoxide and titanium(IV)butoxide as metal precursors. The resulting material $Nb_{0.1}Ti_{0.9}O_2$ had a BET surface area of 136 $m^2\ g^{-1}$. This is two orders of magnitude larger than the value recorded for the solid-phase high temperature synthesis that only produced a material with a surface area of 1.4 $m^2\ g^{-1}$ [12]. The material obtained by the Garcia synthesis was then doped with PtRu nanoparticles prepared by a colloidal method. A complete monoplanar DMFC was then assembled with this anode catalyst. Tests were performed at 70 °C using PtRu/$Nb_{0.1}Ti_{0.9}O_2$ and commercial PtRu/C catalyst from E-TEK for the sake of comparison. Performance comparison showed that the PtRu/$Nb_{0.1}Ti_{0.9}O_2$ delivered higher current densities per Pt unit mass as compared to the E-TEK PtRu/C at a potential of 0.40 V. This has been attributed to a synergistic effect between the support and the catalyst particles.

Nb–TiO_2 supported Pt catalysts have been prepared as a cathode catalyst layer for PEMFCs. Park and Seol [48] produced Nb-doped TiO_2 nanoparticles by a hydrothermal synthesis performed at 120 °C, employing Ti(IV) isopropoxide and Nb(V) ethoxide as a source of metals. The synthesis was then followed by an annealing at 400 °C in a pure H_2 atmosphere. XRD showed that the as-prepared

Nb–TiO_2 presented prevalently a rutile crystal structure. The electrical conductivity was determined as 0.1 S cm^{-1}. The addition of Pt (40 wt %) to the Nb–TiO_2 support was performed using a conventional borohydride reduction method. The mean Pt particle size was around 3 nm, while the Nb-TiO_2 support was nanostructured at the 10 nm scale. Tests at even noble metal loading showed that the catalysts equipped with the Nb-TiO_2 support performed better for the ORR as compared to those supported by conventional carbon blacks. The enhancement in the performance was attributed to the enhanced dispersion of the Pt nanoparticles. More interestingly, the XANES spectra recorded around the Pt L edge of the supported catalysts showed the occurrence of mixed metal phases suggesting the existence of a strong metal to support interaction. Such an interaction may also account for the enhanced activity of Pt/Nb–TiO_2 for the ORR again highlighting the advantages of exploiting promotion effects in electrocatalysis.

Sol–gel methods have also been employed to prepare Ru-doped TiO_2. Pt was then supported on the Ru-doped TiO_2 by an impregnation-reduction method [49]. The characterization of the support showed the composition of $RuxTi_{1\_}xO_2$ material to be in the range between $x = 0.17$ and $x = 0.73$. The phase behavior of these systems is rather complex. Diffraction studies indicated the system is a solid solution above $x = 0.5$ that splits into two phases below $x = 0.5$. SEM investigation of the Ru–TiO_2 material revealed an average particle size in the range between 200 and 300 nm. The conductivity was found to be strongly dependant upon the material composition and had values typical of semiconductors ($0.1 < \sigma < 100$ S cm^{-1}). The addition of platinum nanoparticles via the borohydride method yielded materials with a platinum content varying between 23 and 40 wt %. The Pt crystallite size was found to be rather small leading to distributions peaking in the range 3.3–5.4 nm depending on the platinum loading. Cyclic voltammetry in a 0.5 M phosphoric acid electrolyte showed that the catalysts obtained this way have an electrochemically active surface area comparable to commercially available Pt/C electrocatalysts. It was shown that the ruthenium content plays an important role in determining the size distribution of the Pt nanoparticles. Particularly for $x = 0.71$, the EASA of the Pt supported on Ru-TiO_2 was approximately twice that of the other oxide compositions and larger than the EASA of simple Pt on Vulcan at the same Pt loading.

A comparison of the performance between titanium oxides and metal doped titanium oxides has been performed by Chen et al. [50] in view of their potential application in regenerative fuel cell devices. Particularly, supports such as Ebonex®, Ti_4O_7 and $Nb_{0.1}Ti_{0.9}O_2$ have been considered. Ti_4O_7 and $Nb_{0.1}Ti_{0.9}O_2$ were synthesized by, respectively, reducing and doping rutile titanium dioxide. As a precursor ultrafine rutile with a BET surface area of 110 m^2 g^{-1} has been used. Reduction to Ti_4O_7 was obtained by heating rutile at 1,050 °C under a hydrogen atmosphere, while $Nb_{0.1}Ti_{0.9}O_2$ was prepared by heating a solid-phase mixture of TiO_2 and NbO_2. In the end materials with conductivity comparable to Ebonex® were obtained. Crystallographic analysis showed that $Nb_{0.1}Ti_{0.9}O_2$ consisted in a pure rutile microcrystalline phase. The BET surface area measurements showed that none of the synthesized materials scored well in terms of surface area. Indeed

2 and 1.4 $m^2\ g^{-1}$ were determined for the synthesized Ti_4O_7 and $Nb_{0.1}Ti_{0.9}O_2$, respectively, while the surface area for the commercial Ebonex® resulted as 1 $m^2\ g^{-1}$. $Pt_4Ru_4Ir_1$ was been selected as the catalyst phase. The metal alloy was prepared by reducing the metal salts onto the three supports by the borohydride reduction method. ORR and OER activities were evaluated in a gas diffusion half cell. The most performing electrode was found to be that equipped with the $Nb_{0.1}Ti_{0.9}O_2$ support, its performance resulted superior over the whole range of the explored potentials. The Ti_4O_7-based catalysts showed activities comparable to that the Ebonex®. In order to study the stability, the authors examined Ebonex® and $Nb_{0.1}Ti_{0.9}O_2$-supported catalysts for a relatively long period of anodic polarization in the gas diffusion half cell, with continuous 0.5 M H_2SO_4 electrolyte flow. The $Nb_{0.1}Ti_{0.9}O_2$-supported catalyst maintained constant current, while a large drop in current was observed for the Ebonex®-supported catalyst. This was particularly evident for the oxygen reduction reaction and points to the cause of deactivation as the partial oxidation of Ebonex®, which results in the formation of a resistive layer with the consequent increase of iRlosses. Under ORR conditions both Ebonex® and Ti_4O_7 tend to be oxidized to the less nonconductive TiO_2. The oxidation preferentially occurs at the catalyst/support/electrolyte three-phase interface. The addition of niobium to the titanium dioxide creates a material much more resistant to oxidation, thus retaining its properties as a good electrical conductor. Such stability toward oxidation guarantees the stability under both oxygen evolution and reduction conditions. For this reason $Nb_{0.1}Ti_{0.9}O_2$ is a good candidate as a support material for both ORR and OER electrodes.

6.2.4 Tungsten Oxides

Tungsten oxides (WO_x), are a class of semiconducting materials usually showing a n-type behavior with a band gap generally between 2.6 and 2.8 eV [51]. Due to the large number of stable valence states for tungsten (+2 to +6), its oxides can exhibit a large variety of stoichiometries, also exhibiting an large variability in its physical and chemical properties. Such variability has been extensively exploited in electrochemical applications [52]. Due to the inherent conductivity and the stability under aggressive conditions tungsten oxides have been proposed as alternative materials to carbon for supporting electrocatalysts. Pt particles on sodium tungsten bronze (Na_xWO_3) have been proposed as electrocatalysts for phosphoric acid fuel cells (PAFCs) [53, 54]. Nevertheless, exploitation of tungsten bronze in PAFCs was generally unsuccessful mainly due to the formation of phosphotungstate complexes that caused catalyst deactivation.

On the contrary, the application of Pt/WO_x to Direct Methanol Fuel Cells [55] has shown promising performance due to the high catalytic activity of Pt/WO_x anodes toward methanol oxidation. The enhancement in activity has been attributed to the formation of tungsten bronze materials that are known to promote the dehydrogenation of methanol. Additionally, the oxophilic nature of tungsten oxides also promotes the removal of adsorbed species such as CO, a common intermediate in methanol electrooxidation [56]. A synergistic effect between Pt and WO_x has been hypothesized and is believed to promote the electrooxidation of the adsorbed CO [57, 58]. Micoud et al. [59] have studied carefully the CO tolerance of platinum catalysts supported on WO_x. The synthesis consisted of an impregnation-reduction method. A platinum salt previously embedded on a commercially available monoclinic WO_3 was first reduced to Pt nanoparticles. The catalyst was then investigated by monitoring the CO electrooxidation currents at a potential of 0.1 V versus RHE. It was shown that the Pt/WO_3 catalyst had large CO oxidation current, certainly superior to that of carbon-supported Pt and even carbon-supported PtRu catalysts.

Another beneficial effect associated with the use of tungsten oxide is the enhancement in the proton transfer process during methanol electrooxidation [60, 61]. This phenomenon has been clearly identified by Park et al. [62] in methanol electrooxidation catalyzed by a PtRu–WO3 catalyst and has been ascribed to the formation of tungsten trioxide hydrates.

As for any other electrocatalyst, the Pt loading on the tungsten oxide has a role of primary importance in defining the electrocatalyc activity. As pointed out in many sections of this book, a reduction in Pt loading that does not lower the electrocatalytic activity is one of the primary objectives in fuel cell related material science. Interestingly it has been shown that for WO_3^- supported Pt catalysts, when the metal loading exceeds 50 wt %, the catalytic activity for methanol electrooxidation [63] could be as high as that of a pure Pt catalyst consisting of spherical Pt particles with particle sizes ranging between 50 and 150 nm.

Tailoring morphologies is an important task in defining the activity of the WO_3-supported Pt catalysts. Ganesan and Lee [64] first synthesized WO_3 microspheres and then added Pt catalyst nanoparticles. The obtained material was found to show higher stability for methanol electrooxidation than commercially available 20 wt % PtRu/Vulcan-XC72 and 20 wt % PtRu/carbon-microspheres catalysts at the same metal loading. Mesoporous WO_3 particles with the addition of Pt nanoparticles have been synthesized by Cui et al. [55]. Remarkably, the mesoporous WO_3 showed a high surface area together with an ordered pore structure The critical dimensions for the nanosized wall thickness resulted in the range between 6 and 7 nm. The material was synthesized with a metal loading of 20 wt % Pt and it was shown that the electrocatalytic activity for methanol oxidation was larger than that of commercial 20 wt % Pt/C. The authors also found that the catalytic activity was in the same range of a 20 wt % PtRu/C catalyst, at least in the 0.5–0.7 V versus NHE potential region. To check the possibility to further enhance the catalytic activity toward methanol oxidation Barczuk et al. [65] deposited PtRu at even metal loadings on WO_3 matrices with different morphologies. The obtained catalysts with both nanoporous and microporous WO_3 matrices as supports had high surface areas and performed well in terms of electrocatlytic activity. An interesting investigation on a PtRu catalyst supported on WO_3 is also reported in Ref. [66]. PtRu supported on a three-dimensionally ordered macroporous (3DOM) WO_3 was

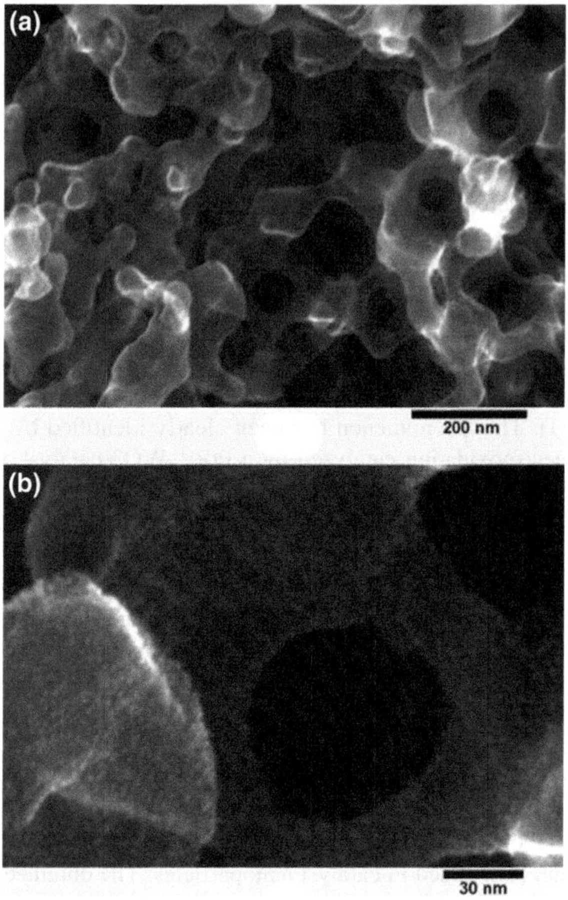

Fig. 6.4 PtRu nanoparticles deposited onto a 3D WO_3 support. **a** low and **b** high magnification SEM images showing the 3D structure of the material and the metal nanoparticle dispersion. Reproduced from Ref. [66] with permission of Elsevier

prepared by a polyol process combined with an ammonia-leaching treatment. Scanning electron microscopy revealed a complex three-dimensional morphology (Fig. 6.4a) with an homogeneous distribution of small PtRu nanoparticles (Fig. 6.4b). Elemental composition and the particle size of the PtRu/WOx catalyst was found to be similar to those of the PtRu catalyst synthesized without the 3DOM WOx template. In addition both the catalysts had a uniform elemental distribution. Linear sweep voltammetry revealed that the PtRu/WOx catalyst had a greater electrocatalytic activity for methanol oxidation than the PtRu catalyst alone at even metal loading. This has been attributed due to synergistic effects with the WOx and the open 3D structure which favors mass transport. Furthermore chronoamperometry (Fig. 6.5) showed that stability was greatly enhanced by the use of 3DOM WO_x

Nanostructuring of tungsten oxides has also been considered. Nanorods of WO_3 have been applied as catalyst supports and it has been demonstrated that the deposition of Pt nanoparticles on 1D WO_3 nanorods provides larger Pt mass

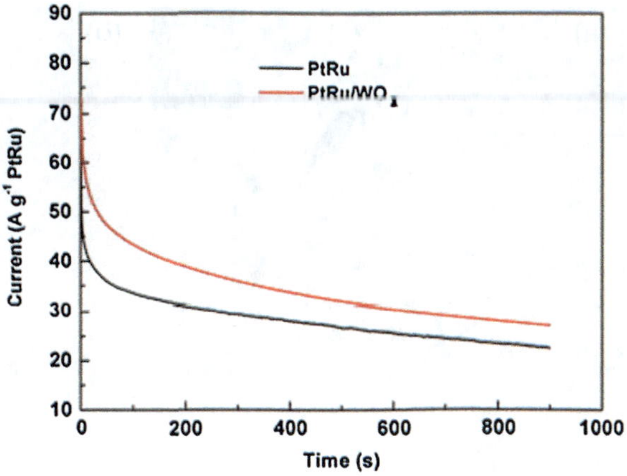

Fig. 6.5 Electrocatalytic activity and stability of PtRu nanoparticles deposited onto a 3D WO_3 support. Chronoamperometry curves at 0.55 V in 0.1 M $HClO_4$ + 0.5 M CH_3OH at 25 °C showing the larger activity and stability of the WO3 supported material as compared to bare Pt/Ru at even metal loading. Reproduced from Ref. [66] with permission of Elsevier

specific surface areas. Furthermore, the strong interaction between the oxide and the metal produce a strong anchoring to the substrate preventing agglomeration and, in the end, better electrocatalyst stability. A simple synthesis of WO_3 nanorods has been proposed by Rajeswari et al. [67]. The synthesis consisted in pyrolyzing surfactant encapsulated tungsten oxide clusters. Nanorods with a length of 130–480 nm and a width of 18–56 nm have been obtained (Fig. 6.6). Platinum nanoparticles were then added by conventional methods. The obtained catalyst has then be tested for methanol oxidation. Activity was compared to that of a commercial 20 % Pt–Ru/C catalyst and it was found that the electrocatalyst had comparable performance to that of the commercial one. A facile template synthesis method has been proposed by Maiyalagan and Viswanathan [68]. They synthesized WO_3 nanorods with a 200 nm length by calcination of phosphotungstic acid previously deposited in the channels of an alumina template. After template dissolution Pt nanoparticles were then added to the WO_3 nanorods to form a Pt/WO_3 catalyst. The electrocatalytic activity was evaluated by depositing an ink with the Pt/WO_3 on a glassy carbon electrode. From electrocatalytic activity assessment the existence of a synergistic effect between the oxide and the metal was shown. The stability in performance was found to be comparable to that of carbon-supported Ptelectrocatalysts (Fig. 6.7).

A synergistic effect has also been recently observed through the comparison between the electrocatalytic performance of Pt/WO_x and Pt/C. It was found that; (1) the addition of WO_x causes a decrease in the charge required for CO_{ads} electrooxidation (2) a partial CO_{ads} electrooxidation between 0.1–0.4 V versus RHE occurs and thus a reduction of the HOR overpotential on CO-poisoned surfaces and (3) a ca. 100 mV negative shift of both the onset and main CO_{ads} electrooxidation peak. These effects have been related to a strong metal–support interaction (SMSI) [69]. SMSI has been demonstrated to induce the spillover of primary oxides from the WO_x support to the adjacent Pt sites [70] and thus blocking those sites (this may result in a decrease of the Pt ECSA). Furthermore

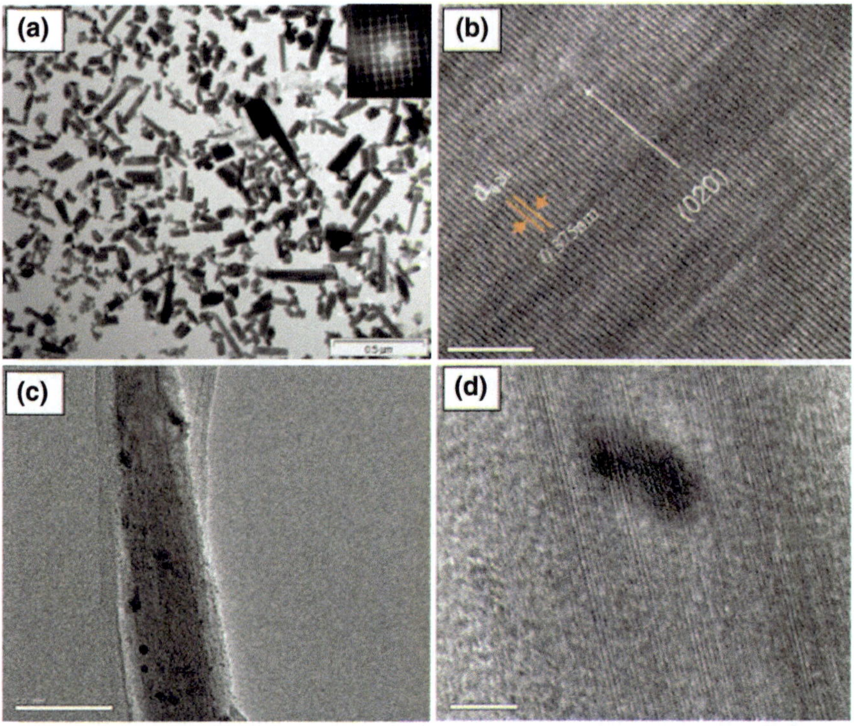

Fig. 6.6 **a** TEM of **WO3** nanorods (inset: an electron diffraction pattern obtained from the WO_3 nanorod), **b** HRTEM image of a WO_3 nanorod, **c** a low magnification HRTEM image of Pt/WO_3 nanorods, and **d** HRTEM image of a Pt nanoparticle on WO_3 nanorod. Reproduced from Ref. [67] with permission of Elsevier

the metal oxides also produce a decrease of the CO fraction on the Pt surface at a potential of 0.1 V versus RHE. This has been hypothesized to be the reason why a negative shift of both the CO electrooxidation on set and peak potentials is observed.

Tungsten oxide-supported Pt also has been found to promote the oxygen reduction reaction. Chhina et al. [52] showed that the oxygen reduction activity of Pt/WO_3 showed high stability with respect to accelerated oxidation tests in deoxygenated 0.5 M H_2SO_4 at 30 and 80 °C, respectively. Interestingly, the activity of commercially available carbon-supported Pt catalyst (Hispec 4,000) dropped quickly after performing tests according to the same accelerated protocol. Tungsten oxides have also been shown to be more thermally stable with respect to Vulcan XC-72R. McLeod and Birss [71] investigated stability employing Pt/WO_x films for methanol electrooxidation. For the film synthesis they used a sol–gel method. Following this procedure they synthesized two types of WO_x, using ethanol or water as the solvent. Both WO_x samples showed high electrical conductivity, a desirable property for fuel cell application as a catalysts support.

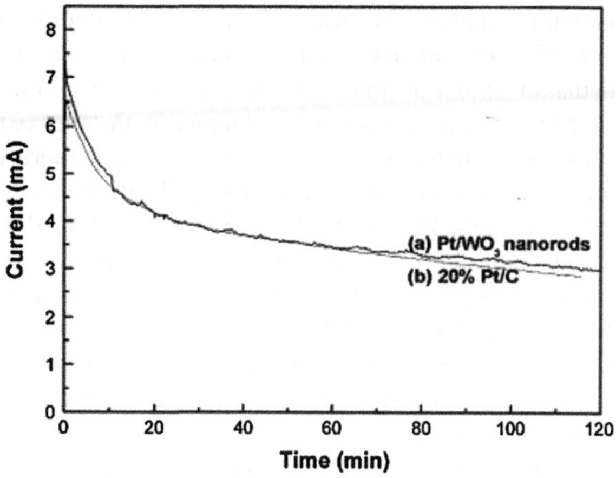

Fig. 6.7 Current density versus time curves at 0.6 V Ag/AgCl. **a** Pt/WO_3 nanorods and **b** Pt/C (Johnson Matthey) measured in 1 M H_2SO_4 + 1 M CH_3OH. Potential stepped from rest potential to 0.6 V versus Reproduced from Ref. [68] with permission of Elsevier

Nevertheless, the film synthesized with ethanol delivered more stable methanol oxidation performance as compared to the sample synthesized in water. The difference in the performance was ascribed to loss of oxide material from the film synthesized in water.

Stability in acidic media of tungsten oxides is limited as dissolution of occurs in acidic media [72]. Attempts to improve the chemical stability of WO_3 in acidic media have been performed. Raghuveer and Viswanathan [73] exploited Ti^{4+} substitution in WO_3 crystals. It was found that the addition resulted effective in preventing WO_3 corrosion. Unfortunately, the addition of titanium ions results in a strong increase of the ohmic resistance.

6.2.5 Other Oxides

Apart from tungsten and titanium a variety of other oxides have been employed as supports for fuel cells and electrolyzer electrocatalysts. Ruthenium oxide (RuO_2) has been considered. This is mainly because of the the bifunctional mechanism in methanol oxidation [72, 73]. The Pt/RuO_2 systems have a strong tendency to form highly mobile hydroxides. Indeed, it is believed that surface ruthenium hydroxides can promote the oxidation of adsorbed CO at potentials lower than bare platinum. Analogous effects have been observed in CeO_2 supported Pd catalysts and exploited in enhancing the energy efficiency of Direct Ethanol Fuel Cells [5]. In alkaline environments, the property of ceria to form highly mobile surface hydroxides results in a net enhancement of the rate of OH adsorption at the surface of Pd. In the end this results in a net enhancement of the rate of ethanol electrooxidation. Hydrated ruthenium oxides have also been synthesized to further exploit the synergistic effect with platinum. Chen et al. [74] proposed a simple method employing a water solution for the synthesis of ruthenium oxide-supported platinum catalysts (Pt/RuO_2 3 xH_2O). A comparison of the electrocatalytic properties for methanol oxidation showed that this catalyst had a reactivity toward methanol oxidation much higher than that of a commercial E-TEK PtRu black catalyst. Similar effects can be obtained with other oxides such as IrO_2 and SnO_2. Silicon dioxide has also been considered as a support for platinum. SiO_2-supported Pt catalyst has been synthesized for application in self-humidifying membranes in PEM fuel cells. Zhu et al. [75] and Wang et al. [76] have developed a novel class of self-humidifying reinforced composite membranes with silicon dioxide as ceramic charge. The fuel cell performance improved by the addition of the strongly hydrophilic Pt/SiO_2 catalyst. The addition of silicon dioxide also resulted in membrane resistance reduction and better anode self-humidification. A reduction in cathode overpotential and consequent increase in the OCV has also been observed. The phenomenon has been attributed to the fine dispersion of the Pt nanoparticles on the SiO_2 support resulting in large Pt surface area. Seger et al. [77] proved that the performance of cathodes employing Pt/SiO_2 were better than a catalyst employing Pt-black. SiO_2 suffers from a relatively low electronic conductivity which limits the potential of its exploitation in practical devices. Possible doping with foreign elements could result in improvements in the conductivity of SiO_2.

Recently, the application of Indium tin oxide (ITO) has also been proposed. ITO is a degenerated n-type semiconductor that shows appreciable electronic conductivity. ITO shows interesting properties such as that of being among the few material showing a good electronic conductivity being at the same time transparent. This has supported its application as a material for photovoltaics and optical devices [78, 79]. For all these reasons ITO is a commercially available material. ITO supported Pt catalysts have been prepared by Chhina et al. [80] and has been proposed as a potential non-carbon catalyst support. A dispersion of platinum with 13 nm crystallite size on nanostructured ITO (38 nm particle size) has been obtained and tested for electrochemical stability. Electrochemical characterization consisted of potential cycling between +0.6 and +1.8 V versus SHE for 100 cycles. It was shown that the catalyst was much more stable than two commercial alternatives, the Hispec 4,000 and the Pt/Vulcan XC-72R. Even the thermal stability was largely superior, indeed TGA showed that Pt/ITO was stable up to 1,000 °C with only 1 wt % loss of ITO while the Hispec 4,000 lost 57 wt % at this temperature.

6.3 Inorganic Metal Carbides and Nitrides

Carbides are compounds composed solely of carbon and a less electronegative element. The literature identifies four categories of carbide: salt-like, covalent compounds, interstitial compounds, and transition metal carbides. The latter category shows the largest potential for application in electrocatalysis, mainly for its CO tolerance, low cost, and stability in acidic and alkaline electrolytes.

Furthermore, some carbides such as WC may exhibit interesting electrocatalytic properties in themselves. In the following section we will review in some detail the application of WC to electrocatalysis. At present WC, for its low cost and large availability is the most widespread carbide support for electrocatalysis. Other materials such as B_4C, SiC and TiC will be briefly discussed in a separate section.

6.3.1 WC

Tungsten carbide exists in two distinct phases, WC and W_2C of which the latter is thermodynamically unstable at temperatures lower than 900 °C [81–83]. The conductivity of tungsten carbide (WC) has been reported to be as high as 10^5 s cm^{-1} [81, 84] and has been proved to have high stability in acidic solutions [85]. Furthermore WC may show catalytic properties similar to that of Pt for a variety of reactions, indicating that WC could become a promising electrocatalyst support in virtue of its peculiar chemical and physical properties [86, 87]. Commercial WC powders produced by the conventional metallurgical route do not show high surface areas thus hampering their application as a support for catalytic nanoparticles. A variety of synthetic strategies for increasing the surface area of WC have been successfully proposed. Carburization of W_2N and WS_2 have been performed by China et al. [84] and Hara et al. [88, 89], respectively. The as synthesized materials had surface areas in the range of 80 m^2 g^{-1}.

A nanostructured Pt/WC catalyst with a Platinum loading of 20 wt % has been synthesized by Jeon et al. [90] for methanol electrooxidation (Fig. 6.8). X-ray diffraction investigations were used to calculate the average Pt particle size which resulted as 7.5 nm. TEM investigations also showed a homogeneous particle size distribution. Electrochemical characterization showed that in the CV the peak potential for PtO reduction increased from 0.72 V for commercial Pt/C to 0.76 V for the synthesized Pt/WC. The combination of CV data together with CO stripping results allowed the authors to conclude that aspill-over proton is transferred from Pt to the WC support. The electrochemically active surface area was calculated from the CV. Particularly, a hydrogen desorption determined surface area of 11.2 and 5.74 m^2/g was calculated for Pt/WC and Pt/C, respectively. It is worth mentioning that the CO electrooxidation peak shifted to less positive potentials, namely from 0.80 V for Pt/C to 0.68 V for Pt/WC. An important role in adsorbed CO oxidation is due to the formation of surface hydroxides. Indeed the reaction of WC with water may result in the formation of surface hydroxide species which are known to help in CO oxidation. Analogous effects have been previously described for alcohol oxidation promotion in a variety of oxide-supported catalysts. The area-specific activity for methanol electrooxidation has been determined and it has been found that the tungsten carbide materials performed better than the carbon-supported ones (144 mA/m^2 for Pt/C to 188 mA/m^2 for Pt/WC).

Fig. 6.8 TEM pictures of the Pt/WC catalyst. Reproduced from Ref. [90] with permission of Elsevier

Elezović et al. [91] synthesized a platinum nanostructured electrocatalyst on nanostructured tungsten for the ORR. This synthetic strategy lead to a large specific surface area of the support (177 $m^2\ g^{-1}$). Platinum (10 wt %) was added using a borohydride reduction method. The STEM characterization (Fig. 6.9) indicated the presence of Pt particles with sizes even lower than 2 nm. It was found that there may exist individual Pt atoms on the WC surface (Fig. 6.9).

The electrochemically active surface area has been assessed by the measurement of the charge of the hydrogen desorption peak and was found to be around 40 $m^2\ g^{-1}$ of Pt.

Assessment of the catalytic properties for the ORR was studied by linear sweep voltammetry at a rotating disk electrode (Fig. 6.10). It was found that the onset potential was shifted positively by 150 mV as compared to Pt/Vulcan XC-72. The Tafel slope of the Pt/WC sample was determined to be -0.105 V dec^{-1}. Table 6.3 summarizes the remarkably high catalytic activity of such materials. Stability has also been found to be remarkably higher than that of the Pt/Vulcan XC72 reference sample.

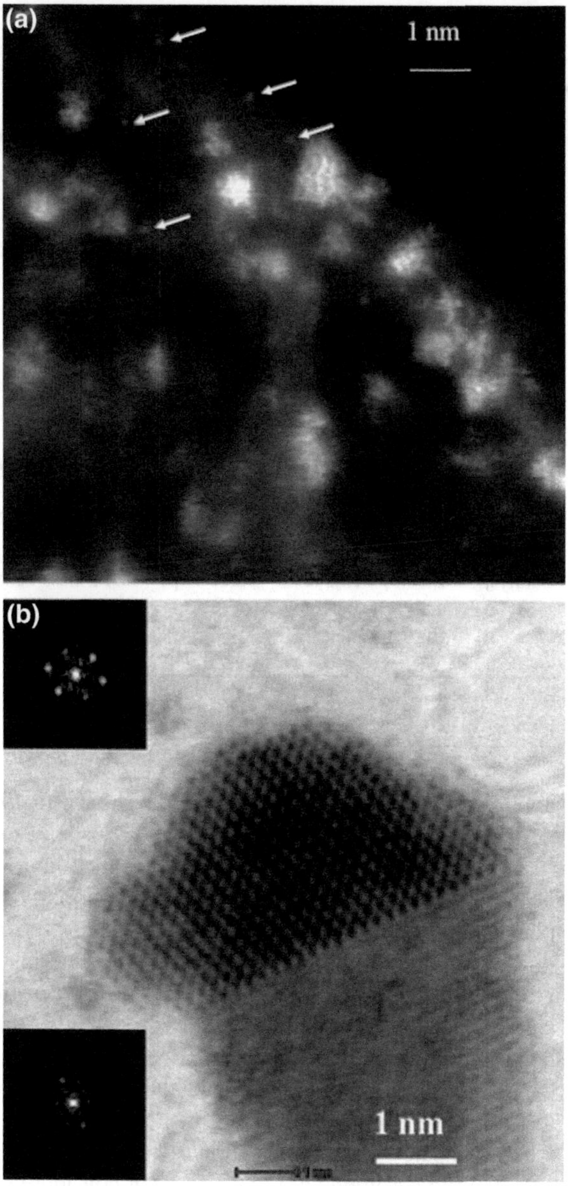

Fig. 6.9 STEM images of the Pt/WC catalyst the smallest particles of Pt lower than 2 nm as clusters of Pt atoms. The *arrows* indicate individual Pt atoms, the indication of Pt particle nucleated on WC support. Reproduced from Ref. [91] with permission of Elsevier

6.3.2 Other Carbides

Boron carbide is known as covalent carbide. It has found many applications in virtue of its mechanical and thermal properties. Indeed B_4C is particularly hard and heat resistant. Furthermore it has an appreciable electronic conductivity. Its

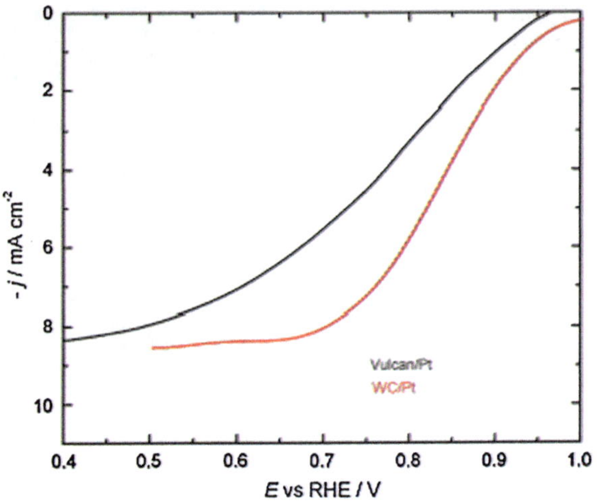

Fig. 6.10 Polarization *curves* (positive-going sweeps) obtained with RDE, for ORR in 0.1 mol dm^{-3} NaOH solution, with a rotation rate 3,040 rpm, at Pt/WC and Pt/Vulcan electrodes. Reproduced with permission from [91]

Table 6.3 Comparison of the catalytic activities for the ORR obtained in 0.1 mol dm^{-3} NaOH at 25 °C for Pt/Vulcan and Pt/WC catalysts

Catalyst	Electrochem. Pt active surface area (m^2/g)	Tafel slope (mV)		Pt mass activity (mA/mg) at $E = 0.85$ V (RHE)	Pt specific activity (mA/cm^2) at $E = 0.85$ V (RHE)
		Low	High		
Pt/Vulcan	46	−65	−135	89	0.212
Pt/WC	40		−105	181	0.453

application as an electrode material dates back to 1963 when it was tested for the first time showing high conductivity and resistance to corrosion and oxidation [92].

Grubb and McKee [93, 94] showed in the middle 1960s that boron carbide can be used as a support for platinum. This property has been exploited mostly in phosphoric acid fuel cells. A further relevant property is that platinum deposited on boron carbide is particularly resistant to sintering, much more than platinum on graphite, suggesting a strong interaction between the metal and the B_4C. Boron carbide also prevents the catalytic hydrogen adsorption which occurs on platinum deposited on graphitic carbon. Application to ammonia/oxygen fuel cells has been explored by McKee et al. [94]. In their investigations, the authors prepared and tested a Pt/Ir catalyst deposited on B_4C. The fuel cell performances were lower when compared to platinum supported on graphite electrodes.

Titanium carbide (TiC) has also been considered as an electrode material [92]. First electrochemical experiments with titanium carbide were related to the electrolytic oxidation of manganese sulfate to MnO_2, where the material has been proved to be stable at current densities exceeding 1,000 A m^{-2}. The use of titanium carbide as a support for fuel cells is still limited. Jalan et al. [95] demonstrated the possibility of using titanium carbide with surface areas in the range of 25–125 m^2/g. Such materials have been prepared by decomposing titanium tetrachloride in a stream of a mixture of hydrogen and hydrocarbons at various temperatures between 500 and 1,250 °C. The resulting TiC had a chain like morphology leading to an open porosity. The open porosity provided two main advantages: (1) maintaining electrical contact and (2) providing fast mass transport, e.g., for oxygen diffusion. The addition of platinum to TiC resulted in an electrochemically active surface area in the range between 20 and 90 m^2 g^{-1}. Authors argued that this support could improve the stability in corrosive environments as well as could lead to enhancements in the electrocatalytic activity.

SiC is another potential candidate as a support material for electrochemical applications. It is highly thermally stable, being able to withstand at temperatures as high as 1,200 °C in strongly oxidizing environments [96]. SiC is also chemically inert and its application as a catalyst support has been proposed [97, 98]. Honji et al. [99] synthesized platinum supported silicon carbide catalysts. Platinum addition has been performed by reduction of hexachloroplatinic acid with methanol as reducing agent performing the reaction in a suspension of SiC powder and surfactants. The electrical conductivity of silicon carbide is relatively low as it ranges, depending on the purity, around 10^{-6} S/cm [100]. Such a value is not adequate for preparing MEAs so the addition of a fraction of carbon black is required. Characterization showed that the platinum particles had a fine dispersion on the SiC substrate. The best polarization performance was found with 23 wt % Pt/SiC, providing 220 mA cm^{-2} at 0.7 V RHE for the oxidation reduction reaction. This value compares well to that of carbon-supported platinum catalysts. Rao et al. [100, 101] prepared nano-SiC-supported platinum nanoparticles. SiC nanoparticles were obtained by treating larger commercial SiC particles in thermal plasma. Electrochemical performance for the ORR was found to be comparable to that of Pt/C (E-TEK) at even metal loading. Even the electrochemical stability has been found to be close to that of commercial Pt/C produced by E-TEK. Krawiec and Kaske [102] as a result of their investigations showed that commercially available silicon carbide does not have a surface area large enough for a successful exploitation as electrocatalyst support. Values ever lower than 25 m^2 g^{-1} have been determined for all the explored commercial SiC powders. On the basis of such findings further research aimed at enhancing the surface area of SiC supports have been proposed and reported in the literature [103–108].

Molybdenum carbide has been explored for application in electrolysis as a platinum free catalyst for use in acidic environments. Molybdenum carbide nanoparticles have been prepared by carburization of ammonium molybdate. The nanoparticles have been obtained both on carbon nanotubes and XC-72R. X-ray diffraction showed that molybdenum was present as Mo_2C. Interestingly X-ray absorption suggested that the Mo2C nanoparticles were inlaid or strongly anchored to the carbon supports. The carbon nanotube-supported Mo_2C has been found to be superior to bulk Mo_2C both in terms of activity and stability toward the hydrogen evolution reaction (HER). A current density of 1 mA cm^{-2} has been measured at 63 mV for nanotube-supported Mo_2C catalysts. The enhanced electrochemical activity has been ascribed to electronic interactions between $Mo_2C$and the carbon nanotubes. Table 6.4 summarizes the main properties for some of the most relevant ceramic support material for electrocatalysis.

Table 6.4 Characteristics of ceramic materials and carbon Vulcan XC-72 supports, and catalytic activity and stability in fuel cell conditions of relative supported catalysts. Reproduced from Ref. [129] with permission of Elsevier

Material	Specific surface area $m^2\ g^{-1}$	Electrical conductivity $S\ cm^{-1}$	Metal dispersion, particle size (d) nm	Catalytic activity of supported metal	Stability in fuel cell conditions	References
Vulcan XC-72	254	4	High d 2.9	High	Low	[128]
Ebonex, Ti_4O_7	1–3	1, 10^3	Low d_{Ti4O7}10–20	Generally low	Conflicting results	[50, 129–131]
M–TiO_2	1–1.4 (high T_{synt}) 136 (low T_{synt})	10^{-1}–10^2	High d_{RuTiO2}3.3–5.4d_{SbTiO2} 3.0	High	High	[45, 47–50]
M–SnO_2	~43 ($RuSnO_2$) 100 ($SbSnO_2$)	10^{-1}–10^3	High d_{SbSnO2}2.5 d_{RuSnO2} 4	High	High	[132–134]
WO_3	18, 86	n.d.	Medium/High d 6.5 d 3–6 (nanorods)	High	High	[55, 64, 67, 68]
$RuO_2{\cdot}xH_2O$	125	1	n.d	Low	n.d.	[74, 135]
S–ZrO_2	80	n.d.	n.d.	Low	High	[136]
Sn–In_2O_3	~22	>10^3	Low d 13	Low	High	[80]
WC	1.6 (T_{synt} > 900 °C), 30–80 (T_{synt} < 900 °C) 176 (microsphere)	10^5	Low/Medium d 6–20 d 2–6 (meso-WC)	Low (low SA) high (high SA)	High	[84, 88–90, 137–141]

6.3.3 Nitrides

Transition metal nitrides have been recently considered as support materials for fuel cells [109–111]. Some of them have been shown to possess electrocatalytic activities that are similar to some noble metals like Pd and Pt. Nevertheless, most of the applications are as catalyst supports [109–111]. To date not so many reports on the application of nitrides as materials for fuel cells are available. The most investigated certainly is titanium nitride (TiN) [92]. The application of nanostructured titanium nitride as a promising material for super capacitors and electrode material in general has been reported in the literature [112–118]. TiN is a worthy subject of investigation mainly because of its high electrical conductivity combined with excellent resistance to oxidation and to corrosion in acid electrolytes [119–121]. Datta and Kumta [112] employed TiN supported PtRu nanoparticles as anode catalyst for methanol oxidation. Avasarala et al. [121] have reported the use of TiN nanoparticles as a catalyst support for Pt anode catalysts for PEMFCs. Investigations with transmission electron microscopes revealed a good dispersion of the Pt/TiN electrocatalyst with an average particle size at 2 nm and with most of the distribution located between 2 and 3 nm. The distribution of Pt on TiN has also been found to be homogeneous by XRD. Electrochemical measurements for the assessment of the ORR activity showed that the synthesized catalyst exhibited a higher catalytic activity as compared to a commercial BASF Pt/C catalyst with a platinum loading of the 20 wt %. TiN like TiO_2 is stable in acidic electrolytes. Avasarala et al. [122] suggested that the nature of the TiN surface is strongly affected by the concentration and the type of acid. Passivation may limit TiN degradation in acidic media and has been observed in perchloric acid as a result of the hydrolysis of the surface layers.

Boron nitride (BN) has also been proposed as a potential candidate for electrocatalyst supports. BN show a structure closely related to that of graphite [123, 124]. As a consequence of its properties, like thermal stability, thermal conductivity, and chemical stability it is a promising candidate for application in systems operating under harsh conditions. A few studies [125, 126] have shown that BN powder has high thermal conductivity ranging between 4 and 6.2 W/cm^3 C as well as thermal stability up to 1,000 °C. More interestingly, relative to application in fuel cells it has been found that BN is stable in acids and base also being resistant to oxidation. Even mechanical properties show enough strength for electrocatalytis application, indicating that the addition of Pt could be interesting in electrochemical applications. Perdigon-Melon et al. [127] have produced high surface area BN powders investigating the effect of the synthetic strategy in the actual area value. It has been found that porous BN synthesized by a sol–gel deposition technique resulted in an excellent support material that then produced a highly dispersed Pt catalyst.

A variety of other nitrides might be considered for electrochemical applications. Molybdenum nitride (Mo_2N) has been reported to be effective in the water gas shift reaction as well as in the methanol steam reforming reactions [142–146]. These materials may also show their own electrocatalytic activity, independently on the addition of catalytic metal nanoparticles.

6.4 Conductive Polymers

As reported by Antolini et al. [147] conductive polymers have been widely considered as materials for supporting electrocatalysts. Indeed polymers may be synthesized or prepared in order to have high surface area to guarantee high metal dispersion, high electric conductivity, and high chemical and electrochemical stability. Furthermore, controlled porosity can be introduced in order to allow fast mass transport of both reactants and products to or from the electrode surface. These properties together indicate the large potential of polymeric materials for application in fuel cells and electrocatalysis. In order to obtain electrical conductivity conjugation is required. Indeed most conducting polymers are conjugated also having heteroatoms in the main chain (heterocyclic polymers). Among them we may site: polyaniline, polypyrrole, polythiophene, and their derivatives (Fig. 6.11). Beyond being electron conductors they may also act as proton conductors. According to this conducting polymers could also offer the opportunity to replace Nafion as ionomers in the preparation of the catalyst layer in MEAs. This aspect may reduce the catalyst layer resistance. More importantly, at least in theory, the use of an electron conducting polymer could require only a two-phase boundary to guarantee the necessary electron and ion transfer during reactions in fuel cells. This architecture would be much more performing in terms of the transportation of electrically charged species than the standard three-phase boundaries of the conventional catalytic layer. Further enhancement of the proton conductivity can also be obtained by mixing proton exchange polymers with the conductive polymers. The most commonly employed proton conducting polymer for this purpose is polystyrene sulphonate. The addition of polystyrene sulphonate to conducting polymers may also result in the formation of a complex three-dimensional architecture. This former point is crucial to enhance the uniformity of the catalytic metal distribution, resulting in more effective noble metal utilization.

The interaction between sulphonated groups and metals is also promising for electrocatalyst stability. Indeed a strong interaction with the substrate, as frequently pointed out in the text, helps to prevent aggregation and hence also the progressive loss in electrochemically active surface area. One of the most performing polymer materials as a support for fuel cell catalysts is nowadays a mix of poly(3,4-ethylenedioxythiophene) and poly(styrenesulfonate) (PEDOT/PSS). Indeed, the activities toward ORR of platinum supported PEDOT/PSS have been found to be comparable to that of platinum supported on carbon blacks [149, 150]. A drawback of the PEDOT/PSS systems is the fact of not being very stable. Stability has not yet reached that of conventional carbon-based electrocatalysts.

As with most of the support materials the catalytic performance of metals supported on conducting polymers is dramatically affected by the deposition method. Addition of metals to conducting polymers has been performed employing a wide variety of methods. Among them electrodeposition is certainly promising.

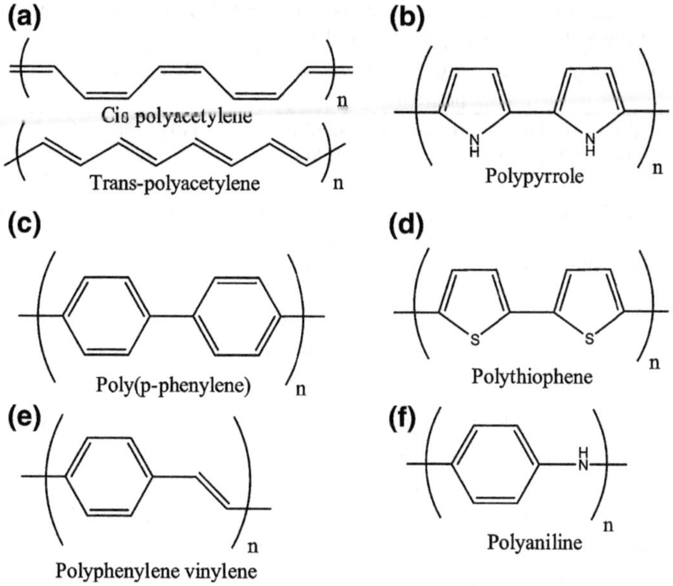

Fig. 6.11 Structure formulas of several conducting polymers showing the extensive conjugation responsible for the electric conductivity. Reproduced from Ref. [148] with permission from Elsevier

Antolini [147] has reported another interesting advantage to the use of conducting polymer supports. In fact unlike carbon and ceramic where the active phase has necessarily to be dispersed at the surface of the support, for conducting polymers the metal can be dispersed inside the support. This leads to the obvious advantage, in principle, of having an added dimension to the system. Whether this fact is really useful for obtaining enhanced electrocatalytic activities is still a debated question. Nowadays, conflicting results concerning the advantage of conducting polymers with metal dispersion inside the support materials as compared to materials with the catalyst on the surface have been reported in the literature [151–156].

Synergistic effects between the active catalyst phase and the conducting polymer have also been reported. Particularly, Polypyrrole (PPy) and Polyaniline (PAni) have been shown to be able to enhance the catalytic activity. This effect has not been ascribed to an increase in metal surface area alone [151, 157–164]. Investigations have shown that promoting effects are especially relevant for PAni supported Pt when oxidizing methanol at low overpotentials [165, 166].

Concerns about the stability of PAni and PPy under fuel cell operating condition have also been raised. While the materials should be in principle remarkably stable, degradation has been observed both in the course of processes for the addition of the catalytic phase and in the fuel cell operation. Degradation derived by the deposition of metal nanoparticles may result in significant reduction of the electronic conductivity of PPy as a result of exposition to the reducing agents used for metal

Table 6.5 Properties of modified conducting polymers (MCPs) and of MCP-supported catalysts. Reproduced from Ref. [147] with permission of Elsevier

Modified conducting polymers (MCPs)	MCP and metal/MCP properties	References
Functionalized-CPs (PEDOT, PoMA, PMPy, PMT, PoT, PAAni)	Better processability (solubility), high stability	[149, 150, 170, 171, 172, 173, 174, 175, 176, 177, 178, 179, 180, 181, 182, 183, 184, 185]
CP/PEP composites (CP/PSS, CP/Nafion) and doped CPs (PAni/DPSA, PPy/DEHS)	High reactant permeability, high ionic conductivity, high metal surface area	[149, 167, 186, 187, 188, 189, 190]
Nanostructured CPs	Higher electronic conductivity, high metal dispersion, high stability	[173, 181, 191, 192, 193, 194, 195].

deposition (formaldehyde, hydrogen, or citrate) and to oxidizing conditions (H_2O_2) frequently encountered at the cathode side especially in automotive application of PEMFC [167]. PAni degradation has also been observed [168] as a result of the exposition to aldehydes. It is worth mentioning that aldehydes are frequently encountered as intermediates or reaction products of alcohol oxidation, especially in acidic conditions. Even the supported catalyst can cause accelerated degradation processes for polymers. This aspect has been investigated in detail in [169]. ORR conditions look especially harmful for conducting polymers. Indeed, degradation of the electronic conductivity of PPy and PAni in the potential range of ORR seriously hampers their exploitation in PEMFC technologies. A list of conducting polymer-based materials extracted from ref. 148 is reported in Table 6.5.

Various investigations [172, 181, 191, 193, 196, 197] have shown that suitable nanostructuring may result in significant improvement of the polymer materials for application is fuel cells. Nanostructured conducting polymers have been tested as catalyst supports for use in low-temperature fuel cells. The physical and chemical properties of such nanostructured conducting polymers resulted in both a net enhancement in the catalytic activity and stability of conducting polymer supported metal nanoparticles [172, 181, 191, 193, 196, 197]. It has been pointed out [198] that structure and morphology tailoring is one of the most promising routes to the successful exploitation of conducting polymers. For example ladder polymers, with their excellent thermal stability, optimal mechanical properties, and high conductivity offer a concrete opportunity for technological exploitation.

6.5 Composite Materials

In order to provide materials with enhanced performance as compared to plain carbon black, ceramic and polymer materials, the possibility of using composite materials as supports for electrocatalysts has been investigated [199]. Among the most investigated composite electrodic materials, a primary role is played by polymer–carbon composites (Table 6.6). Other highly relevant materials are

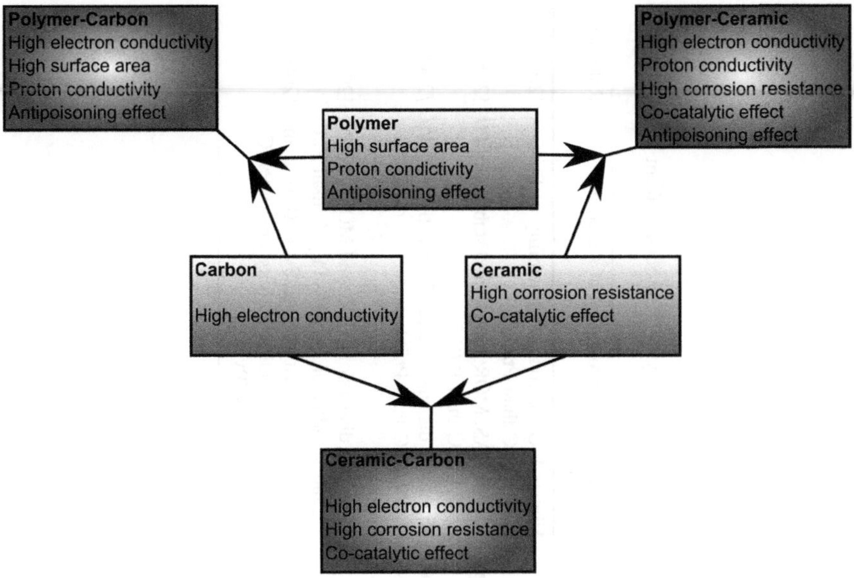

Scheme 6.2 Composite materials as a support for fuel cell materials. Possible combinations

certainly: (i) ceramic–carbon composites, and lately the polymer–ceramic composites. Composite materials offer the unique opportunity to attain desired properties by mixing together materials with different chemical and physical properties. Scheme 6.2 summarizes the various approaches to the design of composite materials for electrochemical applications with regard to their properties. Scheme 6.2 shows the main characteristics of plain and hybrid material supports. It is evident that synergistic properties may originate leading to characteristics that do not belong to either the polymer or the carbon material. It is worth mentioning that high polymeric degree and lower defect density in the polymer structure [200, 201] may be obtained also leading to the formation of a highly porous material structure [202, 203].

Hybrid polymer–carbon materials are classified regarding the most relevant component. Generally, a material is classified as IO if the conductive polymer is added to carbon blacks. This can be done in order to enhance the electrochemical properties of the carbon, such as the accessible surface area and to increase stability and resistance to poisoning. Conversely, when carbon-based materials are added to a polymer matrix a OI material is obtained.

Composite materials are generally synthesized by polymerization of the monomer in the presence of carbonaceous species. Under such conditions carbon may also act as a promoter of polymerization. This is the case of aniline polymerization, which results in a high polymerization degree also showing polymeric structure with low density of defects. Hybrid polymer–carbon materials show several advantages as compared to the single components. These supports have highly accessible surface areas, allowing the deposition of metal nanoparticles

Table 6.6 Characteristics of hybrid polymer–carbon supports and electrocatalytic properties of supported catalysts. Reproduced from Ref. [199] with permission of Elsevier

Carbon type	Support	Support characteristics	Electrocatalytic properties of supported catalyst	References
Carbon back	PAni–C (IO)	Higher anti-poisoning ability than C	Higher MOR activity and stability of Pt/PAni–C than Pt/C	[204]
	PAni–C (IO)	Higher electron and proton conductivities, methanol diffusion coefficient anti-poisoning ability than C	Higher EAS, MOR activity and stability of Pt/PAni–C than Pt/C	[205]
	PAni–C (OI)	Higher polymeric degree and a lower defect density in PAni–C than in PAni. Higher electron conductivity, and lower charge-transfer resistance than PAni	Higher Pt dispersion and MOR activity of Pt/PAni–C than Pt/PAni	[206]
	NSA–PPy–C (IO)	Higher accessible surface area than C	Higher Pt dispersion, HOR and MOR activity and stability of Pt/NSA–PPy–C than Pt/C	[207]
	PPy–C (IO and OI)	The thermal stability of PPy–C decrease with increasing PPy content	Lower ORR activity of Pt/PPy–C (IO)) than Pt/C. Similar ORR activity of Pt/PPy–C (OI) and Pt/PPy	[200]

(continued)

Table 6.6 (continued)

Carbon type	Support	Support characteristics	Electrocatalytic properties of supported catalyst	References
Nanocarbons	PAni–SWNT (OI) and PAni–MWNT (OI)	Higher electron conductivity and porosity than PAni	Higher Pt dispersion and activity for methanol and formic acid oxidation of Pt/PAni–CNT than PAni	[208, 209]
	PAni–SWNT (OI)	Higher polymeric degree and lower defect density than pure PAni. PAni film morphology significantly changed. Higher accessible surface areas, electronic conductivity, and easier charge-transfer at polymer/electrolyte interfaces than PAni	Higher Pt dispersion and activity for methanol and formaldehyde oxidation of Pt/PAni–SWNT than PAni	[202, 210]
	PAni-f-MWNT (OI)	Higher accessible surface areas than PAni	Higher Pt dispersion and activity for formic acid oxidation of Pt/PAni–CNT than PAni	[201]
	PAni-g-MWNT (IO)	Grafting of PANI on MWNT surfaces masks the defect sites in MWNT providing an uniform surface with positively charged sites	Higher Pt dispersion and MOR activity of Au/PAni-g-MWNT than Au/MWNT	[211]
	PPy–MWNT (OI)	Higher accessible surface areas, electronic conductivity, and easier charge-transfer at polymer/electrolyte interfaces than PPy	Higher Pt dispersion and MOR activity of Pt/PPt-MWNT than Pt/PPy	[212]
	PPy–MWNT (IO)	PAni–MWNT show the tubular morphology of MWNT, in which the PPy is coated on each individual MWNT. The average diameter of the nanotubes increases with increasing PPy content in the composite	Higher Pt dispersion and MOR activity of Pt/PPt-MWNT than Pt/MWNT	[212]
	PPy–MWNT (IO)	Change from the hydrophobic surface of MWNT to a hydrophilic PPy–MWNT surface	Higher PtRu dispersion of PtRu/PPt-MWNT than PtRu/MWNT	[213]
	PAANI–MWNT	Higher accessible surface areas, electronic conductivity and easier charge-transfer at polymer/electrolyte interfaces than PAANI	Higher activity and stability for formaldehyde oxidation of Pt/PAANI-SWNT than PAANI	[214]
	Si–MWNT			[215]

(continued)

Table 6.7 Characteristics of hybrid ceramic–carbon supports and electrocatalytic properties of supported catalysts. Reproduced from Ref. [199] with permission of Elsevier

Carbon type	Support	Support characteristics	Electrocatalytic properties of supported catalyst	References
Carbon blacks	TiO_2–C	Improved TiO_2–C dispersion in the aqueous solution of Pt precursor with respect to plain carbon	Modification of Pt electronic structure by Ti. Higher Pt dispersion and ORR activity of Pt/TiO_2–C than Pt/C	[216]
	TiO_2@C	Higher corrosion resistance of the TiO_2@C compared to Vulcan XC-72R.	Higher MOR activity and stability of Pt/TiO_2@C than Pt/C	[217]
Nanocarbons	TiO_2–CNT	CNTs-embedded porous annealed TiO_2substrate. CNTs-embedded TiO_2 substrate facilitates the electron transfer due to its enhanced conductivity	Higher MOR activity of PtRu/TiO_2–CNT than PtRu/TiO_2–C.	[218]
	SnO_2–CNT	Higher electronic conductivity than SnO_2	Higher EOR activity of Pt/SnO_2–CNT than Pt/SnO_2	[219]
	SnO_2–SWNT	Higher thermal stability than SWNT	Higher EOR activity of Pt/SnO_2–SWNT than Pt/SWNT	[220]
	SnO_2@CNT	Higher corrosion resistance than CNT	Same ORR activity, higher stability of Pt/SnO_2@CNT than Pt/CNT	[221]
	MnO_2–CNT	Higher surface area and proton conductivity than CNT	Higher MOR activity of Pt/MnO_2–CNT than Pt/CNT	[222]

Table 6.8 Characteristics of hybrid polymer–ceramic supports and electrocatalytic properties of supported catalysts. Reproduced from Ref. [199] with permission of Elsevier

Support	Support characteristics	Electrocatalytic properties of supported catalyst	References
PAni–V_2O_5	Higher stability of PAni–V_2O_5compared to the individual components	Higher MOR activity and stability of Pt/PAni–V_2O_5than bulk Pt	[223]
PEDOT–V_2O_5	Continuous and relatively homogeneous matrix with a distinct lamellar morphology. No bulk deposition of polymer on the surface of the microcrystallites	Higher MOR activity and stability of Pt/PEDOT–V_2O_5 than Pt/C	[224]
PAni–SnO_2	Lower polymer crystallinity in PAni–SnO_2 than in plain PAni	Higher MOR activity and stability of Pt/PAni–SnO_2than Pt/SnO_2	[225]

smaller in size as compared to the polymer alone. On the other side, a higher electronic conductivity and an easier charge-transfer at the polymer/electrolyte are observed as compared to the single matrix. In the end, the electrocatalytic activity of most of the polymer–carbon composites is enhanced as compared to that of bare polymer (OI materials) or bare carbon (IO materials).

Ceramic–carbon materials are employed to provide co-catalytic properties also leading to an increase in the corrosion resistance. The carbon material contributes significantly to the electronic conductivity of the composite. The properties of some of the most investigated composite carbon ceramic are listed in Table 6.7.

In polymer–ceramic materials (Table 6.8), the required electrical conductivity is provided by the conducting polymer. As in the case of the ceramic-carbon materials, the high corrosion resistance and synergistic catalytic effects are provided by the ceramic component. As in the common carbon inks used for the fabrication of MEAs, the use of an ionomer coating may be required for polymer–carbon materials. The ionomer addition results in the stabilization of the nanoparticles and also enhances the extent of three-phase boundary regions.

It is worth considering that the application of hybrid materials as fuel cell catalyst supports has not yet been fully exploited. The first paper dealing with this appeared in the year 2005. It is expected that in the near future nanotechnology will further improve the properties of such composites with the introduction of new fillers structured at the nanoscale, in other forms, e.g., of nanorods and nanowires [226].

References

1. T. Maiyalagan, B. Viswanathan, U.V. Varadaraju, Nitrogen containing carbon nanotubes as supports for Pt—alternate anodes for fuel cell applications. Electrochem. Commun. **7**, 905 (2005)
2. X. Wang, W. Li, Z. Chen, M. Waje, Y. Yan, Durability investigation of carbon nanotube as catalyst support for proton exchange membrane fuel cell. J. Power Sources **158**, 154 (2006)

3. A. Kongkanand, S. Kuwabata, G. Girishkumar, P. Kamat, Single-wall carbon nanotubes supported platinum nanoparticles with improved electrocatalytic activity for oxygen reduction reaction. Langmuir **22**, 2392 (2006)
4. F.T. Bacon, Fuel cells: Will they soon become a major source of electrical energy? Nature **186**, 589 (1960)
5. V. Bambagioni et al., Energy efficiency enhancement of ethanol electrooxidation on Pd-CeO 2/C in passive and active polymer electrolyte-membrane fuel cells. ChemSusChem **5**, 1266 (2012)
6. A.S. Fialkov, Carbon application in chemical power sources. Russ. J. Electrochem. **36**, 345 (2000)
7. A.S. Fialkov, V.D. Chekanova, New graphitic material, Steklouglerod. Sov Plast, **76** (1973)
8. Y.J. Wang, D.P. Wilkinson, J. Zhang, Noncarbon support materials for polymer electrolyte membrane fuel cell electrocatalysts. Chem. Rev. **111**, 7625 (2011)
9. R.F. Bartholomew, D.R. Frankl, Electrical properties of some titanium oxides. Physical Rev. **187**, 828 (1969).
10. F.C. Walsh, R.G.A. Wills, The continuing development of Magnéli phase titanium sub-oxides and Ebonex® electrodes. Electrochim. Acta **55**, 6342 (2010)
11. K. Kolbrecka, J. Przyluski, Sub-stoichiometric titanium oxides as ceramic electrodes for oxygen evolution-structural aspects of the voltammetric behaviour of TinO2n-1. Electrochim. Acta **39**, 1591 (1994)
12. G. Chen, S.R. Bare, T.E. Mallouk, Development of supported bifunctional electrocatalysts for unitized regenerative fuel cells. J. Electrochem. Soc. **149**, A1092 (2002))
13. T. Ioroi, Z. Siroma, N. Fujiwara, S.I. Yamazaki, K. Yasuda, Sub-stoichiometric titanium oxide-supported platinum electrocatalyst for polymer electrolyte fuel cells. Electrochem. Commun. **7**, 183 (2005)
14. T. Ioroi et al., Stability of corrosion-resistant Magńli-phase Ti4O 7-supported PEMFC catalysts at high potentials. J. Electrochem. Soc. **155**, B321 (2008)
15. L. Adamczyk et al., Effective charge transport in poly (3,4-ethylenedioxythiophene) based hybrid films containing polyoxometallate redox centers. J. Electrochem. Soc. **152**, E98 (2005)
16. E. Slavcheva et al., Electrocatalytic activity of Pt and PtCo deposited on Ebonex by BH reduction. Electrochim. Acta **50**, 5444 (2005)
17. L.M. Vračar, N.V. Krstajić, V.R. Radmilović, M.M. Jakšić, Electrocatalysis by nanoparticles - Oxygen reduction on Ebonex/Pt electrode. J. Electroanal. Chem. **587**, 99 (2006)
18. N.V. Krstajic et al., Advances in interactive supported electrocatalysts for hydrogen and oxygen electrode reactions. Surf. Sci. **601**, 1949 (2007)
19. S.Y. Huang, P. Ganesan, S. Park, B.N. Popov, Development of a titanium dioxide-supported platinum catalyst with ultrahigh stability for polymer electrolyte membrane fuel cell applications. J. Am. Chem. Soc. **131**, 13898 (2009)
20. C.-C. Shih, J.-R. Chang, Pt/C stabilization for catalytic wet-air oxidation: Use of grafted TiO2. J. Catal **240**, 137 (2006)
21. L. Jiang, L. Colmenares, Z. Jusys, G.Q. Sun, R.J. Behm, Ethanol electrooxidation on novel carbon supported Pt/SnOx/C catalysts with varied Pt:Sn ratio. Electrochim. Acta **53**, 377 (2007)
22. K.W. Park, K.S. Ahn, Y.C. Nah, J.H. Choi, Y.E. Sung, Electrocatalytic enhancement of methanol oxidation at Pt-WOx nanophase electrodes and in situ observation of hydrogen spillover using electrochromism. J. Phys. Chem. B **107**, 4352 (2003)
23. D.-S. Kim, E.F.A. Zeid, Y.-T. Kim, Additive treatment effect of TiO_2 as supports for Pt-based electrocatalysts on oxygen reduction reaction activity. Electrochim. Acta **55**, 3628 (2010)
24. D.V. Bavykin, J.M. Friedrich, F.C. Walsh, Protonated titanates and TiO_2 nanostructured materials: synthesis, properties, and applications. Adv. Mater. **18**, 2807 (2006)

25. D. Gong et al., Titanium oxide nanotube arrays prepared by anodic oxidation. J. Mater. Res. **16**, 3331 (2001)
26. M. Wang, D.J. Guo, H.L. Li, High activity of novel Pd/TiO_2 nanotube catalysts for methanol electro-oxidation. J. Solid State Chem. **178**, 1996 (2005)
27. J.M. Macak et al., Self-organized nanotubular TiO_2 matrix as support for dispersed Pt/Ru nanoparticles: Enhancement of the electrocatalytic oxidation of methanol. Electrochem. Commun. **7**, 1417 (2005)
28. S.H. Kang, Y.E. Sung, W.H. Smyrl, The effectiveness of sputtered PtCo catalysts on TiO_2 nanotube arrays for the oxygen reduction reaction. J. Electrochem. Soc. **155**, B1128 (2008)
29. S.H. Kang, T.Y. Jeon, H.S. Kim, Y.E. Sung, W.H. Smyrl, Effect of annealing PtNi nanophases on extended TiO_2 nanotubes for the electrochemical oxygen reduction reaction. J. Electrochem. Soc. **155**, B1058 (2008)
30. T. Kasuga, M. Hiramatsu, A. Hoson, T. Sekino, K. Niihara, Formation of titanium oxide nanotube. Langmuir **14**, 3160 (1998)
31. J.H. Park, S. Kim, A.J. Bard, Novel carbon-doped TiO_2 nanotube arrays with high aspect ratios for efficient solar water splitting. Nano Lett. **6**, 24 (2006)
32. A. Mazare, I. Paramasivam, K. Lee, P. Schmuki, Improved water-splitting behaviour of flame annealed TiO_2 nanotubes. Electrochem. Commun. **13**, 1030 (2011)
33. C. Xu, Y.A. Shaban, W.B. Ingler Jr, S.U.M. Khan, Nanotube enhanced photoresponse of carbon modified (CM)-n-TiO_2 for efficient water splitting. Sol. Energy Mater. Sol. Cells **91**, 938 (2007)
34. Y. Mizukoshi, N. Ohtsu, S. Semboshi, N. Masahashi, Visible light responses of sulfur-doped rutile titanium dioxide photocatalysts fabricated by anodic oxidation. Appl. Catal. B **91**, 152 (2009)
35. S. Hoang, S.P. Berglund, N.T. Hahn, A.J. Bard, C.B. Mullins, Enhancing visible light photo-oxidation of water with TiO_2 nanowire arrays via cotreatment with H_2 and NH_3: synergistic effects between Ti_3 + and N. J. Am. Chem. Soc. **134**, 3659 (2012)
36. S. Hoang, S. Guo, N.T. Hahn, A.J. Bard, C.B. Mullins, Visible light driven photoelectrochemical water oxidation on nitrogen-modified TiO_2 nanowires. Nano Lett. **12**, 26 (2012)
37. A. Ghicov et al., Ion implantation and annealing for an efficient N-doping of TiO_2 nanotubes. Nano Lett. **6**, 1080 (2006)
38. R.P. Vitiello et al., N-Doping of anodic TiO_2 nanotubes using heat treatment in ammonia. Electrochem. Commun. **8**, 544 (2006)
39. G. Wang et al., Hydrogen-treated TiO_2 nanowire arrays for photoelectrochemical water splitting. Nano Lett. **11**, 3026 (2011)
40. H. Wang et al., Photoelectrochemical study of oxygen deficient TiO_2 nanowire arrays with CdS quantum dot sensitization. Nanoscale **4**, 1463 (2012)
41. M. Xu, P. Da, H. Wu, D. Zhao, G. Zheng, Controlled Sn-doping in TiO_2 nanowire photoanodes with enhanced photoelectrochemical conversion. Nano Lett. **12**, 1503 (2012)
42. C. Zhang et al., Supported noble metals on hydrogen-treated TiO_2 nanotube arrays as highly ordered electrodes for fuel cells. ChemSusChem **6**, 659 (2013)
43. Y.X. Chen et al., Electrochemical milling and faceting: Size reduction and catalytic activation of palladium nanoparticles. Angewandte Chem. Int. Ed. **51**, 8500 (2012)
44. D. Morris et al., Photoemission and STM study of the electronic structure of Nb-doped TiO_2. Phys. Rev. B Condens. Matter Mater. Phys. **61**, 13445 (2000)
45. W. Rüdorff, H.H. Luginsland, Z. Anorg, Allg. Chem. **334**, 125 (1964)
46. J. Arbiol et al., Effects of Nb doping on the TiO_2 anatase-to-rutile phase transition. J. Appl. Phys. **92**, 853 (2002)
47. B.L. García, R. Fuentes, J.W. Weidner, Low-temperature synthesis of a PtRu Nb0.1 Ti0.9 O_2 electrocatalyst for methanol oxidation. Electrochem. Solid-State Lett. **10**, B108 (2007)
48. K.W. Park, K.S. Seol, Nb-TiO_2 supported Pt cathode catalyst for polymer electrolyte membrane fuel cells. Electrochem. Commun. **9**, 2256 (2007)

49. O.E. Haas et al., Synthesis and characterisation of RuxTix-1O_2 as a catalyst support for polymer electrolyte fuel cell. J. New Mater. Electrochem. Syst. **11**, 9 (2008)
50. G. Chen, S.R. Bare, T.E. Mallouk, Development of supported bifunctional electrocatalysts for unitized regenerative fuel cells. J. Electrochem. Soc. **149**, A1092 (2002)
51. M.A. Butler, Photoelectrolysis and physical properties of the semiconducting electrode Wo//3. J. Appl. Phys. **48**, 1914 (1977)
52. H. Chhina, S. Campbell, O. Kesler, Ex situ evaluation of tungsten oxide as a catalyst support for PEMFCs. J. Electrochem. Soc. **154**, B533 (2007)
53. S.A. Abbaro, A.C.C. Tseung, D.B. Hibbert, J. Electrochem. Soc. **127**, 1106–110 (1980)
54. A.C.C. Tseung, L.L. Wong, The preparation and characterization of high performance Ag/C oxygen electrocatalysts. J. Appl. Electrochem. **2**, 211 (1972)
55. X. Cui et al., Platinum/mesoporous WO_3 as a carbon-free electrocatalyst with enhanced electrochemical activity for methanol oxidation. J. Phys. Chem. B **112**, 12024 (2008)
56. A.C.C. Tseung, K.Y. Chen, Hydrogen spill-over effect on Pt/WO_3 anode catalysts. Catal. Today **38**, 439 (1997)
57. F. Maillard et al., Is carbon-supported Pt-WOx composite a CO-tolerant material? Electrochim. Acta **52**, 1958 (2007)
58. P.J. Kulesza, L.R. Faulkner, Electrodeposition and characterization of three-dimensional tungsten(VI, V)-oxide films containing spherical Pt microparticles. J. Electrochem. Soc. **136**, 707 (1989)
59. F. Micoud, F. Maillard, A. Gourgaud, M. Chatenet, Unique CO-tolerance of Pt-WOx materials. Electrochem. Commun. **11**, 651 (2009)
60. Y.M. Li, M. Hibino, M. Miyayania, T. Kudo, Proton conductivity of tungsten trioxide hydrates at intermediate temperature. Solid State Ionics **134**, 271 (2000)
62. H. Nakajima, I. Honma, Proton-conducting hybrid solid electrolytes for intermediate temperature fuel cells. Solid State Ionics **148**, 607 (2002)
62. K.W. Park, J.H. Choi, K.S. Ahn, Y.E. Sung, PtRu alloy and PtRu-WO_3 nanocomposite electrodes for methanol electrooxidation fabricated by a sputtering deposition method. J. Phys. Chem. B **108**, 5989 (2004)
63. S. Jayaraman, T.F. Jaramillo, S.H. Baeck, E.W. McFarland, Synthesis and characterization of Pt-WO_3 as methanol oxidation catalysts for fuel cells. J. Phys. Chem. B **109**, 22958 (2005)
64. R. Ganesan, J.S. Lee, An electrocatalyst for methanol oxidation based on tungsten trioxide microspheres and platinum. J. Power Sources **157**, 217 (2006)
65. P.J. Barczuk et al., Enhancement of the electrocatalytic oxidation of methanol at Pt/Ru nanoparticles immobilized in different WO_3 matrices. Electrochem. Solid-State Lett. **9**, E13 (2006)
66. Q. Wang et al., Structure and electrochemical activity of WOx-supported PtRu catalyst using three-dimensionally ordered macroporous WO_3 as the template. J. Power Sources **241**, 728 (2013)
67. J. Rajeswari, B. Viswanathan, T.K. Varadarajan, Tungsten trioxide nanorods as supports for platinum in methanol oxidation. Mater. Chem. Phys. **106**, 168 (2007)
68. T. Maiyalagan, B. Viswanathan, Catalytic activity of platinum/tungsten oxide nanorod electrodes towards electro-oxidation of methanol. J. Power Sources **175**, 789 (2008)
69. S.J. Tauster, S.C. Fung, R.L. Garten, Strong metal-support interactions. Group 8 noble metals supported on titanium dioxide. J. Am. Chem. Soc. **100**, 170 (1978)
70. Q. Fu, T. Wagner, Interaction of nanostructured metal overlayers with oxide surfaces. Surf. Sci. Rep. **62**, 431 (2007)
71. E.J. McLeod, V.I. Birss, Sol-gel derived WOx and WOx/Pt films for direct methanol fuel cell catalyst applications. Electrochim. Acta **51**, 684 (2005)
72. K.Y. Chen, A.C.C. Tseung, Effect of nafion dispersion on the stability of Pt/WO_3 electrodes. J. Electrochem. Soc. **143**, 2703 (1996)

73. V. Raghuveer, B. Viswanathan, Synthesis, characterization and electrochemical studies of Ti-incorporated tungsten trioxides as platinum support for methanol oxidation. J. Power Sources **144**, 1 (2005)
74. Z. Chen et al., Synthesis of hydrous ruthenium oxide supported platinum catalysts for direct methanol fuel cells. Electrochem. Commun. **7**, 593 (2005)
75. X.B. Zhu, H.M. Zhang, Y.M. Liang, Y. Zhang, B.L. Yi, A novel PTFE-reinforced multilayer self-humidifying composite membrane for PEM fuel cells. Electrochem. Solid-State Lett. **9**, A49 (2006)
76. L. Wang et al., Pt/SiO2 catalyst as an addition to Nafion/PTFE self-humidifying composite membrane. J. Power Sources **161**, 61 (2006)
77. B. Seger, A. Kongkanand, K. Vinodgopal, P.V. Kamat, Platinum dispersed on silica nanoparticle as electrocatalyst for PEM fuel cell. J. Electroanal. Chem. **621**, 198 (2008)
78. O.P. Agnihotri, A.K. Sharma, B.K. Gupta, R. Thangaraj, The effect of tin additions on indium oxide selective coatings. J. Phys. D Appl. Phys. **11**, 643 (1978)
79. H.Y. Yu et al., Surface electronic structure of plasma-treated indium tin oxides. Appl. Phys. Lett. **78**, 2595 (2001)
80. H. Chhina, S. Campbell, O. Kesler, An oxidation-resistant indium tin oxide catalyst support for proton exchange membrane fuel cells. J. Power Sources **161**, 893 (2006)
81. S. Shanmugam, D.S. Jacob, A. Gedanken, Solid state synthesis of tungsten carbide nanorods and nanoplatelets by a single-step pyrolysis. J. Phys. Chem. B **109**, 19056 (2005)
82. L. Gao, B.H. Kear, Low temperature carburization of high surface area tungsten powders. Nanostruct. Mater. **5**, 555 (1995)
83. Y. Hatano et al., Solid state reaction between tungsten and amorphous carbon. J. Nucl. Mater. **307–311**, 1339 (2002)
84. H. Chhina, S. Campbell, O. Kesler, Thermal and electrochemical stability of tungsten carbide catalyst supports. J. Power Sources **164**, 431 (2007)
85. H.H. Hwu, J.G. Chen, Potential application of tungsten carbides as electrocatalysts. J. Vacuum Sci. Technol. A Vacuum Surf. Films **21**, 1488 (2003)
86. C.J. Barnett, G.T. Burstein, A.R.J. Kucernak, K.R. Williams, Electrocatalytic activity of some carburised nickel, tungsten and molybdenum compounds. Electrochim. Acta **42**, 2381 (1997)
87. K. Lee, A. Ishihara, S. Mitsushima, N. Kamiya, K.I. Ota, Stability and electrocatalytic activity for oxygen reduction in WC + Ta catalyst. Electrochim. Acta **49**, 3479 (2004)
88. Y. Hara, N. Minami, H. Itagaki, Synthesis and characterization of high-surface area tungsten carbides and application to electrocatalytic hydrogen oxidation. Appl. Catal. A **323**, 86 (2007)
89. Y. Hara, N. Minami, H. Matsumoto, H. Itagaki, New synthesis of tungsten carbide particles and the synergistic effect with Pt metal as a hydrogen oxidation catalyst for fuel cell applications. Appl. Catal. A **332**, 289 (2007)
90. M.K. Jeon, H. Daimon, K.R. Lee, A. Nakahara, S.I. Woo, CO tolerant Pt/WC methanol electro-oxidation catalyst. Electrochem. Commun. **9**, 2692 (2007)
91. N.R. Elezović et al., Pt supported on nano-tungsten carbide as a beneficial catalyst for the oxygen reduction reaction in alkaline solution. Electrochim. Acta **69**, 239 (2012)
92. F. Mazza, S. Trassatti, Tungsten, titanium, and tantalum carbides and titanium nitrides as electrodes i rednox systems. J. Electrochem. Soc. **110**, 847–849 (1963)
93. W.T. Grubb, D.W. McKee, Boron carbide, a new substrate for fuel cell electrocatalysts. Nature **210**, 192 (1966)
94. D.W. McKee, A.J. Scarpellino Jr, I.F. Danzig, M.S. Pak, Improved electrocatalysts for ammonia fuel cell anodes. Electrochem. Soc. J. **116**, 562 (1969)
95. V. Jalan, D.G. Frost, U.S. Patent 4,795,684, (1989)
96. I.K. Sung, Christian, M. Mitchell, D.P. Kim, P.J.A. Kenis, Tailored macroporous SiCN and SiC structures for high-temperature fuel reforming. Adv. Funct. Mater. **15**, 1336 (2005)
97. K.A. Schwetz, in *Silicon Carbide Based Hard Materials*, ed. by R. Riedel. Handbook of Ceramic Hard Materials, vol. 2, (Wiley, Weilheim, 2000) p. 683

98. M.J. Ledoux, C. Pham-Huu, Silicon carbide a novel catalyst support for heterogeneous catalysis. CATTECH **5**, 226 (2001)
99. A. Honji, T. Mori, Y. Hishinuma, K. Kurita, Platinum supported on silicon carbide as fuel cell electrocatalyst. J. Electrochem. Soc. **135**, 917 (1988)
100. C. Venkateswara Rao, S.K. Singh, B. Viswanathan, Electrochemical performance of nano-SiC prepared in thermal plasma. Indian J. Chem. Sect A **47**, 1619 (2008)
101. V. Jalan, E.T. Taylor, D. Frost, B. Morriseau, in *Abstracts of National Fuel Cell Seminar,* (1983) p. 127
102. P. Krawiec, S. Kaskel, Thermal stability of high surface area silicon carbide materials. J. Solid State Chem. **179**, 2281 (2006)
103. M.J. Ledoux, S. Hantzer, C.P. Huu, J. Guille, M.P. Desaneaux, New synthesis and uses of high-specific-surface SiC as a catalytic support that is chemically inert and has high thermal resistance. J. Catal. **114**, 176 (1988)
104. G.Q. Jin, X.Y. Guo, Synthesis and characterization of mesoporous silicon carbide. Microporous Mesoporous Mater. **60**, 207 (2003)
105. J. Parmentier, J. Patarin, J. Dentzer, C. Vix-Guterl, Formation of SiC via carbothermal reduction of a carbon-containing mesoporous MCM-48 silica phase: a new route to produce high surface area SiC. Ceram. Int. **28**, 1 (2002)
106. Z. Liu et al., Low-temperature formation of nanocrystalline β-SiC with high surface area and mesoporosity via reaction of mesoporous carbon and silicon powder. Microporous Mesoporous Mater. **82**, 137 (2005)
107. A.H. Lu, W. Schmidt, W. Kiefer, F. Schüth, High surface area mesoporous SiC synthesized via nanocasting and carbothermal reduction process. J. Mater. Sci. **40**, 5091 (2005)
108. V.G. Pol, S.V. Pol, A. Gedanken, Novel synthesis of high surface area silicon carbide by RAPET (reactions under autogenic pressure at elevated temperature) of organosilanes. Chem. Mater. **17**, 1797 (2005)
109. R.B. Levy, M. Boudart, Platinum-like behavior of tungsten carbide in surface catalysis. Science **181**, 547 (1973)
110. B. Dhandapani, TSt Clair, S.T. Oyama, Simultaneous hydrodesulfurization, hydrodeoxygenation, and hydrogenation with molybdenum carbide. Appl. Catal. A **168**, 219 (1998)
111. E. Furimsky, Metal carbides and nitrides as potential catalysts for hydroprocessing. Appl. Catal. A **240**, 1 (2003)
112. M.K. Datta, P.N. Kumta, Silicon, graphite and resin based hard carbon nanocomposite anodes for lithium ion batteries. J. Power Sources **165** 368 (2007)
113. S.A.G. Evans et al., Electrodeposition of platinum metal on TiN thin films. Electrochem. Commun. **7**, 125 (2005)
114. Y. Wang, H. Yuan, X. Lu, Z. Zhou, D. Xiao, Electroanalysis **18**, 15 (2006)
115. H. Cesiulis, M. Ziomek-Moroz, Electrocrystallization and electrodeposition of silver on titanium nitride. J. Appl. Electrochem. **30**, 1261 (2000)
116. T. Nakayama et al., Use of a titanium nitride for electrochemical inactivation of marine bacteria. Environ. Sci. Technol. **32**, 798 (1998)
117. O.T.M. Musthafa, S. Sampath, High performance platinized titanium nitride catalyst for methanol oxidation. Chem Commun, **67** 1 (2008)
118. J. Giner, L. Swette, Oxygen reduction on titanium nitride in alkaline electrolyte. Nature **211**, 1291 (1966)
119. B. Merzougui, M.K. Carpenter, S. Swathirajan, US Patent App. 11/431,979, (2006)
120. S.T. Oyama (ed.), *The Chemistry of Transition Metal Carbides and Nitrides* (Springer, New York, 1996)
121. B. Avasarala, T. Murray, W. Li, P. Haldar, Titanium nitride nanoparticles based electrocatalysts for proton exchange membrane fuel cells. J. Mater. Chem. **19**, 1803 (2009)
122. B. Avasarala, P. Haldar, Electrochemical oxidation behavior of titanium nitride based electrocatalysts under PEM fuel cell conditions. Electrochim. Acta **55**, 9024 (2010)
123. A.F. Wells, *Structural Inorganic Chemistry* (Oxford University Press, Oxford, 1984)

124. R.T. Paine, C.K. Narula, Synthetic routes to boron nitride. Chem. Rev. **90**, 73 (1990)
125. J.C.S. Wu, Z.A. Lin, J.W. Pan, M.H. Rei, A novel boron nitride supported Pt catalyst for VOC incineration. Appl. Catal. A **219**, 117 (2001)
126. C.A. Lin, J.C.S. Wu, J.W. Pan, C.T. Yeh, J. Catal. **210**, 39 (2002)
127. J.A. Perdigon-Melon, A. Auroux, C. Guimon, B. Bonnetot, Micrometric BN powders used as catalyst support: influence of the precursor on the properties of the BN ceramic. J. Solid State Chem. **177**, 609 (2004)
128. E. Antolini, Carbon supports for low-temperature fuel cell catalysts. Appl. Catal. B **88**, 1 (2009)
129. E. Antolini, E.R. Gonzalez, Ceramic materials as supports for low-temperature fuel cell catalysts. Solid State Ionics **180**, 746 (2009)
130. T. Ioroi, Z. Siroma, N. Fujiwara, S.I. Yamazaki, K. Yasuda, Sub-stoichiometric titanium oxide-supported platinum electrocatalyst for polymer electrolyte fuel cells. Electrochem. Commun. **7**, 183 (2005)
131. T. Ioroi et al., Stability of corrosion-resistant Magńli-phase Ti4O 7-supported PEMFC catalysts at high potentials. J. Electrochem. Soc. **155**, B321 (2008)
132. A.L. Santos, D. Profeti, P. Olivi, Electrooxidation of methanol on Pt microparticles dispersed on SnO_2 thin films. Electrochim. Acta **50**, 2615 (2005)
133. K.S. Lee et al., Electrocatalytic activity and stability of Pt supported on Sb-doped SnO_2 nanoparticles for direct alcohol fuel cells. J. Catal. **258**, 143 (2008)
134. W.J. Zhou et al., Performance comparison of low-temperature direct alcohol fuel cells with different anode catalysts. J. Power Sources **126**, 16 (2004)
135. Z. Jusys et al., Activity of PtRuMeOx (Me = W, Mo or V) catalysts towards methanol oxidation and their characterization. J. Power Sources **105**, 297 (2002)
136. Y. Suzuki, A. Ishihara, S. Mitsushima, N. Kamiya, K.I. Ota, Sulfated-zirconia as a support of Pt catalyst for polymer electrolyte fuel cells. Electrochem. Solid-State Lett. **10**, B105 (2007)
137. S. Zhang et al., The oxidation resistance of tungsten carbide as catalyst support for proton exchange membrane fuel cells. Chin. J. Catal. **28**, 109 (2007)
138. R. Koc, S.K. Kodambaka, Tungsten carbide (WC) synthesis from novel precursors. J. Eur. Ceram. Soc. **20**, 1859 (2000)
139. R. Ganesan, J.S. Lee, Tungsten carbide microspheres as a noble-metal-economic electrocatalyst for methanol oxidation. Angewandte Chem. Int. Ed. **44**, 6557 (2005)
140. R. Ganesan, D.J. Ham, J.S. Lee, Platinized mesoporous tungsten carbide for electrochemical methanol oxidation. Electrochem. Commun. **9**, 2576 (2007)
141. D.J. Ham, Y.K. Kim, S.H. Han, J.S. Lee, Pt/WC as an anode catalyst for PEMFC: activity and CO tolerance. Catal. Today **132**, 117 (2008)
142. J.J. Patt, S.K. Bej, L.T. Thompson, Carbide- and nitride-based fuel processing catalysts. Stud. Surf. Sci. Catal. **147**, 85 (2004)
143. J.J. Patt, *Carbide and Nitride Catalysts for the Water Gas Shift Reaction* (University of Michigan, Michigan, 2003)
144. J. Patt, D.J. Moon, C. Phillips, L. Thompson, Molybdenum carbide catalysts for water-gas shift. Catal. Lett. **65**, 193 (2000)
145. S.K. Bej, E.S. Ranganathan, L.T. Thompson, Abst. Pap. Am. Chem. S. **225**, 327–328 (2003)
146. L.T. Thompson, S.K. Bej, J.J. Patt, C.H. Kim, US Patent App. 10/698,818,(2005)
147. E. Antolini, E.R. Gonzalez, Polymer supports for low-temperature fuel cell catalysts. Appl. Catal. A **365**, 1 (2009)
148. D.J. Walton, Electrically conducting polymers. Mater. Des. **11**, 142 (1990)
149. M.C. Lefebvre, Z. Qi, P.G. Pickup, Electronically conducting proton exchange polymers as catalyst supports for proton exchange membrane fuel cells. Electrocatalysis of oxygen reduction, hydrogen oxidation, and methanol oxidation. J. Electrochem. Soc. **146**, 2054 (1999)
150. J. Shan, P.G. Pickup, Characterization of polymer supported catalysts by cyclic voltammetry and rotating disk voltammetry. Electrochim. Acta **46**, 119 (2000)

151. M. Hepel, The electrocatalytic oxidation of methanol at finely dispersed platinum nanoparticles in polypyrrole films. J. Electrochem. Soc. **145**, 124 (1998)
152. K. Bouzek, K.M. Mangold, K. Jüttner, Platinum distribution and electrocatalytic properties of modified polypyrrole films. Electrochim. Acta **46**, 661 (2000)
153. M. Trueba, S.P. Trasatti, S. Trasatti, Electrocatalytic activity for hydrogen evolution of polypyrrole films modified with noble metal particles. Mater. Chem. Phys. **98**, 165 (2006)
154. F.T.A. Vork, L.J.J. Janssen, E. Barendrecht, Oxidation of hydrogen at platinum-polypyrrole electrodes. Electrochim. Acta **31**, 1569 (1986)
155. F.T.A. Vork, E. Barendrecht, The reduction of dioxygen at polypyrrole-modified electrodes with incorporated Pt particles. Electrochim. Acta **35**, 135 (1990)
156. C.S.C. Bose, K. Rajeshwar, Efficient electrocatalyst assemblies for proton and oxygen reduction: the electrosynthesis and characterization of polypyrrole films containing nanodispersed platinum particles. J. Electroanal. Chem. **333**, 235 (1992)
157. W.T. Napporn, H. Laborde, J.M. Léger, C. Lamy, Electro-oxidation of C1 molecules at Pt-based catalysts highly dispersed into a polymer matrix: Effect of the method of preparation. J. Electroanal. Chem. **404**, 153 (1996)
158. P.O. Esteban, J.M. Leger, C. Lamy, E. Genies, Electrocatalytic oxidation of methanol on platinum dispersed in polyaniline conducting polymers. J. Appl. Electrochem. **19**, 462 (1989)
159. H. Laborde, J.M. Léger, C. Lamy, Electrocatalytic oxidation of methanol and C1 molecules on highly dispersed electrodes part 1: platinum in polyaniline. J. Appl. Electrochem. **24**, 219 (1994)
160. F. Fiçicioğlu, F. Kadirgan, Electrooxidation of methanol on platinum doped polyaniline electrodes: deposition potential and temperature effect. J. Electroanal. Chem. **430**, 179 (1997)
161. K.M. Kost, D.E. Bartak, B. Kazee, T. Kuwana, Electrodeposition of platinum microparticles into polyaniline films with electrocatalytic applications. Anal. Chem. **60**, 2379 (1988)
162. H.J. Salavagione, C. Sanchís, E. Morallón, Friendly conditions synthesis of platinum nanoparticles supported on a conducting polymer: methanol electrooxidation. J. Phys. Chem. C **111**, 12454 (2007)
163. C.C. Chen, C.S.C. Bose, K. Rajeshwar, The reduction of dioxygen and the oxidation of hydrogen at polypyrrole film electrodes containing nanodispersed platinum particles. J. Electroanal. Chem. **350**, 161 (1993)
164. D.J. Strike, N.F. De Rooij, M. Koudelka-Hep, M. Ulmann, J. Augustynski, Electrocatalytic oxidation of methanol on platinum microparticles in polypyrrole. J. Appl. Electrochem. **22**, 922 (1992)
165. A.A. Mikhaylova, E.B. Molodkina, O.A. Khazova, V.S. Bagotzky, Electrocatalytic and adsorption properties of platinum microparticles electrodeposited into polyaniline films. J. Electroanal. Chem. **509**, 119 (2001)
166. B.I. Podlovchenko, Y.M. Maksimov, T.D. Gladysheva, E.A. Kolyadko, Electrocatalytic activity of platinum-polyaniline and palladium-polyaniline systems obtained by cycling the electrode potential. Russ. J. Electrochem. **36**, 731 (2000)
167. Z. Qi, M.C. Lefebvre, P.G. Pickup, Electron and proton transport in gas diffusion electrodes containing electronically conductive proton-exchange polymers. J. Electroanal. Chem. **459**, 9 (1998)
168. C.T. Hable, M.S. Wrighton, Electrocatalytic oxidation of methanol and ethanol: a comparison of platinum-tin and platinum-ruthenium catalyst particles in a conducting polyaniline matrix. Langmuir **9**, 3284 (1993)
169. Y.M. Maksimov, T.D. Gladysheva, B.I. Podlovchenko, Electrochemical behavior of platinum-modified polyaniline films in sulfuric acid solutions of carbon monoxide. Russ. J. Electrochem. **37**, 554 (2001)
170. M.L. Bañón, V. López, P. Ocón, P. Herrasti, Poly(o-toluidine). Deposition of platinum and electrocatalytic applications. Synth. Met. **48**, 355 (1992)

171. D. Profeti, P. Olivi, Methanol electrooxidation on platinum microparticles electrodeposited on poly (o-methoxyaniline) films. Electrochim. Acta **49**, 4979 (2004)
172. C. Sivakumar, Finely dispersed Pt nanoparticles in conducting poly(o-anisidine) nanofibrillar matrix as electrocatalytic material. Electrochim. Acta **52**, 4182 (2007)
173. C. Jiang, X. Lin, Preparation of three-dimensional composite of poly(N-acetylaniline) nanorods/platinum nanoclusters and electrocatalytic oxidation of methanol. J. Power Sources **164**, 49 (2007)
174. P.J. Kulesza et al., Electrocatalytic properties of conducting polymer based composite film containing dispersed platinum microparticles towards oxidation of methanol. Electrochim. Acta **44**, 2131 (1999)
175. A. Yassar, J. Roncali, F. Garnier, Preparation and electroactivity of poly(thiophene) electrodes modified by electrodeposition of palladium particles. J. Electroanal. Chem. **255**, 53 (1988)
176. G. Tourillon, F. Garnier, Inclusion of metallic aggregates in organic conducting polymers. A new catalytic system, [poly(3-methylthiophene)-Ag-Pt], for proton electrochemical reduction. J. Phys. Chem. **88**, 5281 (1984)
177. S. Swathirajan, Y.M. Mikhail, Methanol oxidation on platinum-tin catalysts dispersed on poly(3-methyl)thiophene conducting polymer. J. Electrochem. Soc. **139**, 2105 (1992)
178. A. Galal, N.F. Atta, S.A. Darwish, S.M. Ali, Electrodeposited metals at conducting polymer electrodes. II: study of the oxidation of methanol at poly(3-methylthiophene) modified with Pt-Pd co-catalyst. Top. Catal. **47**, 73 (2008)
179. Z. Cai, C.R. Martin, Electronically conductive polymer fibers with mesoscopic diameters show enhanced electronic conductivities. J. Am. Chem. Soc. **111**, 4138 (1989)
180. B. Rajesh et al., Template synthesis of conducting polymeric nanocones of poly(3-methylthiophene). J. Phys. Chem. B **108**, 10640 (2004)
181. B. Rajesh et al., Pt particles supported on conducting polymeric nanocones as electro-catalysts for methanol oxidation. J. Power Sources **133**, 155 (2004)
182. Z. Qi, P.G. Pickup, Novel supported catalysts: platinum and platinum oxide nanoparticles dispersed on polypyrrole/polystyrenesulfonate particles. Chem. Commun. **1998,** 15 (1998)
183. C. Arbizzani, M. Biso, E. Manferrari, M. Mastragostino, Methanol oxidation by pEDOT-pSS/PtRu in DMFC. J. Power Sources **178**, 584 (2008)
184. S. Patra, N. Munichandraiah, Electrooxidation of methanol on pt-modified conductive polymer PEDOT. Langmuir **25**, 1732 (2009)
186. J.F. Drillet, R. Dittmeyer, K. Jüttner, Activity and long-term stability of PEDOT as Pt catalyst support for the DMFC anode. J. Appl. Electrochem. **37**, 1219 (2007)
186. L.M. Huang, C.H. Chen, T.C. Wen, A. Gopalan, Effect of secondary dopants on electrochemical and spectroelectrochemical properties of polyaniline. Electrochim. Acta **51**, 2756 (2006)
187. L.M. Huang, W.R. Tang, T.C. Wen, Spatially electrodeposited platinum in polyaniline doped with poly(styrene sulfonic acid) for methanol oxidation. J. Power Sources **164**, 519 (2007)
188. S. Kim, S.J. Park, Electroactivity of Pt-Ru/polyaniline composite catalyst-electrodes prepared by electrochemical deposition methods. Solid State Ionics **178**, 1915 (2008)
189. F. Bensebaa et al., Microwave synthesis of polymer-embedded Pt-Ru catalyst for direct methanol fuel cell. J. Phys. Chem. B **109**, 15339 (2005)
190. C. Arbizzani, M. Biso, E. Manferrari, M. Mastragostino, Passive DMFCs with PtRu catalyst on poly(3,4-ethylenedioxythiophene)-polystyrene-4-sulphonate support. J. Power Sources **180**, 41 (2008)
191. B. Rajesh et al., Nanostructured conducting polyaniline tubules as catalyst support for Pt particles for possible fuel cell applications. Electrochem. Solid-State Lett. **7**, A404 (2004)
192. Z. Chen, L. Xu, W. Li, M. Waje, Y. Yan, Polyaniline nanofibre supported platinum nanoelectrocatalysts for direct methanol fuel cells. Nanotechnology **17**, 5254 (2006)

193. H.H. Zhou, S.Q. Jiao, J.H. Chen, W.Z. Wei, Y.F. Kuang, Effects of conductive polyaniline (PANI) preparation and platinum electrodeposition on electroactivity of methanol oxidation. J. Appl. Electrochem. **34**, 455 (2004)
194. B. Rajesh et al., Chem. Commun. **16**, 2022 (2003)
195. J. Li, X. Lin, A composite of polypyrrole nanowire. J. Electrochem. Soc. **154**, B1074 (2007)
196. F.J. Liu, L.M. Huang, T.C. Wen, A. Gopalan, Large-area network of polyaniline nanowires supported platinum nanocatalysts for methanol oxidation. Synth. Met. **157**, 651 (2007)
197. T. Maiyalagan, Electrochemical synthesis, characterization and electro-oxidation of methanol on platinum nanoparticles supported poly(o-phenylenediamine) nanotubes. J. Power Sources **179**, 443 (2008)
198. H. Mizes, E. Conwell, Conduction in ladder polymers. Phys. Rev. B **44**, 3963 (1991)
199. E. Antolini, Composite materials an emerging class of fuel cell catalyst supports. Appl. Catal. B-Environ. **100**, 413 (2010)
200. S. Mokrane, L. Makhloufi, N. Alonso-Vante, Electrochemistry of platinum nanoparticles supported in polypyrrole (PPy)/C composite materials. J. Solid State Electrochem. **12**, 569 (2008)
201. Z.Z. Zhu, Z. Wang, H.L. Li, Functional multi-walled carbon nanotube/polyaniline composite films as supports of platinum for formic acid electrooxidation. Appl. Surf. Sci. **254**, 2934 (2008)
202. Z. Wang, Z.Z. Zhu, J. Shi, H.L. Li, Electrocatalytic oxidation of formaldehyde on platinum well-dispersed into single-wall carbon nanotube/polyaniline composite film. Appl. Surf. Sci. **253**, 8811 (2007)
203. P. Santhosh, A. Gopalan, K.P. Lee, Gold nanoparticles dispersed polyaniline grafted multiwall carbon nanotubes as newer electrocatalysts: preparation and performances for methanol oxidation. J. Catal. **238**, 177 (2006)
204. Y. Xu, X. Peng, H. Zeng, L. Dai, H. Wu, Study of an anti-poisoning catalyst for methanol electro-oxidation based on PAn-C composite carriers. C. R. Chim. **11**, 147 (2008)
205. H. Gharibi, K. Kakaei, M. Zhiani, J. Phys. Chem. C **114**, 3956 (2010)
206. G. Wu, L. Li, J.H. Li, B.Q. Xu, Polyaniline-carbon composite films as supports of Pt and PtRu particles for methanol electrooxidation. Carbon **43**, 2579 (2005)
207. H. Zhao, L. Li, J. Yang, Y. Zhang, Nanostructured polypyrrole/carbon composite as Pt catalyst support for fuel cell applications. J. Power Sources **184**, 375 (2008)
208. J. Shi, Z. Wang, H.L. Li, Electrochemical fabrication of polyaniline/multi-walled carbon nanotube composite films for electrooxidation of methanol. J. Mater. Sci. **42**, 539 (2007)
209. J. Shi, D.J. Guo, Z. Wang, H.L. Li, Electrocatalytic oxidation of formic acid on platinum particles dispersed in SWNT/PANI composite film. J. Solid State Electrochem. **9**, 634 (2005)
210. G. Wu, L. Li, J.H. Li, B.Q. Xu, Methanol electrooxidation on Pt particles dispersed into PANI/SWNT composite films. J. Power Sources **155**, 118 (2006)
211. K. Lee, J. Zhang, H. Wang, D.P. Wilkinson, Progress in the synthesis of carbon nanotube-And nanofiber-supported Pt electrocatalysts for PEM fuel cell catalysis. J. Appl. Electrochem. **36**, 507 (2006)
212. V. Selvaraj, M. Alagar, Pt and Pt-Ru nanoparticles decorated polypyrrole/multiwalled carbon nanotubes and their catalytic activity towards methanol oxidation. Electrochem. Commun. **9**, 1145 (2007)
213. H.B. Bae, J.H. Ryu, B.S. Byun, S.H. Jung, S.H. Choi, Facile synthesis of novel Pt-Ru@PPy-MWNT electrocatalysts for direct methanol fuel cells. Curr. Appl. Phys. **10**, S44 (2010)
214. C. Jiang et al., Preparation of the Pt nanoparticles decorated poly(N-acetylaniline)/MWNTs nanocomposite and its electrocatalytic oxidation toward formaldehyde. Electrochim. Acta **54**, 1134 (2009)
215. Z.C. Wang, Z.M. Ma, H.L. Li, Functional multi-walled carbon nanotube/polysiloxane composite films as supports of PtNi alloy nanoparticles for methanol electro-oxidation. Appl. Surf. Sci. **254**, 6521 (2008)

216. S. Beak, D. Jung, K.S. Nahm, P. Kim, Preparation of highly dispersed Pt on TiO_2-modified carbon for the application to oxygen reduction reaction. Catal. Lett. **134**, 288 (2010)
217. J.M. Lee et al., TiO_2@carbon core-shell nanostructure supports for platinum and their use for methanol electrooxidation. Carbon **48**, 2290 (2010)
218. D. He, L. Yang, S. Kuang, Q. Cai, Fabrication and catalytic properties of Pt and Ru decorated TiO_2{minus 45 degree rule}CNTs catalyst for methanol electrooxidation. Electrochem. Commun. **9**, 2467 (2007)
219. H.L. Pang et al., Preparation of SnO_2-CNTs supported Pt catalysts and their electrocatalytic properties for ethanol oxidation. Electrochim. Acta **54**, 2610 (2009)
220. R.S. Hsu, D. Higgins, Z. Chen, Tin-oxide-coated single-walled carbon nanotube bundles supporting platinum electrocatalysts for direct ethanol fuel cells. Nanotechnology **21** 165705 (2010)
221. C. Du, M. Chen, X. Cao, G. Yin, P. Shi, A novel CNT@SnO_2 core-sheath nanocomposite as a stabilizing support for catalysts of proton exchange membrane fuel cells. Electrochem. Commun. **11**, 496 (2009)
222. M.W. Xu, G.Y. Gao, W.J. Zhou, K.F. Zhang, H.L. Li, Novel Pd/β-MnO_2 nanotubes composites as catalysts for methanol oxidation in alkaline solution. J. Power Sources **175**, 217 (2008)
223. B. Rajesh et al., Pt supported on polyaniline-V_2O_5 nanocomposite as the electrode material for methanol oxidation. Electrochem. Solid-State Lett. **5**, E71 (2002)
224. T. Maiyalagan, B. Viswanathan, Synthesis, characterization and electrocatalytic activity of Pt supported on poly (3,4-ethylenedioxythiophene)-V_2O_5 nanocomposites electrodes for methanol oxidation. Mater. Chem. Phys. **121**, 165 (2010)
225. H. Pang et al., Preparation of polyaniline-tin dioxide composites and their application in methanol electro-oxidation. J. Solid State Electrochem. **14**, 169 (2010)
226. S. Niyogi et al., Chemistry of single-walled carbon nanotubes. Acc. Chem. Res. **35**, 1105 (2002)

Part III
Active Materials

Chapter 7
Supported Metal Nanoparticles

7.1 Key Concepts

The purpose of this chapter is to summarize the main synthetic techniques used to prepare supported metal nanoparticles for use in electrolytic cells (primarily fuel cells and electrolyzers). Each method provides a strategy which has the main goal of controlling particle size, alloy composition and catalyst distribution over the support material. As a consequence the reader is presented here with a general overview of the most common synthetic methods. Some of the more established procedures are now used industrially to produce large quantities of materials. This does not mean they are superior to the others, depending on the end application of the catalyst and the instrumentation available, one method may be advantageous over another. Therefore, the reader is advised to consider the intrinsic advantages and disadvantages of each method when selecting which to use. More recently methods have been developed to control structure on an atomic scale by the formation of surface defects such as twins or stacking faults that can lead to dramatic increases in activity. This will be covered in detail elsewhere.

7.2 Metal Nanoparticle Synthetic Techniques

In recent years, a number of approaches have been developed for the synthesis of nanoparticle based electrocatalysts. Special emphasis has been given to the achievement of a high degree of control over mean particle size, shape, and dispersion on the support. Of equal importance has been the development of reproducible low cost synthetic approaches that are suitable to be scaled up to large substrate areas and material quantities. One of the main pre-requisites of any synthetic strategy is the production of stable and robust materials that preserve their initial morphologies under cell working conditions over long periods, in particular maintaining their narrow metal particle size distribution (i.e., under exposure to reactants, impurities and under high temperatures and pressures).

A. Lavacchi et al., *Nanotechnology in Electrocatalysis for Energy*,
Nanostructure Science and Technology 170, DOI: 10.1007/978-1-4899-8059-5_7,
© Springer Science+Business Media New York 2013

7.2.1 Low Temperature Chemical Precipitation

The preparation of platinum based catalysts by chemical precipitation at low temperatures is a strategy that has been widely used for some time [1]. Both supported and unsupported catalysts can be made this way, and the process is simple and easily scaled up to larger quantities. The process involves the addition of a reducing agent (such as $NaBH_4$ or hydrazine) to a metal salt solution. Bi or tri metallic alloyed catalysts can be made by the co-precipitation of a solution of two or more metal salts, e.g. Ru and Pt salts. In the case of unsupported catalysts the precursor metal salt is reduced and precipitates out of solution as a solid that can be recovered by filtration. The alloy powders that result can be further treated under H_2 flow at high temperatures. Single and bimetallic Pt and Ru catalysts have been prepared by Bock et al. by adding a reducing agent to solutions of H_2PtCl_6 and $RuCl_3$ [1]. PtRu alloys with up to 46 atom % Ru were prepared and no post heat treatment under H_2 was necessary to obtain complete reduction and alloy formation.

An alternative method involves the use of carbonyl precursors that decompose at low temperatures forming metal particles [2]. The metal carbonyl can be formed by bubbling CO through a solution of a metal salt (e.g. H_2PtCl_6) that results in the formation of a $Pt(CO)_2$ precipitate that can be filtered, washed and dried [3]. To form supported catalysts, a carbon black can be added prior to the reduction of the metal-carbonyls. Commonly, the unsupported particles grow more rapidly than supported particles; typically they are at least twice the size of supported particles. Platinum alloys can be made by first forming the bimetallic carbonyl complex via the following route:

$$\left[Pt_3(CO)_6\right]_{10}^{2-} + M_yX_n \rightarrow \left[Pt_xM_y(CO)_6\right]_{10}^{2-}$$

where M=Sn, Ni, Cr, Co and X=Cl^- followed by reduction using hydrogen gas. As an example, Pt–Sn particles (atomic ratio 3:1) with an average size of 2.4 ± 1 nm have been made using this method [2]. Supported, Pt–Ru/C catalysts have been prepared by dissolving the Pt and Ru carbonyl precursors in o-Xylene. Vulcan XC-72 was added and the mixture was refluxed at 143 °C for 24 h to form the carbon supported nano-sized Pt–Ru catalyst (Fig. 7.1) [3].

An adaption of the chemical precipitation method initially reported by Haruta et al. involves the dissolution of the metal precursor salt in a suspension of the support material [4]. Adjustment of the pH (i.e., 5–10) results in the complete precipitation of the metal hydroxide e.g. $Au(OH)_3$, which is deposited onto the surface of the support. The supported hydroxide thus formed is subsequently treated at high temperature (calcined) and reduced to the elemental metal. This can also be achieved under a reducing H_2 atmosphere.

The simplicity of this method allows it to be used also as a system for the screening of multi-metal electrocatalyst combinations using a combinatorial approach. Chen used this method to prepare electrode arrays containing 715

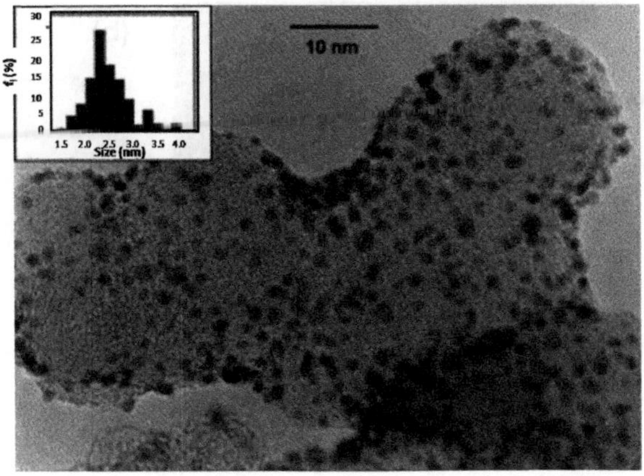

Fig. 7.1 HR–TEM image of PtRu catalyst prepared via the carbonyl route. The dark features correspond to the PtRu catalyst particles. *Inset* Size histogram for the primary particles. Reprinted from Ref. [3] with permission from Elsevier

unique combinations of five different elements (Pt, Ru, Os, Ir, and Rh) [5]. These were prepared by dispersing aqueous solutions of five metal salts ($RhCl_3$, H_2PtCl_6, $RuCl_3$, $OsCl_3$ and $IrBr_3$) onto a Teflon-coated Toray carbon sheet, using a robotic plotter. The completed array contained the same total number of moles of metal at each spot. A 40-fold molar excess of 5 % aqueous sodium borohydride was added to each spot, and the reduced array was thoroughly washed. Each spot was analyzed for activity for both OER and ORR using florescence detection. The ternary catalyst $Pt_{4.5}Ru_4Ir_{0.5}$ was found to be the most active material.

In general chemical precipitation, deposition–precipitation methods produce a broad nanoparticle size distribution and it is difficult to tune the particle size for a particular application owing to the poor control of size that also affects the dispersion and size of the metal nanoparticle with increasing metal loadings. Particle agglomeration is also quite a common phenomena.

7.2.2 Impregnation

Due to the simplicity of preparation, the impregnation method is one of the most commonly used techniques to synthesize electrocatalyst materials. High surface area carbon blacks are impregnated with catalyst precursors by mixing the two in an aqueous suspension [6]. Other methods of impregnating a substrate involve depositing an aliquot of solution containing the catalyst precursor onto the substrate and allowing it to air-dry. This is known as the insipient wetness method. An example for the preparation of a 0.3 wt% Au/C catalyst proceeds as follows [7]; the required amount of solid precursor $HAuCl_4$ was dissolved in a volume of aqueous media, corresponding to the pore volume of the support. The impregnation solution was then added drop wise to the support with intensive mixing. After addition of the solution was complete, the support was only slightly wet. This catalyst precursor material was dried for 16 h at 80 °C and subsequently reduced

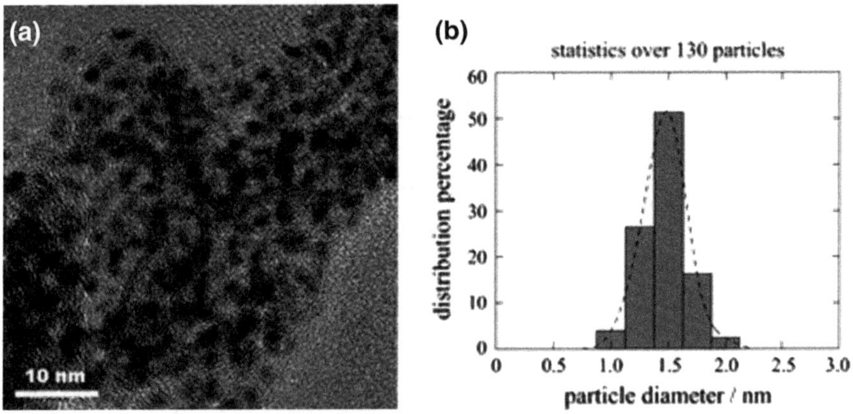

Fig. 7.2 HRTEM image (**a**) and statistics histogram of the particle size distribution (**b**) for PtRu/C (40 wt% Pt + 20 wt% Ru). Reprinted with permission from Ref. [10]. Copyright (2003) American Chemical Society

in the gas phase using 5 % hydrogen in nitrogen at 250 °C for 2 h. Following the impregnation step, a reduction step is required to reduce the catalyst precursor to its metallic state. As reduction occurs after the impregnation step, the nature of the support plays a crucial role in controlling particle size [8]. The porosity of the carbon black support can effectively control the catalyst nanoparticle size and dispersion. Many studies indicate that synthetic conditions, such as the nature of the metal precursors used, the reduction method and the heating temperature, are also crucial in the impregnation process [9]. Common liquid phase reducing agents used include $Na_2S_2O_3$, $NaBH_4$, $Na_4S_2O_5$, N_2H_4 and formic acid. Reduction can be carried out in the gas phase using a flowing hydrogen stream as a reductive agent under elevated temperatures.

Using a simple impregnation method Yang et al. [10] prepared highly dispersed PtRu/C electrocatalysts with metal loadings as high as 60 wt% with a narrow size distribution (1.5 ± 0.5 nm, as shown in Fig. 7.2).

To avoid chlorine poisoning, metal sulfate salts, metal carbonyl complexes and metal nitrate salts are used [3]. Metal carbonyl complexes are particularly attractive as precursors since they can easily be made by direct oxidation of the metal chloride salt with carbon monoxide. Additionally, an external reducing agent is not required, as nanoparticles can then be formed by thermal decomposition of the metal carbonyl complexes impregnated onto carbon supports [3]. Experimental factors, such as the reduction temperature, can significantly influence the morphology of the catalyst. For example the organometalic complex $(\eta\text{-}C_2H_4)(Cl)Pt(\mu Cl)_2Ru(Cl)(\eta^3,\eta^3\text{-}C_{10}H_{16})$ supported on carbon and then reduced under either argon or hydrogen at 400 °C gave 2.5 ± 1 nm particles while reduction at 650 °C gave 5.3 nm particles [11]. The impregnation technique is a comparatively environmentally friendly method as reduction reactions occur either at low temperature or at room temperature, minimizing energy consumption, and organic solvents are avoided by using aqueous media. The disadvantages of the impregnation technique are generally related to using liquid solutions as a processing medium as in solution particles can easily agglomerate and as a consequence it is difficult to control nanoparticle size and distribution.

Binary and ternary electrocatalysts can also be readily prepared by the impregnation method. For example Tayal and co-workers synthesized the following binary and ternary Pt based catalysts with the addition of Re, Pt–Re/C (20:20), Pt–Sn/C (20:20), Pt–Re–Sn/C (20:10:10) and Pt–Re–Sn/C (20:5:15). These electrocatalysts were prepared from their precursors by the co-impregnation-reduction method and the activity towards the electro-oxidation of ethanol in DEFCs was studied [12]. A propanol suspension of the catalyst precursor metal salts was added to a suspension of Vulcan XC-72 in propanol. Impregnation was obtained by ultrasonication of this material. The isolated precursor material was then reduced under H_2 at high temperature to obtain the desired bimetallic and trimetallic catalysts.

7.2.3 Colloidal

The colloidal method of generating nanostructured electrocatalysts is a technique similar to the chemical precipitation method. However, it involves the added benefit of a capping agent that allows for size control thus preventing agglomeration of the catalyst particles. Any molecule that remains adsorbed onto the metal particle surface has the potential to act as a capping agent. The experimental procedure is as simple as combining the metal salt, a reducing agent, and a capping agent together and mixing. Bimetallic and other colloidal catalysts can be prepared by co-reduction. If H_2 gas is the reducing agent, it is typically bubbled through the solution. To synthesize supported catalysts a carbon support is added to the mixture either before or after the formation of the catalyst particles. If a fast reducing agent is used the reactions can be completed within less than 30 min. Conversely, slow reducing agents, like H_2 at room temperature, require longer reaction times and are generally allowed to react overnight or for at least 12 h. The synthesis of bi-metallic carbon supported Pt–Ru/C catalysts of high active surface area (ca. 80 m^2g^{-1}) is known as the Watanabe method [13]. The first step involves the formation of colloidal "PtO_2" from platinic acid (H_2PtCl_6) with the reduction of $NaHSO_3$ and oxidative decomposition with H_2O_2 at a fixed pH of 5:

$$H_2PtCl_6 + 3NaHSO_3 + 2H_2O \rightarrow H_3Pt(SO_3)2OH + Na_2SO_4 + NaCl + 5HCl$$
$$H_3Pt(SO_3)_2OH + 3H_2O_2 \rightarrow PtO_2 + 3H_2O + 2H_2SO_4$$

Subsequently, $RuCl_3$ was added and the formation of a brown-colored colloid "RuO_2" was suggested to have taken place as follows:

$$RuCl_3 + 3/2H_2O_2 \rightarrow RuO_2 + 3HCl + 1/2O_2$$

Carbon black was added to the solution and H_2 was bubbled through it for complete reduction and formation of the carbon-supported PtRu alloys. PtRu alloy formation was proposed to have taken place, as TEM images of the formed nanosized particles indicated a face centered cubic structure (fcc) rather than hexagonally closed packed (hcp) structure typical for metallic Ru. The colloidal synthesis method has also been successfully tailored to allow for particle size and composition control, as well as shape control for Pt-based nanoparticles. Yoshitake et al. used the colloid method for the synthesis of Pt catalysts for use as PEM fuel cell cathodes using single-wall carbon nanohorns (SWNHs) as the support [14]. In the preparation, $NaHSO_3$ and H_2O_2 were added to a H_2PtCl_6 solution to form a Pt oxide colloid solution where the Pt oxide colloid was adsorbed on the SWNH surface. After eliminating Cl, Na and S ions, the samples were dried and reduced by H_2 gas.

7.2.4 Microemulsions

The use of microemulsions or reversed micelles to synthesize nano-sized metal particles has been a strategy widely explored for more than 20 years [15]. In recent years this approach has been applied specifically to the synthesis of nanostructured metal electrocatalysts (e.g. Pt or PtRu) for use in electrolytic devices such as fuel cells [16, 17]. Microemulsions can be described as homogeneous like combinations of water, oils and/or surfactants (often in the presence of an alcohol or amine based compounds). Thus, a solid support like carbon black is impregnated with a microemulsion containing a dissolved metal salt precursor, in a similar way to that of the previously described traditional chemical impregnation. Metal nanoparticles obtained using this methodology have a more controllable, narrow crystallite distribution as compared to those obtained through the traditional impregnation, co-precipitation and precipitation-deposition methods. This has been attributed to the confined location of a limited amount of metal salt in the micelles that are subsequently taken up upon interaction with the support. The microemulsion-support interaction can be enhanced by increasing the hydrophobicity of the support (e.g. silylation of hydroxyl-rich surfaces), making it more chemically compatible with the microemulsion during the deposition step. At high water concentrations an emulsion consists of small oil droplets surrounded by surfactants in a continuous water phase, and vice versa, at high oil concentrations an emulsion consists of small water droplets surrounded by surfactants in a continuous oil phase. Most metal precursors are inorganic salts that are soluble in water, not in oil, therefore, emulsions with high oil concentrations and low water concentrations are used almost exclusively. Reverse micelles are water-in-oil droplets stabilized by a surfactant. A most commonly used surfactant is sodium 2-bis(2-ethylhexyl)sulfosuccinate (Na(AOT)). Alkylthiols (C_nSH),-amines

(C_nNH_2) and -isocyanides are examples of other stabilizers that have also been used to make microemulsions. The droplets are displaced randomly in solution and are subject to Brownian motion. They exchange their water content and re-form into distinct micelles. The size of the water-in-oil droplets increases with increasing water content. The droplet size, i.e., resulting particle size, can also be influenced by the chain length of the stabilizer. In order to obtain catalyst nanoparticles, the metal salt is reduced by adding a reducing agent into the microemulsion system (e.g. N_2H_4, HCHO or $NaBH_4$). Another approach is to mix the microemulsion system that contains a reducing agent with a microemulsion system that contains the metal salt [8, 18]. Once the nanostructured catalysts are formed they can be deposited onto a support, which is done by adding a solvent like tetrahydrofuran (THF) in conjunction with the support powder to the microemulsion [18]. The solvent destabilizes the microemulsion by competing with the surfactant to adsorb onto the particles, and in the destabilized system the particles will adsorb onto the support. It has also been shown that Pt and PtRu catalysts formed using the microemulsion technique can also be directly deposited onto carbon blacks such as Vulcan XC72 by simply mixing and stirring the emulsion with the support, i.e., without the need of an emulsion-breaking solvent [16, 17]. After deposition, residual surfactant molecules are typically removed by heat treatment [8]. Emulsion systems are very sensitive to temperature and therefore the oil and surfactant must be carefully selected [18]. Recently, it has been shown that size control with the microemulsion method appears more complicated than preliminary research indicated. Originally, the water droplets containing dissolved metal salts stabilized by a surfactant were often described as a “micro-reactor” or “nano-cage” [19]. Since the reduction is confined within the nanoscale microemulsion it was originally believed that catalyst size is exclusively controlled by the size of the water droplets in the microemulsion [8]. However, it has been found that particle size is a more complicated phenomenon involving both the nature of the reducing agent as well as the size of the water droplets in the microemulsion [18]. Small particles are formed by fast nucleation caused by a fast or efficient reducing agent [18]. Hydrazine is an example of a fast reducing agent, while dihydrogen gas is a relatively slow reducing agent. The size of the water droplet is influenced by the ratio of water to surfactant and the surfactant concentration (at fixed water/oil ratios). As the ratio of water to surfactant increases, the size of the water droplet increases and consequently, the catalyst size also increases. However, a maximum particle size is reached and further increasing the water-to-surfactant ratio has no effect on catalyst size [16, 17, 20]. For example, in a microemulsion system of water/n-heptane with the surfactant sodium dioctyl sulfosuccinate the Pt–Ru particle size increases from 2.4 (0.1) to 3.2 (0.1) nm when the ratio of water to sodium dioctyl sulfosuccinate increases from 4 to 8 [17]. However, increasing the water-to-dioctyl-sulfosuccinate ratio to 10 does not result in any further increase in catalyst size [17]. Droplet size can also be controlled by varying the surfactant concentration while keeping the concentrations of water and oil constant. For example, increasing the surfactant concentration, with constant water and oil concentrations, increases the number of droplets. As a result, droplet size decreases, resulting in fewer metal ions per droplet and a consequently decreased particle size. Droplet size, however, does not directly or absolutely control particle size. For example, it has been estimated that there are five $PtCl_6^{2-}$ ions in each water droplet in the microemulsion system of water in hexane, stabilized by the surfactant pentaethyleneglycol dodecylether [18]. However, TEM analysis indicated that the platinum catalyst formed had an average diameter of 3.5 nm, which corresponds to 100–1,500 metal atoms (depending on the shape of the particle).

This example illustrates that the final particle is not formed inside the droplet. It is likely that the formation of catalysts via the microemulsion process proceeds by two steps: (1) nucleation of the metal catalyst inside the droplet, followed by (2) aggregation of multiple nuclei via collision and coalescence of droplets to form the final nanostructured catalysts [18].

Wang and co-workers have recently reported another interesting approach using this methodology employing a water-liquid CO_2 (as oil phase) microemulsion stabilized by sodium bis(2-ethylhexyl)sulfosuccinate as surfactant and hexane [21]. In this way, Pd, Rh and Pd–Rh nanoparticles with sizes ranging from 2 to 10 nm could be homogeneously deposited on the surface of multiwalled carbon nanotubes (CNT). Escudero and co-workers have prepared electrocatalysts based on Pt, Pt–Ru and Pt–Pd with a low precious metal loading (0.37–0.05 mg cm^{-2}) using a similar microemulsion method and showed that the MEAs with the electrocatalysts prepared by microemulsion have a performance comparable to that of the MEAs with commercial electrocatalysts [22].

More recently, the microemulsion approach has been applied to more complex systems. For example Yang and coworkers have presented a new approach to synthesizing nanosized Pt–WO_3 using a microwave-assisted microemulsion methodology (Fig. 7.3) [23]. Amorphous WO_3 nanoparticles of 1.0 nm in size were first (Step 1) deposited onto carbon from an alkaline tungstate containing microemulsion via mixing with an acid containing microemulsion under controlled microwave exposure. Platinum was subsequently deposited (Step 2) onto the carbon supported WO_3 nanoparticles by reducing H_2PtCl_6 in a microemulsion under controlled microwave exposure forming nanoparticles with an average size of 2.5 nm. Microwave heating is a good alternative to traditional approaches offering faster, more uniform control of heating. The synthesis of Pt–WO_3 particles with control of the Pt:W ratio and size distribution at the nanoparticle level has been traditionally difficult. The Pt–WO_3 nanoparticles synthesized in this approach show a narrow size distribution and are well dispersed on the carbon support.

The microemulsion method has to date been employed to obtain a number of metallic (Pt, Pb, Fe, Cd, Ag, Au, Cu, Ni, and Co) [24–27] and bimetallic nanoparticles (Co–Ni, Cu–Ni, Au–Ag, Au–Pd, and Au–Pt) [24, 28–32]. In a recent example carbon supported AuM/C (M=Fe, Co, Ni, Cu and Zn) electrocatalysts have been prepared using a water-in-oil microemulsion method [33–35]. These

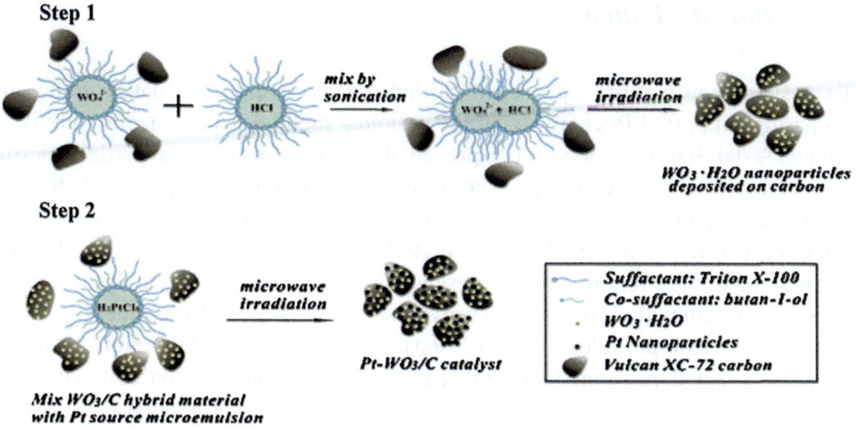

Fig. 7.3 The preparation process for making a Pt–WO_3/C electrocatalyst through the microwave-assisted microemulsion method. Reprinted from Ref. [23] with permission from Elsevier

electrocatalysts were then utilized for the electrochemical oxidation of borohydride. All of the supported bimetallic nanoparticles were found to be spherical with average diameters in the range of 3–9 nm and uniformly distributed on the surfaces of the carbon support.

The microemulsion method can also be readily applied to the formation of non noble-metal particles. The group of Ahmed and co-workers have applied the microemulsion method to form non noble metal metallic and bimetallic nanoparticles for use as HER and OER electrocatalysts in alkaline water electrolysis [24, 28, 36–38]. These researchers have prepared 1:1 CoNi alloyed nanoparticles for OER and nanostructured coreshell particles of Cu–Ni and Cu–Co for use as both OER and HER electrocatalysts. The two step process involved first the microemulsion mediated formation of the metal nanoparticles (cetyltrimethylammonium bromide (CTAB)/1-butanol as surfactants and isooctane as organic phase) followed by annealing under H_2 at high temperature (500–700 °C). In one case these authors were able to prepare spherical, uniform, and highly monodisperse nanoparticles of Fe(75)Co(25), Fe(67)Co(33), Fe(50)Co(50), and Fe(33)Co(67) with an average size of 20, 25, 10, and 40 nm, respectively, and the OER/HER activity of Fe(33)Co(67) was shown to be 100 times higher with respect to the other FeCo alloys [37].

In summary the microemulsion method has been shown to be a technique which allows for a very narrow size distribution of metal particles, with an average size smaller than that of conventional electrocatalysts prepared by impregnation for example. One drawback of this method is that, catalyst fabrication is costly since expensive surfactants and oils are needed.

7.2.5 Polyol Method

Extensive investigations have been carried out to develop alternative routes for preparing supported Pt catalysts by the colloidal method using various stabilizing agents. A stabilizing agent is used to prevent the aggregation of metal particles during the nucleation and growth steps. The intrinsic problem with the process is that the stabilizing organic material remains on the surface of metal colloids and has to be removed before the use of the metal particles in electrocatalysis. Removal of the organic material is required as its presence hinders access to the catalytic sites. In general, the removal of the stabilizer involves heat treatment. Consequently, due to sintering of metal particles, the phase separation and the distribution of metal particles are effected, resulting in lowered catalytic performance. Preparation via the so called polyol process is preferred due to several advantages. The polyol process is a technique in which a polyalcohol such as ethylene glycol is used as both solvent and reducing agent [39, 40]. The procedure used by Boch and coworkers to form Pt–Ru/C electrocatalysts is as follows; firstly $PtCl_4$ and $RuCl_3$ were dissolved in ethylene glycol containing NaOH. The solution was heated under reflux to 160 °C for 3 h, and then cooled. Dark brown solutions containing the PtRu colloids were formed in this manner. The colloidal solutions were then mixed with carbon black for up to 24 h, resulting in the deposition of the PtRu colloids onto the carbon substrates.

A unique property of the polyol process is that it does not require any type of polymer stabilizer. In the polyol process using ethylene glycol, metal ions are reduced to form a metal colloid and ethylene glycol is oxidized to glycolic acid. After reduction the glycolate anion acts as a stabilizer by adsorbing onto the metal colloids [39]. It has been found that the concentration of glycolate anion, which is a function of pH, plays an important role in controlling Pt particle size and loading on carbon [40]. Furthermore removal of these organics from the metal surface by heat treatment below 160 °C has been reported, which is low enough to avoid negative effects associated with heat treatment.

Ternary electrocatalysts can also be readily prepared by the polyol method. In a recent example carbon supported Pt–Ru–Ni and Pt–Sn–Ni catalysts were prepared and studied for the electro-oxidation of ethanol in DEFCs [41]. These electrocatalysts were prepared by heating mixtures of the metal precursor salts and Vulcan XC-72–140 °C for 3 h in an ethylene glycol water mixture. After preparation the materials were employed in DEFCs. When the compound $Pt_{75}Ru_{15}Ni_{10}/C$ was used as an anode catalyst, the current density obtained in the fuel cell was greater than that of all other investigated catalysts.

7.2.6 Microwave Assisted Polyol

Microwaves are electromagnetic waves. Dielectric materials in a microwave field will be heated by the amount proportional to the dielectric loss tangent, which defines the ability of a material to convert electromagnetic energy into heat energy at a given frequency [42]. The greatest advantage of microwave irradiation is that it can heat a substance uniformly through a glass or plastic reaction container, leading to a more homogeneous nucleation and shorter crystallization time compared with those for conventional heating. This is beneficial to the formation of uniform metal colloids. In one of the first examples of the microwave assisted polyol method, polymer-stabilized platinum colloids with nearly uniform spherical shape were prepared as follows; an ethylene glycol solution of NaOH was added drop wise to an aqueous solution of $H_2PtCl_6 \cdot 6H_2O$ and polyvinylpyrrolidone (PVP) dissolved in 24 mL of ethylene glycol. The beaker containing the mixture was placed in the center of a domestic 2,450 MHz microwave oven. After just 30 s of microwave irradiation at the maximum power output of 750 W, the solution changed from orange to black and a PVP-stabilized Pt colloid was formed. The authors concluded that microwave dielectric heating is superior to the conventional heating modes in the preparation of monodispersed nanosized platinum particles [42]. Microwave effects can be divided into two categories: thermal effects and nonthermal effects. The former are caused by the temperature regime which can be created by microwave dielectric heating; the latter are caused by the inherent characteristics of microwaves other than the thermal effect. It now seems to be generally accepted that the thermal effect caused by microwave dielectric heating is the main contributing factor to most phenomena observed.

Several examples of microwave-assisted deposition of metal nanoparticles on supports have since been reported, mainly employing solutions of metal salts as precursors. Microwave irradiation has several advantages over conventional heating methods, including short reaction times, small particle sizes, narrow size distributions and high purity. El-Shall and co-workers have extensively investigated the use of microwaves for the preparation of a range of supported metal nanoparticles including Au and Pd [43, 44]. They have also prepared capped Au and Pd nanoparticles on metal oxides using polyethylene glycol (PEG) and poly(N-vinyl-2-pyrrolidone) as protective polymers prior to microwave-heating to further stabilize the nanoparticles from agglomeration. In this way, the obtained metal nanoparticles were better dispersed and had a narrower particle size distribution, which in turn increased their activity for the investigated application (e.g. oxidation of CO). They claimed that fast and uniform heating (due to high dielectric constants of PEG and PVP) achieved under microwave irradiation allows a quicker reduction of the metal precursor on the support.

Campelo and co-workers have recently reported the preparation of a range of metallic nanoparticles on an ordered mesoporous silica SBA-12 structure [45]. The metallic Au, Ag and Pd nanoparticles were prepared in a very short time (<2 min) under microwave irradiation of a solution of the metal salt precursor in ethanol/water or ethanol/acetone mixtures without the need of additional reducing agent. The microwave protocol afforded dispersed and relatively small metal nanoparticles (2, 3.8 and 11.3 nm average particle size for Au, Ag and Pd, respectively; which were highly active catalysts for oxidation reactions. The time of microwave

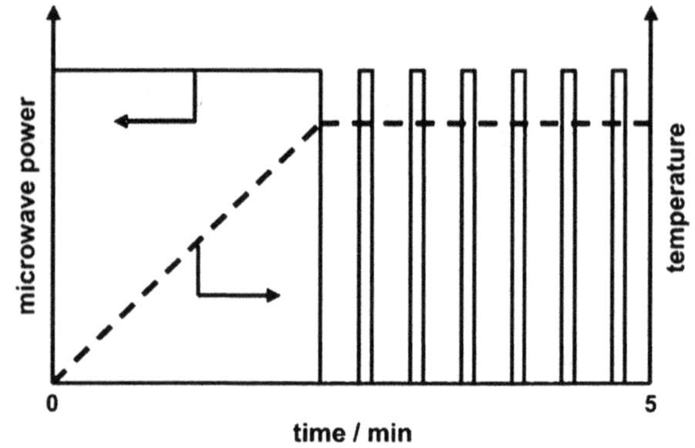

Fig. 7.4 Scheme of the microwave sequence and of the temperature change as a function of time in course of the microwave assisted polyol synthesis procedure. Reprinted Ref. [57] with permission from Elsevier

irradiation is a critical parameter in the preparation of these materials as longer reaction times lead to substantial particle agglomeration.

Pt and Pt–Ru nanoparticles supported on Vulcan XC-72 carbon and CNT have also been prepared by a microwave-assisted polyol process [46, 47]. And in a more recent article high surface area carbon-supported Pt, PtRh, and PtSn catalysts were synthesized by the microwave-assisted polyol technique [48]. The synthetic method involved mixing equal volumes of 0.05 M water solutions of the required metal precursor salts (H_2PtCl_6 alone or together with $SnCl_2$ or $RhCl_3$) with EG and NaOH. The prepared solutions were heated in a microwave oven at 700 W for 60 s (Pt) or 90 s (bimetallic). After microwave heating, the resulting colloidal solutions were uniformly mixed with a water suspension of carbon black (Vulcan XC-72) and 2M H_2SO_4. The microwave assisted polyol has also been used to prepare PtZn/C [49], Pt–CeO_2/C [50], PtCo/C [51], PtSn/C [52], PtPd/C [53] PtPb/C [54] and Pd/C [55] on carbon, MWCNTs and other supports such as TiC–C and TiO_2–C. This methodology has, in general, difficulties with respect to controlling the particle size and distribution of the metal nanoparticle on the support due to the high microwave radiation absorption by ethylene glycol, that leads to the rapid increase of the reaction mixture temperature up to that of reflux. Therefore, it is difficult with this method to control and separate the nucleation and growth steps. It has been found more recently for Pt/C and PtRu/C electrocatalysts that using a pulse microwave method high catalyst surface area can be obtained using a short duration of microwave pulses [56, 57]. The shorter pulse duration (0.1 s) leads to a longer time under controlled microwave irradiation before reaching the maximum temperature set point. This limits nanoparticle agglomeration that forms larger particles, and hence leads to higher active surface area of the Pt/C catalysts. The synthesis of catalysts was performed under continuous microwave irradiation at a power of 1,600 W until reaching the desired reaction temperature, and then microwave pulses were applied to maintain it. Figure 7.4 shows the scheme representing the microwave sequence and corresponding solvent temperature profile during the synthesis process by pulsed microwaves [57].

7.2.7 Electrodeposition

Electrochemical deposition is a technique that has been widely used to deposit Pt and Pt-based nanoparticles on a wide variety of substrates, including glassy carbon [58], highly ordered pyrolytic graphite (HOPG) [59, 60], carbon black inside nafion [61], CNT [62], and PAN-based carbon fibers [63]. Electrochemical deposition occurs at the interface of an electronically conductive substrate and an electrolyte solution containing the salt of the metal to be deposited [64, 65]. Deposition occurs by the application to the substrate (the working electrode) a potential sufficient to reduce the metal salt to its zero valent state. Typically platinum nanoparticles are electrodeposited onto a carbon based electrode using a potentiostatic method from a $H_2PtCl_6 + 0.5$ M H_2SO_4 aqueous solution. The deposition potential used is around −0.25 V (SCE). The amount of deposited platinum is controlled by the deposition charge. There are five stages to electrochemical deposition of metals (1) transport of metal ions in solution to the electrode surface, (2) electron transfer (3) formation of metal ad-atoms via adsorption (4) nucleation and growth, two or three dimensional, of metal particles and (5) growth of the three dimensional bulk metal phase. If the growth process is stopped after the fourth step, then nanosized catalysts are produced. If however the particles are allowed to grow, then the end result is the formation of metal films. 2D and 3D growth can also be controlled by utilizing either underpotential deposition (UPD) or overpotential deposition (OPD), respectively; UPD refers to deposition initiated at a potential more positive than the reversible potential M/M^{z+}, while OPD occurs at a potential more negative than the reversible potential.

In a procedure recently developed by Taylor and co-workers, Pt was first electrodeposited onto Nafion™ coated carbon supports in a plating bath [66]. In this way, the Pt ions diffused through the surface of the Nafion™ coating into the carbon support surface to form Pt nanoparticles. In a similar study small Pt particles with diameters less than 4–5 nm have been deposited on the carbon surface to form a Pt/C catalyst in which it is suggested that the ionic path channels inside the Nafion™ coating could serve as the diffusion path for Pt ions [67]. In this study the Pt particles size obtained was 2–3.5 nm at a Pt loading of less than 0.05 $mg.cm^{-2}$. A number of papers also discuss the deposition of Ru nanoparticles onto Pt single crystals, mainly Pt(111) [68, 69].

7.2.8 Pulse Electrodeposition

Noble metal nanoparticles prepared by the electrodeposition method usually have metal particles with large particle size (10s–100s of nm). It is therefore a challenge to synthesize noble metal particles by electrodeposition that have a high dispersion and small particle size. The application of galvanostatic/potential pulses with durations in the range of several ms to 100s of ms favors the formation of

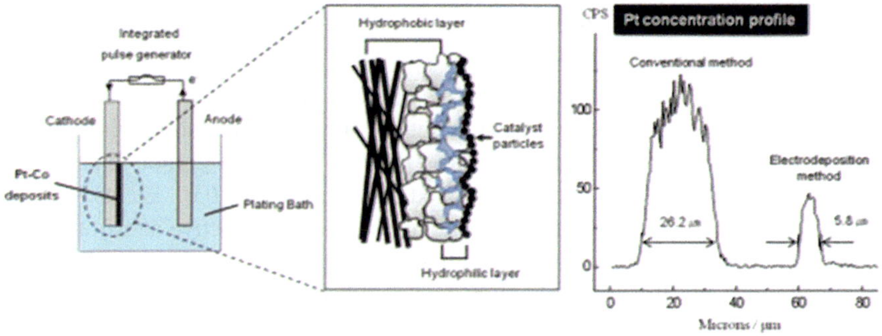

Fig. 7.5 Catalyst layer formation by PED of Pt–Co alloy onto Nafion^{TM} bonded carbon electrode. Reprinted from Ref. [71] with permission from Elsevier

nucleation sites and thus a high dispersion of the deposited metal. In a recent example pulse electrodeposition (PED) of Pt nanocluster catalysts on a graphene oxide (GO)-carbon paper composite electrode was carried out by Hsieh et al. The PED was carried out at a potential of −0.8 V versus SCE. The deposition and rest periods were set at 0.5 and 5 s, respectively [70]. Low loading PtCo (0.7 mg_{Pt}/cm^{-2}) alloy catalysts have also been prepared on a Nafion^{TM} bonded carbon electrode by a galvanostatic pulse technique [71]. The thickness of the catalyst layer was four times smaller with respect to traditional Pt/C catalyst (Fig. 7.5).

Electrodeposition was performed in a plating bath which contained the solution of K_2PtCl_4 and various concentrations of $CoCl_2$ dissolved in 0.5 M NaCl. The parameters for galvanostatic PED were a peak current density of 300 mA cm^{-2}, an on/off time of 10/100 ms and a total charge density of 1 C cm^{-2}. After the electrodeposition process, the electrodes were heated at 250 °C in H_2 atmosphere for 30 min to remove any organic solvent and fully reduce the catalysts. The same authors used a similar procedure to obtain CO tolerant PtRu/C nafion bonded electrodes [72].

Xiao et al. used an ultrasonic electrodeposition method to prepare AuPt nanoparticles on the surface of a MWCNT ionic liquid (ILS). Nanoparticles of non noble metals can also be readily deposited by PED. Highly ordered TiO_2 nanotube arrays fabricated by anodization were employed by Zhang and coworkers as a substrate and loaded with Ni nanoparticles by PED [73]. A three electrode system was employed (TiO_2/Ti as the working electrode, nickel plate as a counter electrode and Ag/AgCl electrode as reference electrode). The deposition was based upon modulated pulse signals in the microsecond range (Fig. 7.6).

Ni nanoparticles with average size ranging from 19 to 84 nm were obtained depending on the electrodeposition parameters. At constant current off-time (t_{off}) and pulse time of both negative and positive currents, the particle size decreased with increasing amplitude of both negative and positive current.

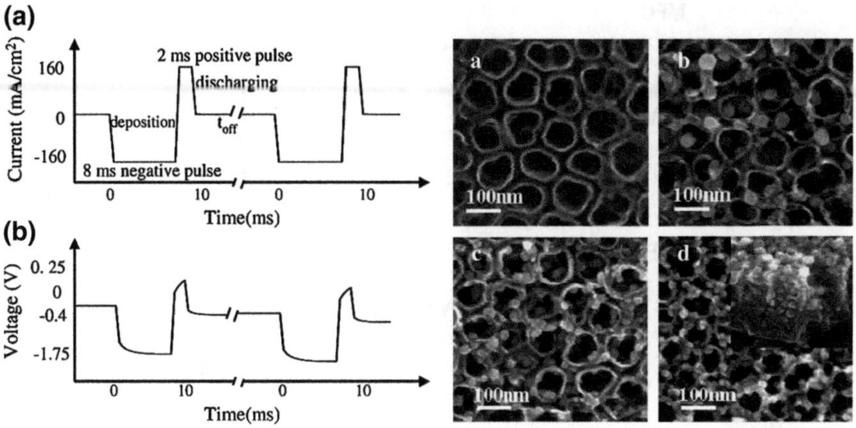

Fig. 7.6 *Left* **a** Current–time and **b** voltage–time *curves* for PED with negative pulse (-160 mA cm^{-2}, 8 ms), positive pulse ($+160$ mA cm^{-2}, 2 ms) and current off-time (1,000 ms) and *Right* FESEM images of **a** TiO_2 nanotube arrays and Ni/TiO_2 nanocomposite fabricated at constant current off-time (1,000 ms), pulse time of both negative and positive current (8, 2 ms) and different current amplitudes: **b** 70 mA cm^{-2}, **c** 110 mA cm^{-2} and **d** 160 mA cm^{-2}. Reprinted from Ref. [73] with permission from Elsevier

7.2.9 Vapor Phase Methods

Chemical vapor deposition (CVD) and atomic layer deposition (ALD) are ideal methods for forming thin metal films. Film growth can potentially be controlled mono-layer by mono-layer. CVD is typically used to form thin films but it can also be used to make nanoparticles. A typical CVD process begins by vaporizing the precursor, an inorganic compound containing the desired metal(s) (Fig. 7.7) [74]. The substrate is placed in a reaction chamber into which the vaporized precursor, mixed with a carrier gas and any other gaseous reagents, is introduced. The precursors diffuse or are carried to and adsorb onto the substrate surface where they decompose thermally, forming the metallic film. The precursor is designed to ensure that any reaction byproducts are gaseous and desorb into the gas phase.

Fuel cell electrocatalysts with improved properties have been prepared by dispersing Pt nanoparticles onto CNT using the CVD method. (Trimethyl) methylcyclopentadienyl platinum ($MeCpPtMe_3$) has been used as the Pt precursor in the CVD process and the CVD conditions have been optimized to obtain small Pt particles. Pt particles synthesized by CVD have a relatively uniform size of approximately 1 nm [74].

ALD is a modified version of metal–organic CVD, which relies on the self limiting chemistry of precursors and the interaction between substrates and precursor molecules [75, 76]. Accordingly, the ALD process shows great promise in the field of catalysis because the film growth is self limiting for each deposition cycle, capable of maximizing the accessible surface area of the catalyst.

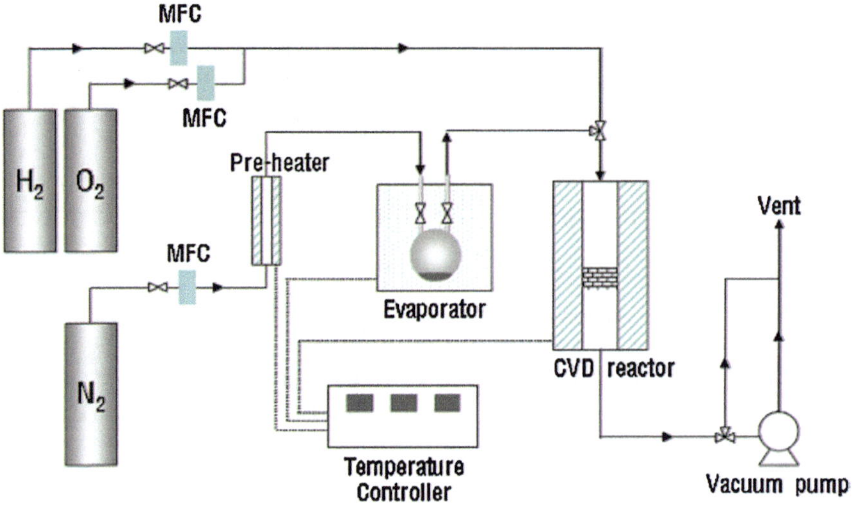

Fig. 7.7 A typical CVD apparatus. Reprinted from Ref. [74] with permission from Elsevier

Hsueh and co-workers used the same Pt precursor ($MeCpPtMe_3$) to deposit platinum nanoparticles on nitric acid-treated multiwalled CNT by ALD at 250 °C [77, 78]. Formation of uniform and well-distributed Pt nanoparticles was achieved (see Fig. 7.8). The size and number of Pt nanoparticles was found to be controlled by the ALD cycle number. In PEMFC tests the membrane electrode assembly made of both anode- and cathode-deposited Pt (0.019 and 0.044 mg cm^{-2}, respectively) after 100 cycles of ALD had 11 times higher specific power density than that made of commercial E-Tek electrodes containing 0.5 mg cm^{-2} of Pt. The ALD method has also been used to prepare Pt nanocatalysts supported on GO nanosheets [79].

Metal alloys can be fabricated by CVD when a heterometallic precursor is used [80]. Using a single source precursor (that remains coordinated in the vapor phase) allows for precise control of the ratio of the two metals. Pt–Ru particles 2 nm in diameter have been formed by vapor deposition using a mixture of commercially available single metal precursors, namely platinum(II)-acetylacetonate and ruthenium(III)-acetylacetonate [81].

The precursors were adsorbed onto carbon black by sublimation and subsequently decomposed at 320 °C in H_2 or N_2. While particle size was virtually independent of sublimation temperature, the Pt:Ru ratio decreased as sublimation temperature increased from 170 to 240 °C. A maximal Pt:Ru ratio, equaling the Pt:Ru ratio of the precursor salts, was reached at 220 °C. The composition is affected by the sublimation temperature, as the vapor pressures of the two precursors are influenced by the sublimation temperature. This example illustrates how the vapor pressure of the precursors can act as a limitation. To avoid this more recently an alternative method for the formation of Ru–Pt/C catalysts with different Ru/Pt ratios has been developed using the selective CVD of Ru onto

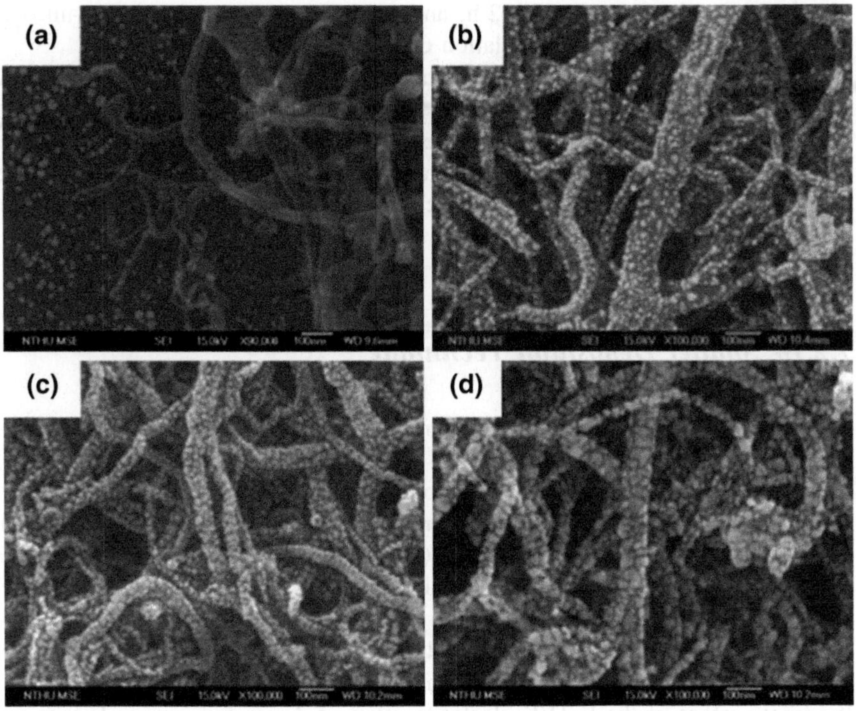

Fig. 7.8 SEM images of Pt nanoparticles deposited on **a** pristine CNTs with 100 cycles of ALD and CNTs acid-treated with **b** 100, **c** 200, and **d** 300 cycles of ALD. Reprinted from Ref. [78] with permission from Elsevier

a Pt/C surface [82]. The CVD catalyst shows an improved CO tolerance because Ru is preferentially deposited as nano-scale particles on the Pt surface and, consequently, the number of Pt particles that are in close contact with the added Ru is greater in the CVD catalyst.

Co was deposited on the surface of a 10 wt% Pt/C catalyst using an atmospheric CVD apparatus (as shown in Fig. 7.7) [83]. $CoCp(CO)_2$ which is a red liquid with a vapor pressure of 0.5 Torr at room temperature, was used as the Co precursor in the CVD process. The $CoCp(CO)_2$ vapor obtained at room temperature was introduced into the reactor containing the pre-reduced Pt/C catalyst in a flowing hydrogen stream diluted with nitrogen ($H_2/N_2 = 1$) for different periods. The catalyst containing the adsorbed Co precursor was then heated to 300 °C for 1 h in a flowing hydrogen–nitrogen stream in order for the Co precursor to be decomposed on the catalyst surface.

The same authors prepared Cr-modified Pt/C catalysts by the CVD of Cr on Pt/C [84]. In the CVD process $Cr(CO)_6$, was injected to a hydrogen–nitrogen stream ($H_2/N_2 = 1$) flowing through a reactor, which contained pre-reduced Pt/C, at 55 C for 1 h. The catalyst containing the adsorbed Cr precursor was heated to

150 °C in flowing nitrogen for 2 h, and finally treated in a hydrogen–nitrogen stream for 1 h for the decarbonylation of the Cr precursor.

A CO-tolerant $PtRu_xSn_y/C$ electrocatalyst, with an optimal x/y ratio of 0.8/0.2, was prepared by selectively depositing Sn using CVD on the metallic surface of $PtRu_{0.8}/C$ [85]. The CO tolerance of the catalyst was greater when Sn was added by CVD than by a conventional precipitation method because most of the Sn after addition by CVD was located in the vicinity of the Pt and Ru surfaces, on which CO molecules were strongly adsorbed.

7.2.10 Sputter Deposition Technique

The sputter deposition method has been recently applied to the preparation of PEM fuel cell catalysts. Mukerjee et al. [86] found that the performance of a fuel cell electrode was improved fourfold after sputter deposition of a thin film of platinum (50 nm thick and 0.05 mg cm^{-2} loading) on the front surface of an electrode containing a supported electrocatalyst (20 % Pt/C, 0.4 mg cm^{-2} loading). This study compared the electrode kinetic parameters, electrochemically active surface areas, activation energies and reaction orders for the oxygen reduction reaction (ORR) for the sputtered and unsputtered electrodes in proton exchange membrane fuel cells as a function of temperature and pressure. The Pt film was sputtered onto the front surface of the E-TEK electrode using an Argon ion source. Hirano and co-workers have prepared a low Pt loading catalyst layer (0.1 mg cm^{-2}) on an uncatalyzed E-TEK electrode also using the sputter deposition technique [87]. When compared to a 32 μm thick E-TEK electrode (0.4 mg_{Pt} cm^{-2}), the sputtered electrode showed only slightly less performance in the low current density region but better performance in the high current region probably due to low mass transport losses of the thinner electrode. More recently, this technique has been further employed to prepare CNT-supported Pt catalysts by several research groups. Chen et al. first fabricated CNTs on the carbon cloth onto which the Pt particles were sputtered [88]. A bias voltage of −100 V was found to be an optimum condition for Pt sputtering deposition. The employed sputtering current and time were 10 mA and 30 s, respectively. For comparison Pt particles were deposited on the same CNTs using an electroless method. TEM results showed that the sputtering method generated highly uniform Pt nanoparticles compared to those produced by the electroless method. The Pt particle sizes deposited by the sputtering method were very uniformly distributed at 2 nm, while those deposited by the electroless method had a particle size range of 2–5 nm on the same CNTs. Sun et al. deposited Pt nanoparticles on nitrogen containing CNTs (CN_x NT) [89]. The CN_x NTs were grown on a Si substrate through microwave-plasma-enhanced chemical vapor deposition (MPECVD) using CH_4, N_2 and H_2 gases. For the Pt deposition, a DC sputtering technique was employed. Highly dispersed Pt nanoparticles were formed with an average diameter of 2 nm on the CN_x NTs while a continuous Pt thin film was observed on the bare Si substrate. Their results suggest that the sputter deposition is a very valuable technique to deposit small and uniform Pt nanoparticles. This method can also generate a thinner catalyst layer that could give a higher fuel cell cathode performance and, at the same time, reduce the Pt loading considerably. However, with respect to the electrode mass production, the sputter deposition technique may face some technical challenges.

7.2.11 Sonochemistry and Sonoelectrochemistry

The use of ultrasound to produce high performing and efficient electrocatalysts is a very promising technique that has been only recently developed [90, 91]. Ultrasound is defined as a sound wave with a frequency above 16 kHz. The main cause for most of the observed effects of ultrasound on surfaces and chemical reactions is recognized as being due to the "cavitation" effect which occurs as a secondary effect when an ultrasonic wave passes through a liquid medium. Cavitation is a phenomenon where microbubbles are formed which tend to implode and collapse violently in a liquid leading to the formation of high velocity jets of liquid. The cavitation bubble collapse leads to a near adiabatic heating of the vapor that is inside the bubble, creating the so-called "hot-spot" in the fluid. High temperatures and pressures are produced and water vapor is pyrolised into hydrogen radicals (H·) and hydroxyl radicals (OH·), known as water sonolysis.

Recently, the use of ultrasound to reduce metallic ions in aqueous solutions by using special ultrasonic devices (probes and baths) has been developed. The effect of surfactants and the presence of alcohols are also important in controlling the process. Caruso et al. investigated the sonochemical reduction of Pt(IV) in the presence of various alcohols to form Pt nanoparticles (2.6 nm) [92]. They attributed the Pt nanoparticle synthesis to the water sonolysis process induced by cavitation where the primary H· and OH· radicals formed are scavenged by interfacial adsorbed alcohol molecules which subsequently diffuse away from the cavitation bubble and react with Pt(IV) in the bulk solution. A number of researchers have produced a wide range of mono- and bi-noble metal electrocatalysts at various ultrasonic frequencies and powers in several surfactants and alcohols. For example it is possible to produce Pt nanoparticles below 1 nm at 20 kHz in a solution of platinum salt containing PVP, EG and citrate [93].

Recently, increasing efforts have been employed in the use of the sonoelectrochemical method to produce noble metal and fuel cell electrocatalysts. The sonoelectrochemical method involves either depositing a metal under continuous electrical current and ultrasound or producing nanosized metals at various currents and ultrasonic pulses (a few ns) at a vibrating electrode. A mechanism has been proposed whereby metallic ions are reduced by a short current pulse to produce metallic nanoparticles on the sonoelectrode surface, which are then dislodged by the ultrasonic pulse [91].

Wang and co-workers have recently prepared Pt–Ru/C electrocatalysts (up to 60 wt%) by ultrasound (40 kHz; 70 W) and mechanical stirring [94]. The Pt–Ru nanoparticles (2–4 nm) thus formed were found to be homogeneously dispersed on the Vulcan XC-72R surface. Electrochemical and DMFC experiments showed that compared to standard electrocatalysts the Pt–Ru/C electrocatalyst prepared by sonication had enhanced electrocatalytic activity for methanol electrooxidation and better performance in DMFCs. They concluded that ultrasound provides a uniform environment for the nucleation and growth of metal nanoparticles on the carbon surface and simultaneously hinders Pt–Ru nanoparticle agglomeration. The ultrasound technique has been also applied to the preparation of other bimetallic carbon supported electrocatalysts with various applications e.g. Pd–Sn/C (ORR alkaline), Pd/C Au/C, Pd–Au/C (MOR and ORR alkaline) [95–97].

More recently, the sonication method has been combined with the polyol technique to produce electrocatalysts such as Pt/MWCNTs and Pt/TiO_2 [98, 99]. No stabilizer or surfactants are necessary in this way. The insertion of Pt into mesoporous TiO_2 was performed by Gedanken et al. using the polyol reduction method in the presence of ultrasound (20 kHz, 100 W cm^{-2}) [99]. The resulting Pt supported on mesoporous TiO_2 exhibited threefold higher ORR activity than Pt supported on Vulcan XR-72.

7.2.12 Spray Pyrolisis

The experimental procedure of spray pyrolysis is simple. First, an aqueous solution containing the metal precursor is atomized into a carrier gas that is passed through a furnace. Second, the atomized precursor solution deposits onto a substrate, where it reacts and forms the final product [100]. The process has many advantages compared to other metal-forming techniques [101]: (1) it is very easy to dope films or form alloys in any proportion by manipulating the spray solution; (2) neither high-purity targets and substrates nor vacuum set-ups are required; (3) deposition rates and therefore film thickness can easily be controlled by using the spray parameters; (4) moderate operation temperatures (100–500 °C) allow for deposition on temperature-sensitive substrates and ensure that the overall process is less energy intensive; (5) the technique has relatively limited environmental impact since aqueous precursor solutions can be used; and (6) the process is scalable, with production rates as high as 1.1 kg h^{-1} [102]. Ultrasonic nebulizers [101] can be used to form micrometer- and submicrometer-sized droplets. Droplets formed by ultrasonic waves have very small sizes and size distributions. To deposit the droplets onto a substrate the aerosol is transported to the heated substrate, where the solvent vaporizes. A heterogeneous reaction occurs that leads to the formation of thin solid films. The spray jet can be scanned continuously to coat a large area. To form nanoparticles the aerosol is pyrolyzed. To improve the deposition efficiency, the ratio of atoms effectively deposited to those supplied, a corona discharge is used to control the transport of aerosol droplets towards the substrate [101]. Electrostatic spray pyrolysis is another method used to control the transportation of the atomized precursor solution from the atomizer to the substrate. It is accomplished by applying a positive voltage, up to 12.5 kV, to the spray nozzle, forming a positively charged spray. Electrostatic forces guide the spray to the hot substrate where pyrolysis takes place. Spray pyrolysis can be used to form nanoparticles. PtRu/C catalysts have been formed by spray pyrolysis using H_2PtCl_6 and $RuCl_3$ as the precursors, dissolved in an aqueous solution containing carbon black and various molecular lengths of PEG [100]. Once atomized, the droplets of precursor solution in the carrier gas were passed through a quartz tube that was heated to 180 °C by a tube furnace. The solvent evaporation and precursor decomposition resulted in nanoparticle deposition on the carbon black.

7.2.13 Supercritical Fluids

7.2.13.1 Supercritical Deposition Technique

Supercritical deposition is an alternative and promising way to prepare electrocatalysts. This process involves the dissolution of a metallic precursor in a supercritical fluid (SCF) and the exposure of a porous support to the solution. After adsorption of the precursor on the support, the metallic precursor is converted to its metal form by chemical or thermal reduction. Using a SCF as the processing medium for synthesis of electrocatalysts has many advantages which are directly related to the special properties of the SCFs. Properties of a SCF are different from those of ordinary liquids and gases and are tunable simply by changing the pressure and temperature. In particular, density and viscosity change drastically at conditions close to the critical point. Since fluid densities can approach or even exceed those of liquids, various SCFs are good solvents for a wide range of organic and organometallic compounds. Compared with conventional liquid solvents, high diffusivities in SCFs combined with their low viscosities result in enhanced mass transfer characteristics. The low surface tension of SCFs permits better penetration and wetting of pores than liquid solvents do. Among the SCFs, supercritical carbon dioxide ($scCO_2$), readily accessible with a Tc of 31 BC and a Pc of 7.38 MPa, is particularly attractive since it is abundant, inexpensive, nonflammable, nontoxic, environmentally benign and leaves no residue on the treated medium [103]. This promising catalyst preparation technique results in small particle sizes and homogeneous dispersions [104, 105]. An additional advantage of this technique is the ability to thermodynamically control the metal loading [106]. Bayrakçeken et al. used supercritical deposition technique to prepare Pt-based electrocatalysts for polymer electrolyte membrane fuel cells as follows [107]. Carbon supports, Vulcan XC-72R, BP2000 and MWCNT were impregnated with Pt using the $scCO_2$ deposition technique. Prior to impregnation all carbon supports were heat treated at 150 °C for 4 h under a N_2 atmosphere. In this synthesis, dimethyl(cyclooctadiene)platinum(II) ($PtMe_2COD$) (99.9 %) was used as the Pt precursor. The heat-treated carbon support was placed into a pouch made of a filter paper and placed into the vessel together with a certain amount of $PtMe_2COD$ precursor, and a stirring bar. The vessel was sealed and heated to 70 °C using a circulating heater/cooler apparatus and then charged slowly with carbon dioxide to a pressure of 24.2 MPa. These conditions were maintained for a period of 6 h. After allowing the vessel to cool, the pouch was removed and the impregnated carbon support was weighed to determine the amount of precursor adsorbed. The carbon support was then placed in an alumina process tube in a tube furnace, and the precursor was reduced thermally under flowing N_2 for 4 h at 150 °C. The electrochemical surface area (ESA) of the prepared Pt/Vulcan and Pt/MWCNT catalysts were about three times larger than that of a commercial E-TEK catalyst with similar (10 wt% Pt) loading. By using this method Pt nanoparticles, about 1–2 nm in diameter, were dispersed uniformly on the carbon supports. The Pt/ Vulcan catalyst prepared by supercritical deposition showed the best performance for the ORR, substantially higher than a commercial Pt/C catalyst.

7.2.14 High Energy Ball Milling

The high energy ball milling technique also known as mechanical alloying, is a solid state method for the synthesis of metallic powders or ceramics. In the ball milling process, powder particles are submitted to high energy impacts from balls inside an air tight container (crucible). During milling, materials are subjected to intense mechanical deformations, and solid state reactions are induced at the atomic level. The powder particles are repetitively flattened, fractured and cold welded until a balance between fragmentation (fracture) and agglomeration (cold-welding) is established [108, 109]. Through this process, a structure refinement, a particle size reduction and a homogeneous alloy formation are obtained. The grinding and mixing of materials can be undertaken with or without the presence of a liquid. The process can also be undertaken under inert atmosphere (Ar or N_2) or in the presence of a reducing gas like H_2. The milling is done in a rotating cylinder or conical mill using balls (steel or ceramic) and typically a powdered catalyst is prepared. Ball-milling is a technique that has the potential to extend the solubility limit of one element into the other, alloy difficult elements, make alloys of defined grain size down to the nanometer range, and synthesize new crystalline and quasi-crystalline phases.

Denis and co-workers have used high energy ball milling to produce unsupported CO tolerant Pt and Pt–Ru anode catalysts for PEM fuel cells [110–112]. PtRu alloys of different composition (Pt/Ru 0.18–3.00) were obtained by milling together the precursor materials, e.g. Pt and Ru metal powders, a dispersing agent, and often a process control agent. Typically, 6 g of the metal powders and WC balls were loaded in a WC vial. The ball to powder weight ratio was always around 4/1.

Milling times of 40 h were routinely used. Typical process control agents (PCAs) used in ball-milling are organic compounds such as stearic acid, hexane, oxalic acid, and polyvinyl pyridine. The PCAs act as surface active agents, interfering with the solid by lowering the surface tension and preventing the catalyst from sticking to the balls. Organic PCAs typically decompose during ball milling and are incorporated into the catalyst as carbides and oxides. As a consequence various inorganic PCAs have been tested. Initially Al was used and most of the Al was leached out after milling using 1 M NaOH solutions. However, PtAl alloys could also be formed that were shown to have a beneficial effect on the catalytic activity. The most promising PCAs were found to be NaF and MgH_2; the last has the dual function of acting as a PCA as well as a dispersing agent. MgH_2 partially decomposes during milling, releasing H_2, and with the possibility of forming metallic Mg that may be incorporated into the catalyst lattice. The presence of MgH_2 was found to be necessary to obtain a specific surface area of between 50 and 75 m^2g^{-1} [113]. After ball-milling, the powdered samples contain "impurities", namely the dispersing and process controlling agent, and typically also parts of the mill equipment (i.e., from the vials and balls). Tungsten carbide (WC) balls and vials are often used in ball-mill equipment to make fuel cell catalysts and as a consequence the products generally contain WC as impurity.

More recently, other researchers have adapted the method to other Pt based catalysts for fuel cells [114–116]. Pt–Co alloys of three compositions ($Pt_{25}Co_{75}$, $Pt_{50}Co_{50}$ and $Pt_{75}Co_{25}$) were prepared by Enzo et al. [116]. For $Pt_{75}Co_{25}$ the average crystallite size was reduced down to about 15 nm after 12 h milling. Further milling lead to an increase in crystallite size. Mechanical alloying of Pt and Co powders was undertaken by Farhat et al. with various amounts of Vulcan XC-72R carbon black were used as a PCA. Binary Pt–Co and tertiary Pt–Co–C materials were produced [114]. Carbon as PCA proved very effective in preventing excessive agglomeration and reducing WC jar and ball wear. An average particle size of 33 nm and a surface area of 43 m^2g^{-1} was obtained after acid leaching of a large amount of the Co from the ball milled catalyst.

7.3 Commercial Supported Nanoparticles for Electrocatalysis

Currently, commercially available electrocatalysts are principally based upon Pt and PtRu black materials or the same metal nanoparticles supported on carbon black. The principle suppliers include Johnson Matthey (Alfa Aesar a division of JM, sells small quantities for R&D). The BASF chemical company does not supply catalyst materials but only produces and supplies PEMFC MEAs. Cabot corporation also develops electrocatalyst powders for fuel cell applications through a patented spray pyrolisis process flow method originally developed by Superior Micropowders. Although the type and content of the catalysts that they offer is not stipulated. A Japanese company The Tanaka Kikinzoku Group which specializes in precious metal products also commercializes Vulcan carbon supported Pt (40–50 wt%) and PtRu (50–58 wt%) electrocatalysts for use in PEMFCs and DMFCs. Some performance data is given for the Pt/C and PtRu/C materials under standard fuel cell testing conditions.

Johnson Matthey's HiSPec™ range of fuel cell catalysts offer Pt/C (40–87 wt%) and PtRu/C (Pt:Ru 2:1) 70 wt% catalysts for use in PEMFCs and DMFCs. They also sell readymade HiSPec™ MEAs for use in fuel cells although the catalyst content is not stipulated. No performance data is given. Often HiSPec™ catalysts and MEAs have been used as benchmark materials in electrocatalyst development. The performance of new electrocatalytic materials is often compared directly to these commercially available catalysts measured under the same conditions [117–119], thus providing a *pseudo* industry standard to which new materials can be benchmarked. As the reader may notice when reviewing the scientific literature, often new catalytic materials are compared with *in-house* prepared simple carbon supported metal nanoparticles. Care has to be taken when doing this as the difference and consequently improvements can be exaggerated if the standard material is actually poor performing.

References

1. C. Bock, B. MacDougall, Y. LePage, J. Electrochem. Soc. **151**, A1269–A1278 (2004)
2. N. Alonso-Vante, Fuel Cells **6**, 182–189 (2006)
3. A.J. Dickinson, L.P.L. Carrette, J.A. Collins, K.A. Friedrich, U. Stimming, Electrochim. Acta **47**, 3733–3739 (2002)
4. M. Haruta, S. Tsubota, T. Kobayashi, H. Kageyama, M.J. Genet, B. Delmon, J. Catal. **144**, 175–192 (1993)
5. G.Y. Chen, D.A. Delafuente, S. Sarangapani, T.E. Mallouk, Catal. Today **67**, 341–355 (2001)
6. O.V. Cherstiouk, P.A. Simonov, V.I. Zaikovskii, E.R. Savinova, J. Electroanal. Chem. **554**, 241–251 (2003)
7. C. Baatz, N. Decker, U. Prusse, J. Catal. **258**, 165–169 (2008)
8. H.S. Liu, C.J. Song, L. Zhang, J.J. Zhang, H.J. Wang, D.P. Wilkinson, J. Power Sources **155**, 95–110 (2006)
9. N. Fujiwara, K. Yasuda, T. Ioroi, Z. Siroma, Y. Miyazaki, Electrochim. Acta **47**, 4079–4084 (2002)
10. B. Yang, Q.Y. Lu, Y. Wang, L. Zhuang, J.T. Lu, P.F. Liu, Chem. Mater. **15**, 3552–3557 (2003)
11. E.V. Spinace, A.O. Neto, M. Linardi, J. Power Sources **124**, 426–431 (2003)
12. J. Tayal, B. Rawat, S. Basu, Int. J. Hydrogen Energ. **37**, 4597–4605 (2012)
13. M. Watanabe, M. Uchida, S. Motoo, J. Electroanal. Chem. Interfacial Electrochem. **229**, 395–406 (1987)
14. T. Yoshitake, Y. Shimakawa, S. Kuroshima, H. Kimura, T. Ichihashi, Y. Kubo, D. Kasuya, K. Takahashi, F. Kokai, M. Yudasaka, S. Iijima, Phys. B **323**, 124–126 (2002)
15. M.P. Pileni, Adv. Colloid Interfac. **46**, 139–163 (1993)
16. L. Xiong, A. Manthiram, Solid State Ionics **176**, 385–392 (2005)
17. D.R.M. Godoi, J. Perez, H.M. Villullas, J. Electrochem. Soc. **154**, B474–B479 (2007)

18. S. Eriksson, U. Nylen, S. Rojas, M. Boutonnet, Appl. Catal. a-Gen. **265**, 207–219 (2004)
19. V. Uskokovic, M. Drofenik, Adv. Colloid Interfac. **133**, 23–34 (2007)
20. I. Lisiecki, M.P. Pileni, J. Am. Chem. Soc. **115**, 3887–3896 (1993)
21. J.S.F. Wang, H.B. Pan, C.M. Wai, J. Nanosci. Nanotechnol. **6**, 2025–2030 (2006)
22. M.J. Escudero, E. Hontanon, S. Schwartz, M. Boutonnet, L. Daza, J. Power Sources **106**, 206–214 (2002)
23. C.Z. Yang, N.K. van der Laak, K.Y. Chan, X. Zhang, Electrochim. Acta **75**, 262–272 (2012)
24. J. Ahmed, K.V. Ramanujachary, S.E. Lofland, A. Furiato, G. Gupta, S.M. Shivaprasad, A.K. Ganguli, Colloid Surf. A **331**, 206–212 (2008)
25. I. Capek, Adv. Colloid Interfac. **110**, 49–74 (2004)
26. C. Stubenrauch, T. Wielpuetz, T. Sottmann, C. Roychowdhury, F.J. DiSalvo, Colloid Surf. A **317**, 328–338 (2008)
27. A.K. Ganguli, A. Ganguly, S. Vaidya, Chem. Soc. Rev. **39**, 474–485 (2010)
28. J. Ahmed, S. Sharma, K.V. Ramanujachary, S.E. Lofland, A.K. Ganguli, J. Colloid Interf. Sci. **336**, 814–819 (2009)
29. D.H. Chen, C.J. Chen, J. Mater. Chem. **12**, 1557–1562 (2002)
30. M.L. Wu, D.H. Chen, T.C. Huang, J. Colloid Interf. Sci. **243**, 102–108 (2001)
31. M.L. Wu, D.H. Chen, T.C. Huang, Langmuir **17**, 3877–3883 (2001)
32. M.L. Wu, D.H. Chen, T.C. Huang, Chem. Mater. **13**, 599–606 (2001)
33. P.Y. He, X.Y. Wang, Y.J. Liu, L.H. Yi, X. Liu, Int. J. Hydrog. Energ. **37**, 1254–1262 (2012)
34. P.Y. He, X.Y. Wang, Y.J. Liu, X. Liu, L.H. Yi, Int. J. Hydrog. Energ. **37**, 11984–11993 (2012)
35. L.H. Yi, Y.F. Song, X. Liu, X.Y. Wang, G.S. Zou, P.Y. He, W. Yi, Int. J. Hydrog. Energ. **36**, 15775–15782 (2011)
36. J. Ahmed, A. Ganguly, S. Saha, G. Gupta, P. Trinh, A.M. Mugweru, S.E. Lofland, K.V. Ramanujachary, A.K. Ganguli, J. Phys. Chem. C **115**, 14526–14533 (2011)
37. J. Ahmed, B. Kumar, A.M. Mugweru, P. Trinh, K.V. Ramanujachary, S.E. Lofland, Govind, A.K. Ganguli, J. Phys. Chem. C **114**, 18779–18784 (2010)
38. J. Ahmed, T. Ahmad, K.V. Ramanujachary, S.E. Lofland, A.K. Ganguli, J. Colloid Interf. Sci. **321**, 434–441 (2008)
39. C. Bock, C. Paquet, M. Couillard, G.A. Botton, B.R. MacDougall, J. Am. Chem. Soc. **126**, 8028–8037 (2004)
40. H.S. Oh, J.G. Oh, Y.G. Hong, H.S. Kim, Electrochim. Acta **52**, 7278–7285 (2007)
41. E. Ribadeneira, B.A. Hoyos, J. Power Sources **180**, 238–242 (2008)
42. W.Y. Yu, W.X. Tu, H.F. Liu, Langmuir **15**, 6–9 (1999)
43. G. Glaspell, H.M.A. Hassan, A. Elzatahry, V. Abdalsayed, M.S. El-Shall, Top. Catal. **47**, 22–31 (2008)
44. G. Glaspell, L. Fuoco, M.S. El-Shall, J. Phys. Chem. B **109**, 17350–17355 (2005)
45. J.M. Campelo, T.D. Conesa, M.J. Gracia, M.J. Jurado, R. Luque, J.M. Marinas, A.A. Romero, Green Chem. **10**, 853–858 (2008)
46. Z.L. Liu, L.M. Gan, L. Hong, W.X. Chen, J.Y. Lee, J. Power Sources **139**, 73–78 (2005)
47. Z.L. Liu, X.Y. Ling, X.D. Su, J.Y. Lee, L.M. Gan, J. Power Sources **149**, 1–7 (2005)
48. S. Stevanovic, D. Tripkovic, J. Rogan, K. Popovic, J. Lovic, A. Tripkovic, V.M. Jovanovic, J. Solid State Electr. **16**, 3147–3157 (2012)
49. C.T. Hsieh, W.M. Hung, W.Y. Chen, J.Y. Lin, Int. J. Hydrogen Energ. **36**, 2765–2772 (2011)
50. D.M. Gu, Y.Y. Chu, Z.B. Wang, Z.Z. Jiang, G. Yin, Y. Liu, Appl. Catal. B-Environ. **102**, 9–18 (2011)
51. C.T. Hsieh, W.Y. Chen, I.L. Chen, A.K. Roy, J. Power Sources **199**, 94–102 (2012)
52. C.T. Hsieh, Y.Y. Liu, W.Y. Chen, Y.H. Hsieh, Int. J. Hydrogen Energ. **36**, 15766–15774 (2011)
53. Y.Y. Chu, Z.B. Wang, Z.Z. Jiang, D.M. Gu, G.P. Yin, J. Power Sources **203**, 17–25 (2012)
54. Y.Y. Huang, S.Y. Zheng, X.J. Lin, L.Q. Su, Y.L. Guo, Electrochim. Acta **63**, 346–353 (2012)

55. C.T. Hsieh, Y.Y. Liu, Y.S. Cheng, W.Y. Chen, Electrochim. Acta **56**, 6336–6344 (2011)
56. E. Lebegue, S. Baranton, C. Coutanceau, J. Power Sources **196**, 920–927 (2011)
57. S. Harish, S. Baranton, C. Coutanceau, J. Joseph, J. Power Sources **214**, 33–39 (2012)
58. S. K, W. D, K. T, J. Electroanal. Chem. **223**, 223–224 (1987)
59. J.V. Zoval, J. Lee, S. Gorer, R.M. Penner, J. Phys. Chem. B **102**, 1166–1175 (1998)
60. G.J. Lu, G. Zangari, J. Phys. Chem. B **109**, 7998–8007 (2005)
61. O. Antoine, R. Durand, Electrochem. Solid State **4**, A55–A58 (2001)
62. Z.B. He, J.H. Chen, D.Y. Liu, H. Tang, W. Deng, W.F. Kuang, Mater. Chem. Phys. **85**, 396–401 (2004)
63. G. N, J. D, K. P, J. Electroanal. Chem. Interf. Electrochem. **264**, 235–245 (1989)
64. E. Budevski, G. Staikov, W.J. Lorenz, *Electrochemical Phase Formation and Growth: An Introduction to the Initial Stages of Metal Deposition* (VCH, Weinheim, New York, 1996)
65. R. Greef, R. Peat, L. Peter, D. Pletcher, J. Robinson, *Electrocrystallisation* (Ellis Horook Limited, New York, 1985)
66. E.J. Taylor, E.B. Anderson, N.R.K. Vilambi, J. Electrochem. Soc. **139**, L45–L46 (1992)
67. K. Lee, A. Ishihara, S. Mitsushima, N. Kamiya, K. Ota, J. Electrochem. Soc. **151**, A639–A645 (2004)
68. H. Hoster, T. Iwasita, H. Baumgartner, W. Vielstich, Phys. Chem. Chem. Phys. **3**, 337–346 (2001)
69. F. Maillard, G.Q. Lu, A. Wieckowski, U. Stimming, J. Phys. Chem. B **109**, 16230–16243 (2005)
70. C.T. Hsieh, J.M. Wei, J.S. Lin, W.Y. Chen, Catal. Commun. **16**, 220–224 (2011)
71. S. Woo, I. Kim, J.K. Lee, S. Bong, J. Lee, H. Kim, Electrochim. Acta **56**, 3036–3041 (2011)
72. Y. Ra, J. Lee, I. Kim, S. Bong, H. Kima, J. Power Sources **187**, 363–370 (2009)
73. Y.H. Zhang, Y.N. Yang, P. Xiao, X.N. Zhang, L. Lu, L. Li, Mater. Lett. **63**, 2429–2431 (2009)
74. H. Kim, S.H. Moon, Carbon **49**, 1491–1501 (2011)
75. J.S. King, A. Wittstock, J. Biener, S.O. Kucheyev, Y.M. Wang, T.F. Baumann, S.K. Giri, A.V. Hamza, M. Baeumer, S.F. Bent, Nano Lett. **8**, 2405–2409 (2008)
76. J.H. Shim, X. Jiang, S.F. Bent, F.B. Prinz, J. Electrochem. Soc. **157**, B793–B797 (2010)
77. Y.C. Hsueh, C.C. Wang, C. Liu, C.C. Kei, T.P. Perng, Nanotechnology **23** (2012)
78. Y.C. Hsueh, C.C. Wang, C.C. Kei, Y.H. Lin, C. Liu, T.P. Perng, J. Catal. **294**, 63–68 (2012)
79. T. Shu, S.J. Liao, C.T. Hsieh, A.K. Roy, Y.Y. Liu, D.Y. Tzou, W.Y. Chen, Electrochim. Acta **75**, 101–107 (2012)
80. S.F. Huang, Y. Chi, C.S. Liu, A.J. Carty, K. Mast, C. Bock, B. MacDougall, S.M. Peng, G.H. Lee, Chem. Vapor Depos. **9**, 157–161 (2003)
81. P. Sivakumar, R. Ishak, V. Tricoli, Electrochim. Acta **50**, 3312–3319 (2005)
82. H.T. Kim, H.I. Joh, S.H. Moon, J. Power Sources **195**, 1352–1358 (2010)
83. S.J. Seo, H.I. Joh, H.T. Kim, S.H. Moon, J. Power Sources **163**, 403–408 (2006)
84. S.J. Seo, H.I. Joh, H.T. Kim, S.H. Moon, Electrochim. Acta **52**, 1676–1682 (2006)
85. J.S. Yoo, H.T. Kim, H.I. Joh, H. Kim, S.H. Moon, Int. J. Hydrogen Energ. **36**, 1930–1938 (2011)
86. S. Mukerjee, S. Srinivasan, A.J. Appleby, Electrochim. Acta **38**, 1661–1669 (1993)
87. S. Hirano, J. Kim, S. Srinivasan, Electrochim. Acta **42**, 1587–1593 (1997)
88. C.C. Chen, C.F. Chen, C.H. Hsu, I.H. Li, Diam. Relat. Mater. **14**, 770–773 (2005)
89. C.L. Sun, L.C. Chen, M.C. Su, L.S. Hong, O. Chyan, C.Y. Hsu, K.H. Chen, T.F. Chang, L. Chang, Chem. Mater. **17**, 3749–3753 (2005)
90. P. Sakkas, O. Schneider, S. Martens, P. Thanou, G. Sourkouni, C. Argirusis, J. Appl. Electrochem. **42**, 763–777 (2012)
91. B.G. Pollet, Int. J. Hydrogen Energ. **35**(2010), 11986–12004 (2004)
92. R.A. Caruso, M. Ashokkumar, F. Grieser, Colloid Surf. A **169**, 219–225 (2000)
93. D. Radziuk, H. Mohwald, D. Shchukin, J. Phys. Chem. C **112**, 19257–19262 (2008)
94. X.G. Wang, J.H. Liao, C.P. Liu, W. Xing, T.H. Lu, Electrochem. Commun. **11**, 198–201 (2009)

95. J. Kim, J.E. Park, T. Momma, T. Osaka, Electrochim. Acta **54**, 3412–3418 (2009)
96. M. Nakanishi, H. Takatani, Y. Kobayashi, F. Hori, R. Taniguchi, A. Iwase, R. Oshima, Appl. Surf. Sci. **241**, 209–212 (2005)
97. Y. Mizukoshi, T. Fujimoto, Y. Nagata, R. Oshima, Y. Maeda, J. Phys. Chem. B **104**, 6028–6032 (2000)
98. Z.C. Wang, D.D. Zhao, G.Y. Zhao, H.L. Li, J. Solid State Electr. **13**, 371–376 (2009)
99. S. Shanmugam, A. Gedanken, J. Phys. Chem. C **113**, 18707–18712 (2009)
100. X.Z. Xue, C.P. Liu, W. Xing, T.H. Lu, J. Electrochem. Soc. **153**, E79–E84 (2006)
101. P.S. Patil, Mater. Chem. Phys. **59**, 185–198 (1999)
102. R. Mueller, L. Madler, S.E. Pratsinis, Chem. Eng. Sci. **58**, 1969–1976 (2003)
103. Y. Zhang, C. Erkey, J Supercrit Fluid **38**, 252–267 (2006)
104. A. Bayrakceken, U. Kitkamthorn, M. Aindow, C. Erkey, Scripta Mater. **56**, 101–103 (2007)
105. Y. Zhang, D.F. Kang, C. Saquing, M. Aindow, C. Erkey, Ind. Eng. Chem. Res. **44**, 4161–4164 (2005)
106. C.D. Saquing, T.T. Cheng, M. Aindow, C. Erkey, J. Phys. Chem. B **108**, 7716–7722 (2004)
107. A. Bayrakceken, A. Smirnova, U. Kitkamthorn, M. Aindow, L. Turker, I. Eroglu, C. Erkey, J. Power Sources **179**, 532–540 (2008)
108. L. Lü, M.O. Lai, *Mechanical alloying* (Kluwer Academic Publishers, Boston, 1998)
109. C. Suryanarayana, Prog. Mater Sci. **46**, 1–184 (2001)
110. M.C. Denis, G. Lalande, D. Guay, J.P. Dodelet, R. Schulz, J. Appl. Electrochem. **29**, 951–960 (1999)
111. G. Lalande, M.C. Denis, D. Guay, J.P. Dodelet, R. Schulz, J. Alloy. Compd. **292**, 301–310 (1999)
112. M.C. Denis, P. Gouerec, D. Guay, J.P. Dodelet, G. Lalande, R. Schulz, J. Appl. Electrochem. **30**, 1243–1253 (2000)
113. M.C. Denis, M. Lefevre, D. Guay, J.P. Dodelet, Electrochim. Acta **53**, 5142–5154 (2008)
114. Z. Farhat, A. Alfantazi, Mat. Sci. Eng. A-Struct. **476**, 169–173 (2008)
115. P. Pharkya, A. Alfantazi, Z. Farhat, J. Fuel Cell Sci. Tech. **2**, 171–178 (2005)
116. M. Lucariello, N. Penazzi, E. Arca, G. Mulas, S. Enzo, Mater. Chem. Phys. **114**, 227–234 (2009)
117. B. Liu, L.P. Zheng, S.J. Liao, J.H. Zeng, J. Power Sources **210**, 54–59 (2012)
118. B. Millington, S.F. Du, B.G. Pollet, J. Power Sources **196**, 9013–9017 (2011)
119. Z.Q. Jiang, Z.J. Jiang, Y.D. Meng, Appl. Surf. Sci. **257**, 2923–2928 (2011)

Chapter 8
Shape and Structure-Controlled Metal Nanoparticles

8.1 Key Concepts

It will become clear to the reader after a perusal of this book, that the size and shape of metal nanoparticles are the main characteristics that determine their electrocatalytic properties. Generally, electrocatalytic properties are also directly related to the proportion of surface atoms with respect to the total number of atoms in the nanoparticle. The smaller the particle becomes the more surface atoms there are. For particles with diameters less of than 10 nm the proportion of surface atoms is large (about 20 % for d = 5 nm with a total number of around 6,000 atoms) and about 50 % for d = 2 nm (total number of about 300 atoms). With a diameter of 1 nm, it can be considered that all atoms of the particles are surface atoms [1].

However, the unique electrochemical properties of nanocrystals are not only determined by the large proportion of surface atoms but also by their crystallographic arrangement at the particles surface. Characterization of both the size and surface domains of metal nanoparticles is essential if we are to be able to synthesis well-shaped nanocrystals with well-defined desired properties.

Metallic nanoparticle formation starts with the nucleation step. The formation of the stable solid crystal phase occurs via the generation of solid seeds. The classical homogeneous nucleation approach [2] assumes that the growth of a cluster is thermodynamically governed by the change in Gibbs free energy, which can be decomposed into two terms. The bulk energy term of the cluster formation is negative and varies as the volume of the particle, whereas the surface free energy term is positive and varies with the surface area of the particle. To obtain stable particles, the bulk energy has to be much larger than the energy involved at the solid–liquid interface. Under thermodynamic equilibrium conditions, the equilibrium shape of the macroscopic crystal is unique. The minimal energy is obtained for a polyhedron with central distance between faces being proportional to their surface energy. In the case of face centered cubic metals (fcc) such as platinum, palladium, or gold, the equilibrium shape with lowest energy is a truncated octahedron, with eight hexagonal (111) faces and six square (100) faces.

A. Lavacchi et al., *Nanotechnology in Electrocatalysis for Energy*,
Nanostructure Science and Technology 170, DOI: 10.1007/978-1-4899-8059-5_8,
© Springer Science+Business Media New York 2013

The cuboctahedron, which has eight triangle (111) faces and six (100) faces is less stable than the truncated octahedron.

For reactions that are sensitive to the surface structure, the crystallographic structure of the catalytic site is most important. Theoretical analysis on ideal geometric structures (octahedrons and truncated octahedrons) by Van Herdeveld and Hartog demonstrated the change in the ratios of the different crystallographic surface sites as a function of particle size [3]. In the case of these ideal structures, the proportion of edge and corner atoms with respect to the total surface atoms was shown to increase with decreasing particle size. Therefore, understanding the topography of surface sites on a nanoscale, such as the presence of terraces, steps, kinks, adatoms and vacancies, and their effects on electrocatalytic properties is the key to designing nanosized functional materials for use in electrolytic devices for energy applications [4–6].

8.2 Identification of High-Index Facets

The performance of nanocrystals used as catalysts depends strongly on the surface structure of facets enclosing the crystals. Thermodynamics usually ensures that crystal facets evolve to have the lowest surface energy during the crystal growth process. For a pure metal, the surface energy relies on coordination numbers (CNs) of surface atoms as well as their density. For example, it increases in the order of $\gamma\{111\} < \gamma\{100\} < \gamma\{110\} < \gamma\{hkl\}$ on a face-centered cubic (fcc) metal, where {hkl} represents high-index planes with at least one Miller index larger than 1 [7, 8]. For a metal oxide, the surface energy increases with increasing density of dangling bonds. High-energy surfaces that have an open surface structure also possess enhanced catalytic properties. Extensive studies in surface science have shown that Pt high-index planes with open surface structure exhibit much higher reactivity than that of (111) or (100) low-index planes, because high-index planes have a large density of low-coordinated atoms situated on steps and kinks, with high reactivity required for high catalytic activity [9, 10]. More importantly, on high-index planes, there exist short-range steric sites (such as "chair" sites) that consist of the combination of several (typical 5–6) step and terrace atoms [11, 12]. Due to the synergistic effect between step and terrace atoms, steric sites usually serve as highly active catalytic sites.

Face-centered cubic (fcc) metals have their atoms close-packed in an ABCABC staking sequence {111}. Although the coordination number (CN) of atoms inside the fcc nanocrystal is 12, the CN of surface atoms can vary with the surface structure. For fcc metals (such as Pt, Pd, and Au), a unit stereographic triangle (Fig. 8.1a) [13] is often used to illustrate the coordinates of different crystal planes [12, 14]. In the triangle, the polyhedral crystals (i.e., octahedron, rhombic dodecahedron) enclosed by the three basal facets (i.e., {111}, {100}, {110}, respectively) are located at the three vertexes. Other planes lying in the sidelines and inside the triangle are high-index planes. The three sidelines represent the

[001], $[01\bar{1}]$, and $[1\bar{1}0]$ crystallographic zones, and corresponding Miller indices can be expressed as {hk0}, {hkk}, and {hhl} ($h \neq k \neq l \neq 0$). The planes inside the triangle can be expressed as {hkl}. Figure 8.1a also illustrates atomic arrangement models of several typical planes. Atomic arrangements are largely correlated to the Miller indices. The (111) and (100) planes are atomically *flat* with closely packed surface atoms, and the coordination numbers (CNs) of surface atoms are 9 and 8, respectively. The (110) plane and high-index planes have open surface structures with low-coordinated step or kink atoms. Step atoms on {hkk}, {hhl}, and (110) planes have the same CNs of 7, while kink atoms on {hk0} and {hkl} planes have the lowest CNs of 6. Surfaces that possess a high density of atoms with CNs of 6 and 7 have also a very high catalytic activity [12, 15].

Examples of single-crystalline polyhedral nanocrystals of fcc metals that have been successfully synthesized are (i) tetrahexahedron (THH) [16–19], (ii) concave cube [20], (iii) truncated ditetragonal prism [17, 21] enclosed by {hk0}, trisoctahedron by {hhl} [22], octapod by {hkk} [23], and concave hexoctahedron by {hkl} [24]. The presence of twinned boundaries can significantly enrich the morphology of nanocrystals enclosed by a specific surface. For example, as shown in Fig. 8.1b, although single-crystalline polyhedra enclosed by {111} are limited to octahedron, tetrahedron and truncated tetrahedron, a large variety of {111} enclosed polyhedra having twinned boundaries have also been prepared. These twinned polyhedra include single-twinned trigonal bipyramid/truncated bipyramid, fivefold twinned decahedron, and multiply twinned icosahedron. Additionally, the shape of nanocrystals of fcc metals can be further enriched by having mixed facets (Fig. 8.1c). Examples of this are the cuboctahedron, truncated octahedron, five-twinned nanorod/nanowire, and hexagonal plate that are enclosed by a mixture of {111} and {100} facets. The concave tetrahedron and concave bipyramid structures exhibit both {111} and {110} facets.

With the knowledge that the presence of high-index facets leads to higher electrocatalytic activity the challenge for synthetic chemists is to reproduce the desired structural motifs on the nanoscale in metals of interest to energy applications (i.e., Pt, Pd, and their alloys). To better control the shape of noble metal nanocrystals, the following two important issues have to be considered: (1) how to control the surface structure of the nanocrystals and (2) how to induce the formation of twinned boundaries or defects in the nanocrystals. We will now examine several examples of specific interest to energy-related electrocatalysis.

8.3 Surface Structure Effects in Electrocatalysis

Continued research effort is being applied in studying the intrinsic relationship between surface structure and electrocatalytic properties by using metal single crystal electrodes with well-defined surface structures. On the other hand, the catalytic performance of the nanocrystals can also be finely turned by their shape,

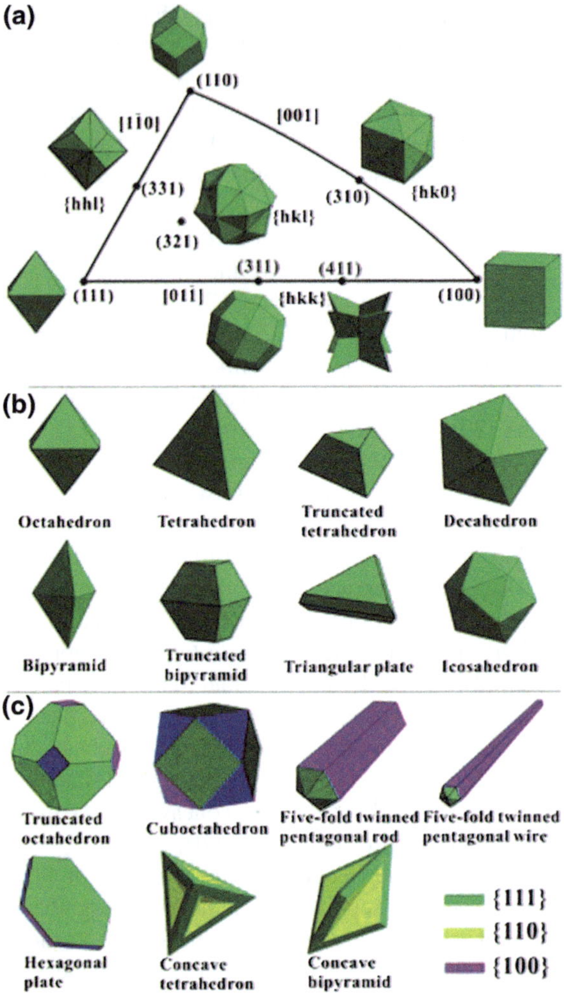

Fig. 8.1 Typical morphologies of fcc metal nanocrystals. **a** Stereographic triangle single-crystalline polyhedral of a fcc metal enclosed by various types of crystal planes. **b** Polyhedra enclosed by {111} facets. **c** Polyhedra enclosed by mixed facets. Reprinted from reference 10 with permission from John Wiley and Sons

which determines surface atomic arrangement and coordination [25, 26]. And this research mainly focuses on the precise control of the surface structure of the catalyst. As particle dimensions reduce toward the nanoscale, the surface-to-volume ratio proportionally increases and small-size effects associated with nanoparticles become more pronounced. Understanding the nanoscale topography of surface sites, such as terraces, steps, kinks, adatoms and vacancies, and their effects on catalytic and other physicochemical properties is the key to designing nanoscale functional materials by nanotechnology.

8.3.1 The Oxidation of Small Organic Molecules

The electrooxidation of small organic molecules (SOMs) is the most important field where these materials are applied. At the anode of direct fuel cells (DFCs) (see also Chap. 2) SOMs undergo electrooxidation. The most common fuels utilized are methanol, ethanol, ethylene glycol, glycerol, and formic acid. The principle catalysts used are Pt, Pt alloy, and Pd-based electrocatalysts. These reactions are very sensitive to surface structure since they involve cleavage of strong chemical bonds e.g., C–C, C–H, and C–O.

It is generally accepted that the electrooxidation of SOMs on Pt surfaces proceeds via a duel pathway mechanism; (i) reactive intermediates (direct pathway) and (ii) poisoning intermediates (indirect pathway). On Pt poisoning intermediates are generally adsorbed CO (CO_{ad}) species, which are derived from the dissociative adsorption of SOMs. Such species can only be removed electrochemically at electrode potentials above 0.6 V (RHE), where oxygen-containing species are generated on Pt surfaces. High-index planes can catalyze both the direct and indirect pathways. Therefore at relatively high potentials, high-index planes exhibit higher activity than that of the basal planes. In the case of the electrooxidation of methanol on Pt basal planes the catalytic activity is in the order of Pt(110) > Pt(100) > Pt(111) [27]. Introducing atomic steps on the surfaces generally promotes the overall oxidation rate, although the formation of poisoning CO species derived from dissociative adsorption of methanol on the step sites may also increase [28, 29]. Koper et al. [29–31]. have studied methanol oxidation on Pt single crystals planes lying in the [1$\bar{1}$0] zone and found that the catalytic activity of the investigated Pt single crystal planes is in the order of Pt(111) < Pt(110) < Pt(554) (= 9(111) × (110)) < Pt(553) (= 4(111) × (110)). Although the stepped surfaces are deactivated faster, they still remain with a much higher activity than Pt(111). Tripkovic et al. [32] studied methanol oxidation on Pt stepped surfaces belonging to the [01$\bar{1}$] zone and demonstrated that Pt(755) had the highest activity, while the catalytic activity decreased by further increasing the density of step sites due to poisoning, i.e., Pt(755) (= 6(111) × (100)) > Pt(211) (= 3(111) × (100)) > Pt(311) (= 2(111) × (100)). Density functional theory (DFT) calculations also indicated that the initial steps of C–H and O–H cleavage of methanol oxidation are more favorable at the Pt step sites [33].

For the electrooxidation of formic acid, the catalytic activity of Pt single-crystal basal planes has been ranked in an order of Pt(110) > Pt(111) > Pt(100) [34]. Sun et al. studied formic acid oxidation on Pt stepped surfaces lying in the [001] zone by CV i.e., Pt(110), Pt(210), Pt(310), and Pt(410) (0.1 M HCOOH, 0.5 M H_2SO_4) [12, 35]. On the return potential sweep, two peaks were seen at 0.80 and 0.40 V (RHE) correspond to formic acid oxidation on (110) steps and (100) terraces. The activity of (110) and (100) were quite different depending on the surface structure. The activity of the (110) step sites is increased by 2 times on Pt(210) and increased by 1.5 times on Pt(310), while it decreased by about 70 % on Pt(410). This effect has been attributed to the particular surface sites consisting of (110) steps and (100) terraces. On Pt(210), the height of (110) steps and the width of (100) terraces are comparative, resulting in a high density of six-atom chair-sites. Consequently, a very high catalytic activity toward formic acid oxidation is shown by Pt(210). Adzic et al. [36]. found that formic acid oxidation increased by introducing atomic steps on Pt(100) and Pt(110) surfaces but decreased by introducing steps on the Pt(111) surface. Hoshi et al. [37, 38]. investigated formic acid electrooxidation on Pd basal planes and Pd(S)-[n(100) $\times$ (111)] ($n = 2 - 9$) stepped surfaces. The catalytic activity of the basal planes was shown to be in the order of Pd(110) < Pd(111) < Pd(100). The highest activity was observed for the high-index plane Pd(911) ($n = 5$).

8.3.2 Electrooxidation of CO

Adsorbed CO (CO_{ad}) is the primary catalyst poisoning species that results from then electrooxidation of SOMs. As a consequence, the study of electrooxidation of CO_{ad} has been of great interest in the development of fuel cell anode catalysts. It has been found that Pt high-index planes in addition to having high activity for the electrooxidation of SOMs, also exhibit high catalytic activity toward CO oxidation [39, 40]. The oxidation of CO_{ad} on stepped surfaces of Pt; Pt(15,15,14), Pt(554), Pt(553) and on basal planes of Pt(111) and Pt(110) in 0.5 M H_2SO_4 has been studied by Lebedeva et al. [9, 41]. It was shown that the overpotential required for the oxidation of a saturated CO adlayer, as well as a CO submonolayer, is increased in the sequence of Pt(553) < Pt(554) < Pt(111). The rate of oxidation was proportional to the step fraction on the Pt surfaces, confirming that the active sites for CO_{ad} oxidation consist of step atoms. Garcia et al. [42] illustrated that, in alkaline solutions, the activity for CO_{ad} oxidation follows the order of kinks (defects) > steps > terraces. The same authors[43] recently showed that CO oxidizes first at step and defect sites, but the product of the reaction, carbonate, remains adsorbed at or near the active site, blocking subsequent oxidation of other CO molecules. Consequently, new CO molecules will be oxidized at the next most active and available oxidation sites. The carbonate does not remain adsorbed on the Pt(111) surface and therefore CO can diffuse over that surface without limitations. Mikita et al. [44] studied the electrooxidation of CO_{ad} on series of stepped surfaces (i.e., Pt(210), Pt(510), and Pt(910)) and found that Pt(210) exhibits the highest activity, and the lowest onset potential. The high activity of Pt(210) was explained by the highest density of kink atoms. High activity for the oxidation of bulk (not adsorbed) CO was also shown by the Pt high-index surfaces [44].

8.3.3 Oxygen Reduction

The Oxygen Reduction Reaction (ORR) has also been studied on Pt basal planes. The catalytic activity was determined in the order of Pt(110) > Pt(100) > Pt(111) [45].

Another study of the structure sensitivity of ORR on Pt high-index planes lying in the [01$\bar{1}$] and [1$\bar{1}$0] zones revealed that the catalytic activity increased linearly with increasing density of step atoms [46, 47]. In the [01$\bar{1}$] zone the Pt(211) surface gave the highest activity while in the [1$\bar{1}$0] zone Pt(331) showed the highest activity. These authors also highlighted a very important factor in determining the structure sensitivity of the ORR on Pt surfaces. They demonstrated that in aqueous solution oxygen must compete for the adsorption sites with other anions present as well as the formation of surface oxides. The main structural effects on ORR are the anion adsorption and oxide formation on the electrode, whereas the oxygen adsorption energy on the different sites actually plays a secondary role in determining the electrocatalytic activity of the electrode. In Chap. 9, the reader will learn that it is now generally accepted that ORR activity and stability is greatly enhanced on smooth Pt surfaces, where the binding of oxygen containing intermediates that block the ORR is weakened thus enhancing the ORR.

8.3.4 Effects of Surface Structure on Selectivity in Higher Alcohol Electrooxidation

Higher alcohols such as ethanol, ethylene glycol and glycerol are promising alternatives to methanol in DAFCs due to their lower toxicity and their availability from biomass [48–52]. However, the electrooxidation of these alcohols in DAFCs is at present only partial, as it involves the cleavage of the C–C bond. In the case of ethanol the main products are acetaldehyde and acetic acid, and CO_2 production contributes to less than 2 % of the total current even on the most effective electrocatalysts (e.g. Pt–Sn alloys) [53]. As a consequence, the fuel efficiency of an ethanol fed DAFC is actually very low. Higher conversion to CO_2 has been demonstrated working with very dilute ethanol solutions in acidic DEFCs (Faradaic efficiency 64 % for 0.1 M ethanol at 80 °C) [54].

Studies of hydrocarbon reactions (e.g., hydrogenolysis and isomerization) on Pt single-crystal planes have revealed that the product selectivity highly depends on surface atomic arrangements and the presence of step atoms, especially kink atoms, which can promote the breaking of the C–C bond. This knowledge is very helpful in designing catalysts toward electrooxidation of SOMs containing C–C bonds. Tarnowski et al. [55] studied the effects of step atoms on the selectivity of ethanol electrooxidation by using Pt(533), Pt(755), and Pt(111) electrodes. The yield of CO_2 was increased by introducing (100) step on (111) terrace. At the same time the yield of acetic acid on the Pt(533) electrode was only approx. 25–30 % of that on the Pt(111). Sun and co-workers [56] studied the electrooxidation of isopropanol and found that Pt(610) was the most active surface for yielding CO_2. The order of activity for producing CO_2 was Pt(610) > Pt(111) > Pt(100) > Pt(211) > Pt(110). These results demonstrate that high-index surfaces with a high density of step and kink atoms do significantly promote complete electrooxidation of fuels containing C–C bonds.

8.4 Common Strategies and Synthetic Methods

What follows is a summary of the major synthetic strategies that have been developed to obtain nanocrystals with well-defined high-index facets.

8.4.1 Small Adsorbate-Assisted Facet Control of Pt and Pd Nanocrystals

A number of small molecules or ions have been found to have a major impact on facet evolution during the synthesis of noble metal nanocrystals. Compared with surfactants or polymers used in traditional nanocrystal synthesis, using small strong adsorbates has allowed better understanding of how the facet evolution is achieved from a chemical perspective. The strong binding of polymeric capping agents on the surface of nanocrystals creates the difficulty of cleaning them from the surface, which is essential before they can be used in catalyst applications. Often the cleaning process alters significantly the surface structure of the nanocrystals that have been freshly prepared. As an alternative, the use of small strong adsorbates avoids these problems as they do not form a thick impenetrable chemisorbed layer and hence can be more easily removed. At the same time, the adsorption of small molecules can be readily studied by various techniques thus helping to understand and design the nanocrystal surface structure. The range of molecules studied has greatly expanded in recent years to include NO_2 [57], HS^-/S^{2-} [58] and $C_2O_4{}^{2-}$ [59, 60]. The most important examples will be discussed here with the purpose of demonstrating how small adsorbates can be used to induce the formation of well-defined nanocrystals.

8.4.1.1 Carbon Monoxide

Carbon monoxide is well known as a common poisoning agent of Pd and Pt catalysts due to the strong adsorption of CO on their surfaces which prevents other molecules from accessing the catalytic surfaces. CO molecules do behave differently in the controlled synthesis of Pd and Pt nanostructures [13]. CO prefers to adsorb on the Pd{111} surface that facilitates the growth of ultrathin Pd nanosheets and tetrapod/tetrahedral nanocrystals having Pd{111} as the main exposure surface. The preferential adsorption of CO on Pt{100} induces the formation of Pt nanocubes [61, 62].

Researchers have extensively studied the adsorption of CO on various surfaces of Pd under vacuum conditions and have shown that CO exhibits strong adsorption. [71–73]CO molecules adsorbed on Pd can be readily removed by applying a high temperature or an electrochemical oxidation potential, so it is simple to clean the surface of the as-prepared CO-capped Pd nanocrystals.

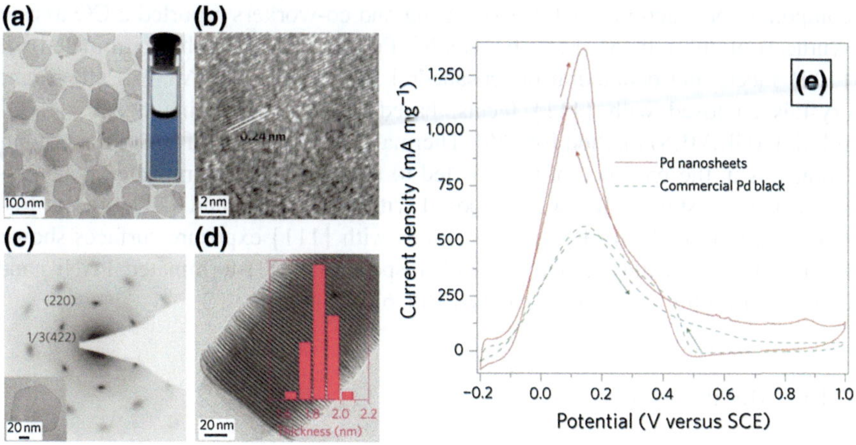

Fig. 8.2 *Left* (**a–d**) Ultrathin palladium nanosheets synthesized in the presence of PVP, halide, and CO gas in DMF. **e** Comparison of formic acid electrooxidation activity of palladium nanosheets and palladium black (0.5 M H_2SO_4). Reprinted by permission from Macmillan Publishers Ltd: Nature Nanotechnology, 2011, 6, pp 28–32, copyright (2011)

Very recently, when introduced into the synthesis of Pd nanocrystals, CO was demonstrated as an excellent adsorbate to facilitate the formation of Pd nanocrystals with Pd{111} as their exposure surfaces. Huang et al. [63] introduced pure CO gas into the synthesis of Pd nanocrystals in the presence of a halide salt and PVP. After 3 h at 100 °C, monodisperse, ultrathin Pd nanosheets (thickness 1.8 nm) were obtained (Fig. 8.2a–d). The Pd nanosheets were shaped in well-defined hexagon with {111}ure facets as their basal planes and {100} facets as their six side planes.

Experiments showed that the basal {111} surfaces of the Pd nanosheets were absorbed by CO, and the side {100} surfaces were bonded by halide. On the upper and lower {111} facets the CO bonded strongly while the halide binding on the sides was weaker. This meant that the nanocrystals grew preferentially on the sides creating the nanosheet structure. A direct result of the high surface area of the nanosheets was a high activity for the electrooxidation of formic acid (2.5 times that of a commercial palladium black catalyst) (Fig. 8.2e) [63].

CO has also been used in the shape-controlled synthesis of Pt nanocrystals [13]. Zheng et al. concluded that the selective binding of CO on Pt{100} was the principle reason for the preferential formation of Pt nanocubes in the presence of CO gas. Both supported and unsupported Pt nanocubes were prepared in the presence of CO [62]. The as-prepared Pt nanocubes were surfactant-free and could be surface-cleaned after being exposed in air for several hours. The as prepared supported Pt catalysts contained clean Pt{100} facets and thus exhibited enhanced electrocatalytic activity for methanol electrooxidation in acidic media. Gaseous CO was also successfully used to synthesize Pt_3M nanoalloys (M = Fe, Co, Ni) with exposed {100} and even {111} facets depending on the transition metal component or reaction kinetics [64]. Yang and co-workers reported a CO assisted synthesis of cubic Pt–M (M = Fe, Co, Ni, Pd) alloyed nanocubes terminated with {100} facets and octahedral or icosahedral Pt–M (M = Ni, Au, Pd) alloy nanocrystals enclosed with {111} facets, based on a gas reducing agent in liquid solution (GRAILS) method [65, 66]. Thermal decomposition of both $Pt(acac)_2$ and $Co(acac)_2$ in the presence of CO gas led to the formation of truncated octahedral Pt_3Co nanocrystals which were exposed with both {100} and {111} facets [64]. Icosahedra or octahedra Pt_3Ni nanocrystals with {111} exposure surfaces showed enhanced electrocatalytic activity with respect to {100}-terminated Pt_3Ni nanocubes or a highly active Pt/C catalyst [65–67].

8.4.1.2 Halide Anions

It is well known that halide anions chemisorb strongly on crystalline Pd surfaces in the order of $Cl^- < Br^- < I^-$ [68, 69]. Xiong and co-workers in 2007 first used halide anions in the shape controlled synthesis of Pd nanocrystals [70]. Pd nanobars and nanocubes enclosed by {111} facets were successfully synthesized using a mixture of KBr, PVP, and Na_2PdCl_4 in a polyol-based synthesis. The use of KCl led to the formation of cubooctahedra bound by a mixture of {111} and {100} facets. The use of KI led to poorly defined nanocrystals most likely due to the much stronger chemisorptions of I^- on the surface. The ORR activities of both the nanocubes and the octahedra were studied. The {100}-terminated Pd nanocubes had about an order of magnitude higher activity with respect to the {111}-enclosed Pd octahedral. To avoid the effect of using Cl^- ions, Zheng and co-workers used another Pd salt $Pd(acac)_2$ in the synthesis. In the presence of I^- Pd nanocubes were produced [71]. The formation of Pd nanocubes bounded by {100} facets in the presence of I^- can be expected due to the stronger adsorption of I^- on Pd{100}. With either Cl^- or Br^- multiply twinned nanoparticles were produced.

Polyol synthesis in the presence of halide anions has also yielded Rh, Pt or Pd nanocubes [72]. When I^- was used as the shape control agent, {100}-facet-enclosed PtPd nanocubes were produced. With Cl^-{111}-facet enclosed PtPd octahedral/tetrahedral were obtained.

8.4.1.3 Amines

Pt concave nanocrystals with high-index {411} facets and a unique octapod morphology (Fig. 8.3) have been prepared by Huang et al. using a solvothermal synthesis by introducing methylamine into a DMF mixture containing PVP and hexachloroplatinic (IV) acid [23]. The octapod nanocrystals had 24 kite-like high-index {411} facets. When these researchers replaced methylamine with other amines (i.e., ethylamine, butylamine, 4-methylpiperidine, trimethylamine) or NH_3 concave octapod nanocrystals were also obtained. The creation of high-index {411} facets in the presence of amines was due to the stabilizing effect of the

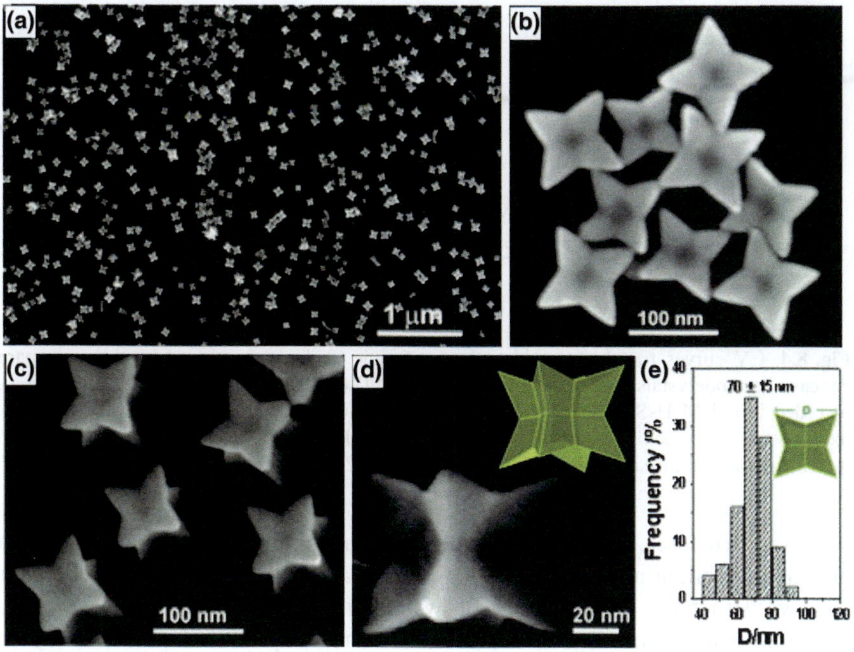

Fig. 8.3 Typical **a** large-area, **b** enlarged, and **c** tilted SEM images of the as-prepared concave Pt nanocrystals. **d** High-magnification SEM image of a single concave Pt nanocrystal. The *top-right inset* shows an ideal geometrical model of the concave Pt nanocrystal with the same orientation as the nanocrystal in the SEM image. **e** Size distribution of the as-prepared concave Pt nanocrystals. Reprinted with permission from Ref. [23]. Copyright (2011) American Chemical Society

strong selective coordination of amines on the seven-coordinated Pt sites. Another advantage to using amines is that although under neutral and basic conditions they bind strongly to Pt, under acidic conditions they are readily protonated and removed from the Pt surface. So after acid treatments, the Pt octapod nanocrystals displayed excellent activities in the electrocatalytic oxidation of formic acid (Fig. 8.4). The specific activity of the concave Pt nanocrystals normalized by electrochemically active surface area (ECSA) was 2.3 and 5.6 times greater than those of commercial Pt black and Pt/C, respectively. The increase in activity is due to the presence of active sites with low CNs on the surface of Pt{411} facets.

8.4.1.4 Formaldehyde

Formaldehyde has been widely utilized as both a strong reducing agent as well as a facet control agent. The formation of concave tetrahedral/trigonal bipyramidal Pd nanocrystals bounded with {111} and {110} facets was achieved by Huang et al. [73]. The synthesis involved a mixture of $Pd(acac)_2$, PVP, and formaldehyde in benzyl alcohol. Replacement of formaldehyde with benzaldehyde also lead to

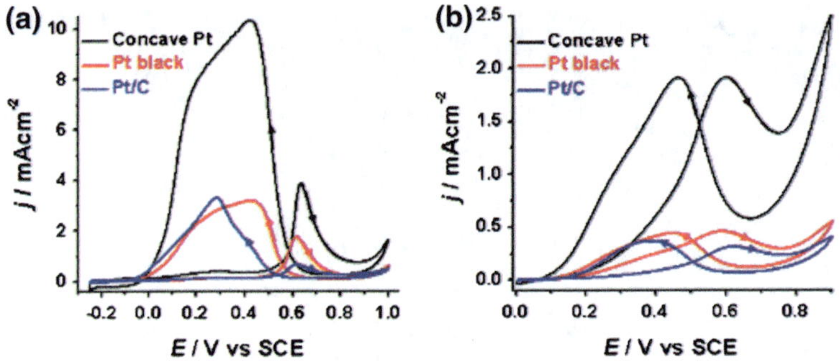

Fig. 8.4 CV curves for electro-oxidation of **a** formic acid and **b** ethanol by the as-prepared concave Pt nanocrystals, commercial Pt black, and Pt/C (E-TEK). The formic acid oxidation was recorded in 0.5 M H_2SO_4, 0.25 M HCOOH solution at a scan rate of 50 mV/s. The ethanol oxidation was recorded in 0.1 M $HClO_4$ + 0.1 M CH_3CH_2OH solution at a scan rate of 50 mV/s. Reprinted with permission from Ref. [23]. Copyright (2011) American Chemical Society

concave structures. Without formaldehyde a mixture of morphologies was obtained highlighting the importance of formaldehyde in the formation of the concave structures. Additionally, it was found that the degree of concavity increased with increasing the amount of formaldehyde used. The activity for formic acid oxidation also increased with increased proportion of {110} facets. Other researchers have used formaldehyde in the selective formation of various Pd nanocrystals including icosa-, deca-, octa-, tetrahedral, and triangular plate-like Pd nanocrystals (enclosed with {111} facets) [74] and truncated cubes, cuboctahedra, truncated octahedral and octahedra [75]. In another recent article, Huang et al. showed that well-defined concave tetrahedral Pd nanocrystals can also be prepared using only CO as reducing agent. It could be possible that the origin of shape control using formaldehyde could result from the decomposition of formaldehyde to CO and H_2, a reaction that commonly occurs on Pd [76].

8.4.2 Facet Control by Electrochemical Methods

The group of Sun and co-workers in 2007 made a breakthrough in the synthesis of Pt nanocrystals enclosed with high-index facets through the development of a novel electrochemical method [18, 77]. The process involves a electrochemical square-wave potential deposition process on the glassy carbon (GC) substrate in a solution containing 0.1 M H_2SO_4 and 30 mM ascorbic acid. Nearly all of the growing Pt nanocrystals on the GC surface are tetrahexahedral (THH Pt NCs). The as-prepared Pt nanocrystals show good agreement with a geometrical model of THH (Fig. 8.5).

Miller indices of exposed surfaces on the THH Pt NCs were identified as mainly {730} facets through the measurement of plane angles between two adjacent facets

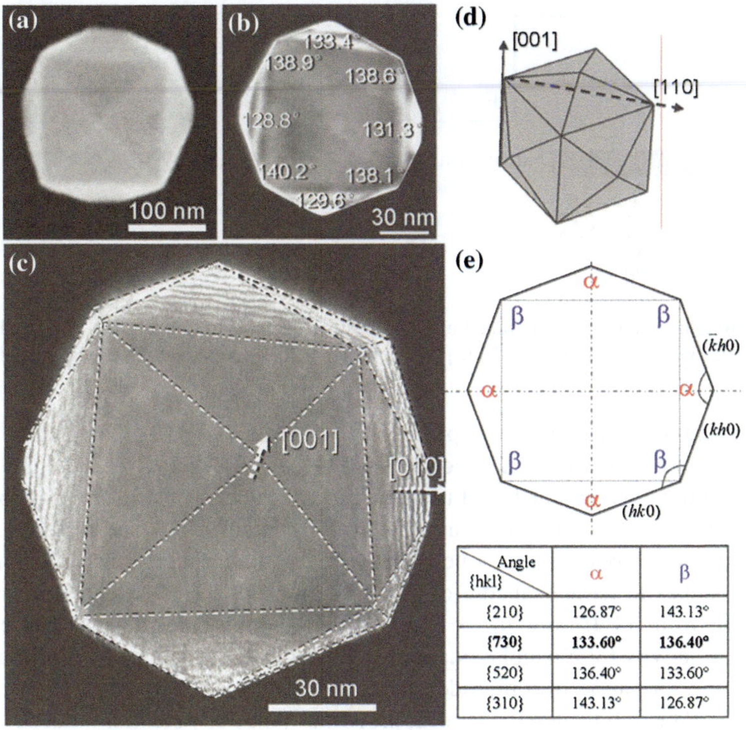

Angle / {hkl}	α	β
{210}	126.87°	143.13°
{730}	**133.60°**	**136.40°**
{520}	136.40°	133.60°
{310}	143.13°	126.87°

Fig. 8.5 **a** SEM image of a THH Pt nanocrystal and **b**, **c** two dark-field TEM images with incident electron beam along and approx 10° away from the [001] direction, respectively. **d** Three-dimensional model of the THH Pt nanocrystal. **e** A two-dimensional projection draw of the THH nanocrystal and a table of the faceted angles calculated for the different surface planes. Reprinted with permission from Ref. [77]. Copyright (2007) American Institute of Physics

parallel to the [001] zone axis in a TEM image. The Pt (730) plane is periodically composed of two (210) microfacets followed by one (310) microfacet, and has a density of step atoms as high as 5.1×10^{14} cm^{-2} (i.e., 43 % of surface atoms are step atoms). More importantly, all surface atoms on the THH Pt NCs are arranged in such a way that they form active sites for catalysis. Therefore, THH Pt NCs exhibit high catalytic activity.

It has been demonstrated that for formic acid electrooxidation, the catalytic activity of THH Pt NCs is 1.6–4.0 times higher than that of polycrystalline Pt nanospheres, and 2.0–3.1 times larger than that of commercial Pt/C catalyst from E-TEK Co., Ltd. For ethanol electrooxidation, the enhancement factor of the catalytic activity obtained on the THH Pt NCs varies from 2.0 to 4.3 relative to that of Pt nanospheres, and 2.5–4.6 relative to commercial Pt/C catalyst [18].

Sun and co-workers then applied their electrochemically shape controlled method to obtain high-index facet Pt NCs supported on carbon black with a size (2–10 nm) comparable to that of standard commercial Pt/C catalysts [78]. Rather

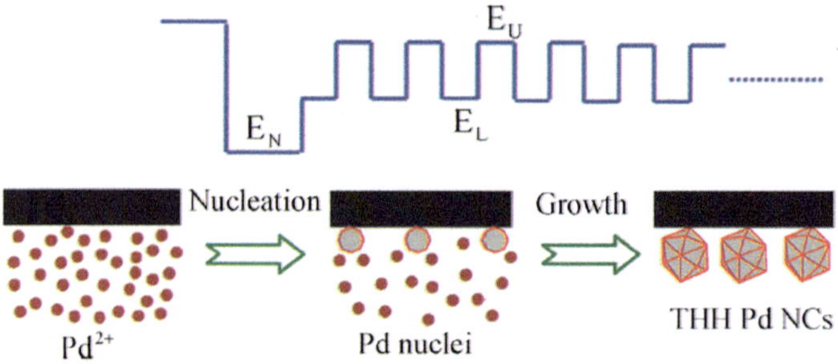

Fig. 8.6 Illustration of programmed electrodeposition method for preparation of THH Pd NCs. Reprinted with permission from Ref. [19]. Copyright (2010) American Chemical Society

than using Pt nanospheres as precursor they used Cs_2PtCl_6. The much higher density of low coordinate atomic steps ({110}, {210}, {310}, {510}, {211}, and {311} steps) thus obtained lead to a doubling of the cleavage of the C–C bond in the electrooxidation of ethanol as evidenced by in situ FTIR spectroscopy.

Square wave potential treatment was used to obtain shape transformation from Pt nanocubes to tetrahexahedra of around 10 nm [79]. The surface structure was changed from {100} low-index facets to {310} high-index facets. The electrocatalytic activity for ethanol oxidation was also greatly enhanced by this treatment, due to the formation of a high density of step atoms on high-index facets.

Since Pd exhibits electrochemical properties (such as potential induced oxygen adsorption/desorption) similar to that of Pt, but less stable and with a hydrogen absorption feature, the electrochemical square-wave potential method may be also used, with some modifications, for the preparation of Pd nanocrystals with high-index facets.

To prepare THH Pd NCs, Tian et al. developed a modified method [19]. As illustrated in Fig. 8.6, THH Pd NCs were directly electrodeposited (from solution) on the GC substrate in 0.2 mM $PdCl_2$ + 0.1 M $HClO_4$ solution by a programmed electrodeposition method.

The SEM image of as-prepared THH Pd NCs is shown in Fig. 8.7a. The yield of the THH Pd NCs was over 80 %, and the other shapes found were agglomerations of imperfect THHs and irregular polyhedra. The average size of THH Pd NCs was 61 nm. The exposed facets on the THH Pd NCs are also mainly {730} facets, as determined by HRTEM and SAED. The {210} and {310} steps can be discerned on the border atoms in the HRTEM image, as marked in Fig. 8.7b, c. Owing to their high density of surface active sites, the THH Pd NCs exhibited 4–6 times higher catalytic activity per unit surface area when compared to commercial Pd black catalyst (Johnson Matthey, Inc.) for ethanol electrooxidation in alkaline solutions (Fig. 8.7d). Besides, the THH Pd NCs also exhibits a high stability. After 1,000 potential cycles, 75.0–95.5 % of the initial catalytic activity was maintained.

A clever adaption of the square wave method has been recently reported called *Electrochemical Milling and Faceting* (ECMF) [80]. Large Pd NPs (35 nm) with

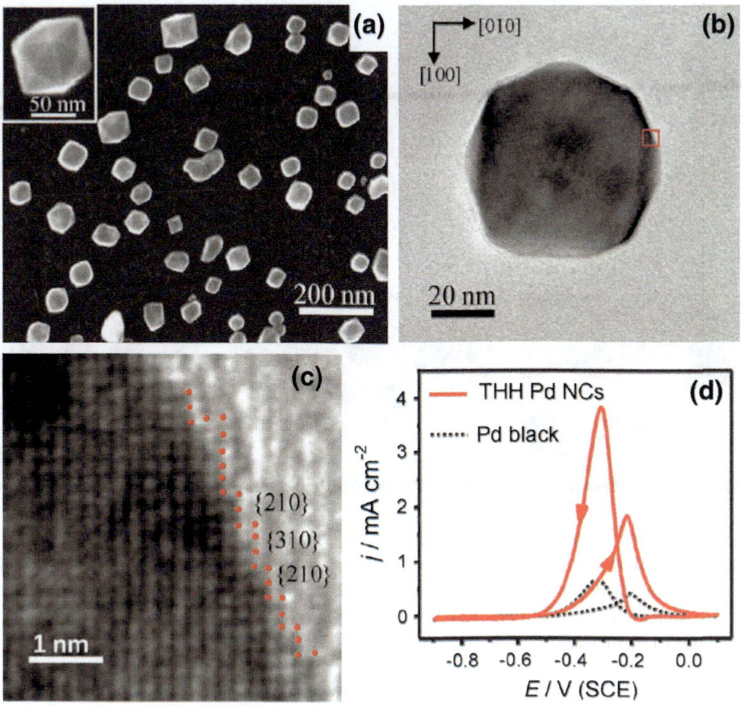

Fig. 8.7 **a** SEM image of THH Pd NCs. The *inset* is a high magnification SEM image. **b** TEM image of a THH Pd NC recorded along the [001] direction. **c** HRTEM image recorded from the *boxed area* in (**b**), showing some {210} and {310} steps that have been marked by *red dots*. **d** Cyclic voltammograms of THH Pd NCs (*solid line*) and Pd black catalyst (*dashed line*) at 10 mV s^{-1} in 0.1 M ethanol + 0.1 M NaOH. Reprinted with permission from Ref. [19]. Copyright (2011) American Chemical Society

low-index facets were first supported on a TiO_2 nanotube array (Fig. 8.8A). A two-step square wave electrochemical treatment was then applied. In the first "heavy" step a palladium oxidation was applied at 4.55 V (vs. RHE) for 180 s, followed by the reduction of the Pd oxides at −1.95 V (vs. RHE) for 180 s (Fig. 8.8B). This was followed by a milder treatment with a frequency of 0.025 Hz for 3 h between +3.35 and −0.75 V (vs. RHE) (Fig. 8.8C). The overall treatment resulted not only in a net reduction in mean particle size to 7 nm particles (milling) but also the formation of HIFs, multiple twins and a high density of step atoms (Faceting). Analysis showed the presence of high-index facets {210} and {410} along the ⟨100⟩ direction, and {211} and {311} facets along the < 110 > direction, respectively (Fig. 8.9). Cyclic voltammetry using an ethanol-containing electrolyte showed a peak current density of 201 $mAcm^{-2}$ (Fig. 8.10, curve 3), corresponding to a normalized mass-specific activity for Pd of 8,965 Ag^{-1}. A value remarkably higher than those reported in the literature determined under comparable conditions (e.g., 3,600 Ag^{-1}) [81]. The onset potential for the oxidation of ethanol

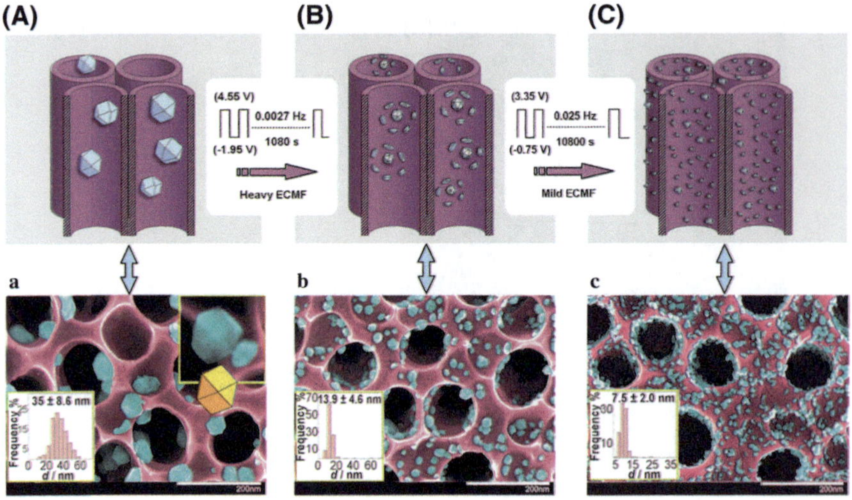

Fig. 8.8 **A** TNTAs with as-deposited Pd and **a** the corresponding SEM image. **B** TNTAs with Pd after heavy ECMF and **b** the corresponding SEM image. **C** TNTAs with Pd after heavy and mild ECFM and **c** the corresponding SEM image. False coloring of the SEM images shows Pd NPs (*light blue*) and TNTA support (*violet*). The white scale bars in (**a**–**c**) are 200 nm. Reprinted from Ref. [80] with permission from John Wiley and Sons

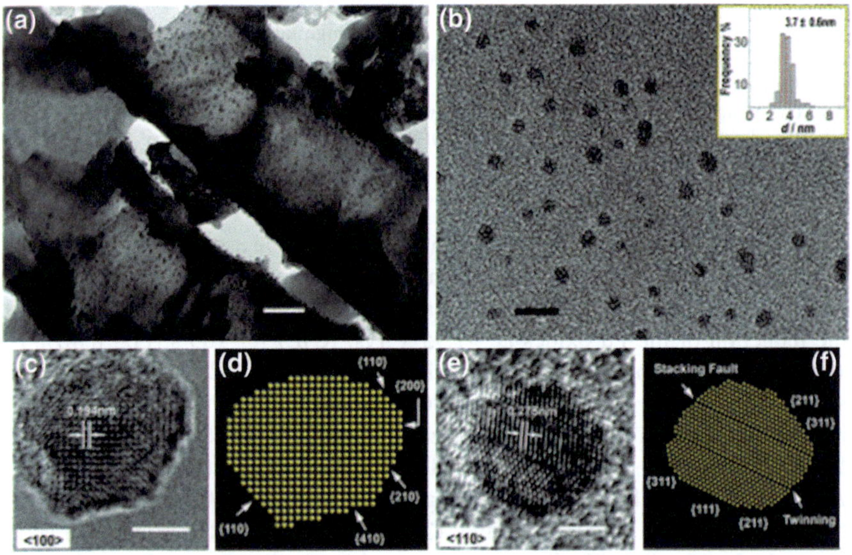

Fig. 8.9 ⟨⟩**a** TEM image of the Pd-loaded TNTA electrode after heavy and mild ECMF (scale bar = 50 nm). **b** Pd nanoparticles found in the electrolyte after heavy and mild ECMF (scale bar = 35 nm). **c** HRTEM image (scale bar = 2 nm) and **d** atomic models with face assignment of the TNTA-supported Pd nanoparticle along the ⟨100⟩ direction. **e** HRTEM image (scale bar = 2 nm) and f) face assignment of the TNTA-supported Pd nanoparticles along the ⟨110⟩ direction. Reprinted from Ref. [80] with permission from John Wiley and Sons

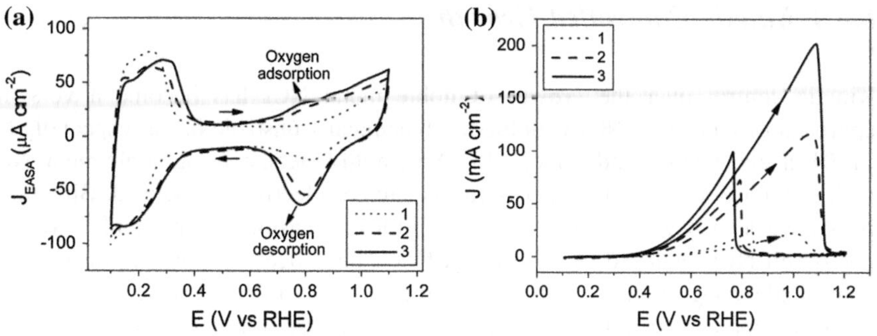

Fig. 8.10 Cyclic voltammograms of TNTAs with deposited Pd recorded in **a** 0.1 M $HClO_4$ and **b** 2 M KOH with 10 wt % EtOH. Scan rate: 50 mV s^{-1}. Curve 1: TNTA-Pd as deposited. Curve 2: TNTA-Pd after heavy ECMF. Curve 3: TNTA-Pd after heavy and mild ECMF. Reprinted from Ref. [80] with permission from Jon Wiley and Sons

shifted 0.17 V more negative than the potential obtained for the as-deposited sample. The same authors applied an analogous treatment to a polycrystalline Pd electrode and obtained an increase in activity for alcohol electrooxidation which was attributed to a combination of increased electro active surface area and an increased concentration of low coordination palladium atoms [82].

8.4.3 UPD

UPD of impurity metals can provide an alternative strategy to control the shape evolution and exposed facets of noble metallic nanostructures [17]. For example, hexoctahedral (HOH) AuPd alloy NCs enclosed by {431} high-index facets were synthesized with a Cu UPD-assisted process [83]. In the synthesis, $AuCl_4^-$ was first preferentially reduced because of its high reduction potential. A Cu UPD monolayer then deposits on the Au surface. As the reduction potential of $PdCl_4^{2-}$ is higher than the potential of $Cu^{2+/0}$ (0.341 V), $PdCl_4^{2-}$ may be reduced to Pd on the Au surface by galvanic replacement, and Cu atoms are reoxidized to Cu^{2+} ions. Simultaneously, the reduction of $AuCl_4^-$ may occur, and the Au–Pd alloy ultimately forms. Besides the UPD effect, the surfactant octadecyl trimethyl ammonium chloride (OATC) and co-solvent EG also contributed to the formation of {431} high-index facets. As there are high-density atomic steps and kinks on {hkl} high-index surfaces, the HOH Au–Pd alloys as expected exhibited high activity for formic acid electrooxidation. UPD has also been used to deposit a highly active monolayer on to high-index faceted metal nanocrystals. Lu et al. [84] reported the use of Au TDPs enclosed by 12 high-index {310} facets as substrate and nanofacet activator to deposit a Pt monolayer by Cu UPD and subsequent galvanic replacement reaction. The resultant Au(TDP)-Pt(ML) exhibited an enhanced catalytic activity compared to pure TDP-Au.

8.4.4 Kinetic Controlled Growth

Kinetic control over the growth of noble metal NCs has become a versatile approach to synthesize NCs with thermodynamically unstable facets, especially for {110} facets and high-index facets [85, 86]. With kinetic control, noble metal NCs with different types of high energy facets can be selectively produced. Bimetallic NCs have also been prepared in a few limited occasions. In particular Au@Pd core–shell NCs enclosed by high-index facets have been synthesized through the controlled over-growth of Pd shells onto the surfaces of Au NC seeds [85–88]. In a recent example, Au@Pd core-shell NCs enclosed by different high-index facets were synthesized with concave TOH Au NC as seeds [86]. Polyhedral Au@Pd NCs with three different classes of high-index facets, including concave TOH NCs with {hhl} facets, concave hexoctahedral (HOH) NCs with {hkl} facets, and tetrahexahedral (THH) NCs with {hk0} facets, were formed by varying Pd/Au ratios and amount of NaBr added. When the Pd/Au ratio was low (1/4), the templating effect of the Au seeds are dominant and this leads to the formation of TOH Au@Pd NCs with similar index facets to the seeds. With a higher concentration of Pd (Pd/Au = 1/2), the faster growth rate of the $< 110 >$ directions gradually fill the concave space of the TOH NCs, resulting in the shape transformation from TOH to HOH or THH.

Sang Woo Han and co-workers have recently reported the synthesis of a HOH Au–Pd bimetallic alloy NCs enclosed exclusively by high-index facets {541} [89]. These alloyed Au–Pd NCs that exhibit a unique structure, were realized using a simple one-pot aqueous synthesis that didn't need seed or additional metal ions as structure regulating agent. The HOH Au–Pd alloy NCs were prepared by the co-reduction of Au and Pd precursors using ascorbic acid (AA) in the presence of cetyltrimethylammonium chloreide (CTAC) over 2 h. The resulting NCs had an average size of 114 nm with a HOH structure (polyhedron bounded by 48 triangular high-index {hkl} ($h > k > l > 0$) facets) (Fig. 8.11). To elucidate the growth process a TEM study was undertaken sampling the NCs at various reaction times. Initially, under fast kinetics small NCs with an average size of 30 nm were produced with an Au–Pd alloy HOH structure, as the reaction continued (slow kinetics) the NC size gradually increased but the HOH structure was retained. The HOH Au–Pd NCs exhibited higher catalytic performance toward the electro-oxidation of ethanol than Au–Pd NCs bound by low-index facets (Fig. 8.12).

8.4.5 Seeded Growth

This method involves the use of pre-prepared small nanocrystals with well-defined shape and surface structures. Seeded growth can be used to prepare not only monometallic but also bimetallic nanocrystals (core–shell) with well-defined facets of the external shell metal. Of great importance is the discovery that by tuning the

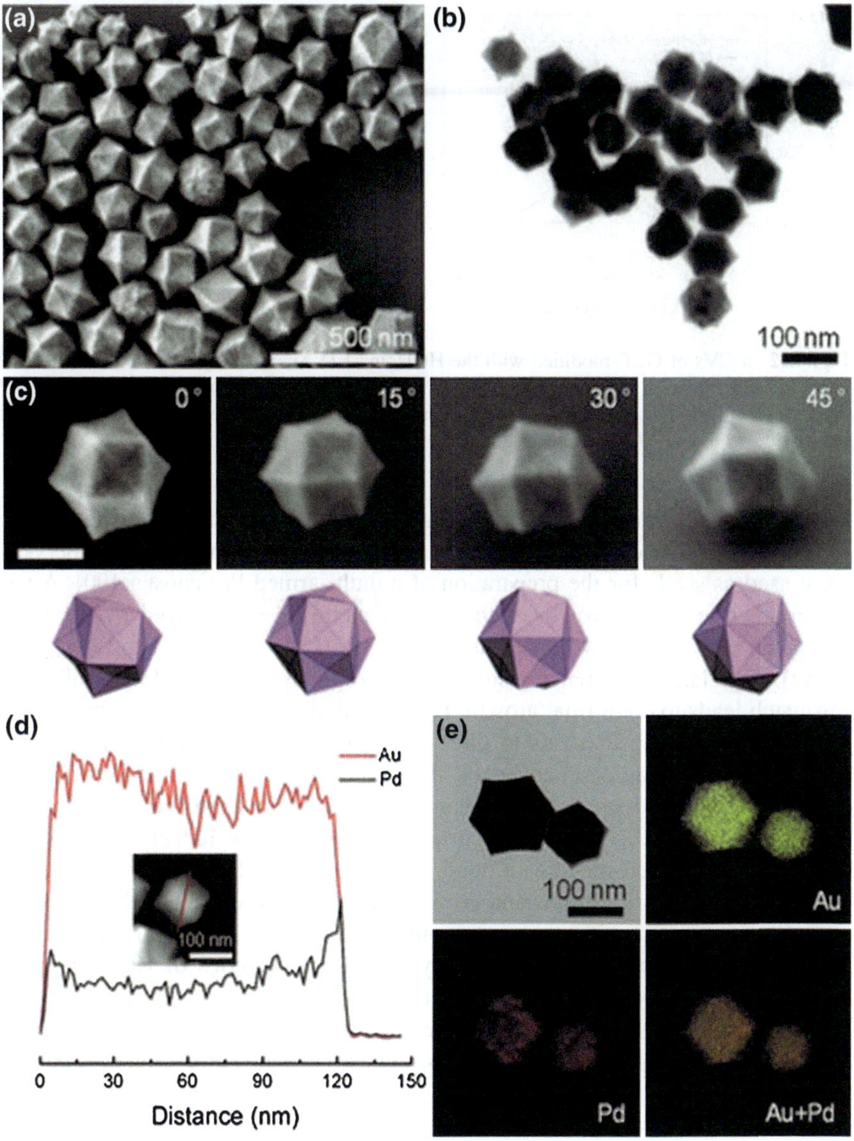

Fig. 8.11 **a** SEM and **b** TEM images of the HOH Au–Pd NCs. **c** SEM images of the NCs tilted from 0° to 45°. The scale bar indicates 100 nm. The corresponding structural models with different orientations are shown below the SEM images. **d** HAADF-STEM image and cross-sectional compositional line profiles of a HOH Au–Pd NC. **e** HAADF-STEM–EDS mapping images of the HOH Au–Pd NCs Reprinted from Ref. [89] with permission from John Wiley and Sons

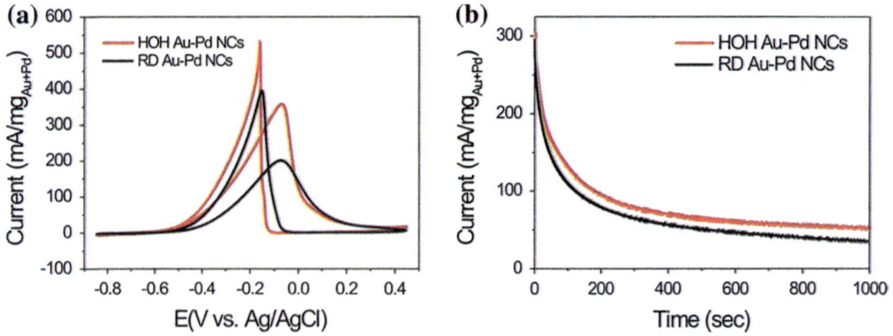

Fig. 8.12 **a** CVs of GCE modified with the HOH and RD Au–Pd alloy NCs obtained in 0.1 M KOH + 0.5 M ethanol at a scan rate of 50 mV s^{-1}. **b** CA curves of the HOH and RD Au–Pd alloy NCs on GCE obtained at −0.1 V versus Ag/AgCl. Reprinted from Ref. [89] with permission from Jon Wiley and Sons

seeded growth conditions, the exposed facets developed on the shell surface can be different from that of the seeds. In a recent example, tetrahedral Pt nanocrystals were used as seeds for the preparation of a multi-armed Pt nanostar [90]. A very important role in the growth of the second metal layer is played by the lattice mismatch between the two different metals (e.g., 0.77 % for Pt/Pd and 4.08 % for Pt/Au). High lattice mismatch results in anisotropic overgrowth while small lattice mismatch leads to conformal growth. Habas et al. demonstrated this concept using highly faceted cubic Pt seeds to direct the epitaxial overgrowth of a secondary metal [57]. With lattice-matched Pd they produced conformal shape-controlled core–shell particles, on the other hand with lattice-mismatched Au, this gave anisotropic growth. It is very important to have a slow reduction rate of the precursor of the shell metal as a rapid reduction rate will break the near equilibrium condition and result in island growth [91–94]. This was shown by Zhang and co-workers using a syringe pump to control the addition rate of precursor to the seed solution, which controlled directly the growth kinetics of Rhodium nanocrystals [95]. Epitaxial overgrowth on seed nanocrystals with well-defined shapes provides a straightforward and effective route to control of exposure surfaces of the obtained core–shell metal heterostructures. Citric acid, which is a mild reducing agent, has been used to obtain slow reduction and hence epitaxial growth of Pt shells on Pd nanoplates [96]. {111}-Bounded Pd@Pt nanoplates with hexagonal and triangular shapes were obtained. The ultimate goal is to obtain metal shells bounded with desired high-index facets. This has been successfully achieved using metal templates with high-index facets [88]. Template-directed epitaxial overgrowth of Pd on THH and TOH Au nanocrystals with high-index facets has been demonstrated by Wang and co-workers [88]. Slow reduction of the Pd precursor with ascorbic acid resulted in the Pd nanoshell retaining the {730} and {221} facets of the THH and TOH Au seed nanocrystals. The thus obtained Au@Pd core–shell nanocrystals possessed a large number of coordinatively unsaturated Pd atoms at steps on their high-index facets.

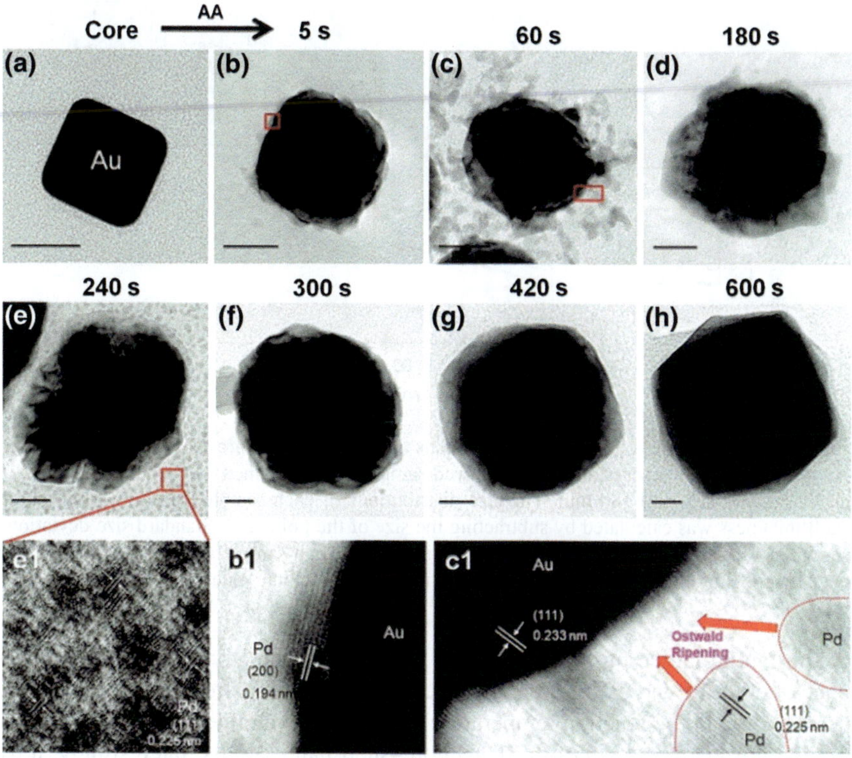

Fig. 8.13 **a** TEM image of a Au nanocube. **b–h** TEM images of the Au–Pd core–shell heterostructures taken at different times after the addition of ascorbic acid to disclose the formation process of the THH nanocrystals. Scale bar is equal to 20 nm for every image. (**b1**) Enlarged TEM image of the *square region* in panel b showing a thin Pd shell. (**c1**) TEM image of the *rectangular region* in panel **c**. Reprinted with permission from Ref. [19]. Copyright (2010) American Chemical Society

It is possible to use the seeded growth method to obtain core-shell nanocrystals that do not adopt the morphologies of the seed nanocrystals. In fact the final morphology of the core-shell nanocrystals is determined by the synthetic conditions including the type of facet specific capping agents used in the overgrowth process [57].

Seeded growth can also be used to form catalytically highly active surfaces from metal seeds with low-index facets. For example, Lu and co-workers [97] have successfully developed a simple method for the fabrication of Au@Pd core–shell nanocrystals with a tetrahexahedral (THH) structure with exposed {730} ure facets. They used low index faceted Au nanocubes as the structure-directing template (Fig. 8.13f, g). While the Au nanocubes were enclosed by {100} facets, the obtained Au@Pd core–shell THHs possessed high-index {730} exposure surfaces.

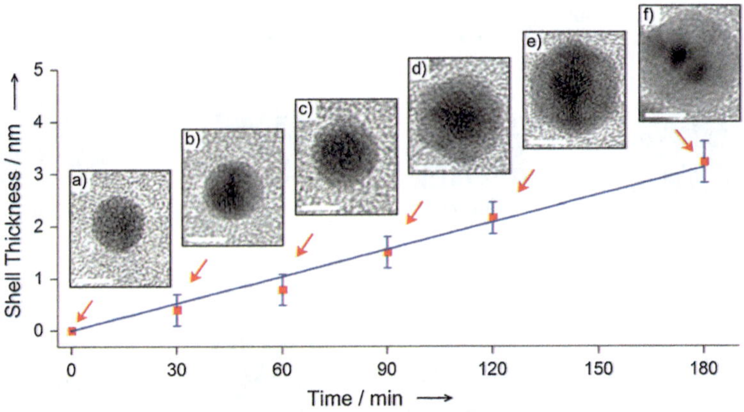

Fig. 8.14 Palladium shell thickness (nm) versus time (min) for the growth of faceted icosahedral Au–Pd core–shell nanocrystals. Time-resolved samples were obtained by quenching the reaction at 30, 60, 90, 120, and 180 min. The size distributions of each sample were measured and the shell thickness was calculated by subtracting the size of the gold seed. Standard size deviation is shown as error bars. Representative HRTEM images of each sample are shown in (**a**–**f**). Scale bars: 5 nm. Reprinted from Ref. [99] with permission from John Wiley and Sons

A substantial lattice mismatch between Au and Pd, oxidative etching in the presence of chloride and oxygen, the use of cetyltrimethylammonium chloride (CTAC) surfactant, and the reaction temperature (30–60 °C) were identified as key factors facilitating the formation of the THH core–shell nanocrystals. Interestingly, novel concave octahedral and octahedral Au@Pd nanocrystals were prepared by lowering the reaction temperature and prolonging the reaction time.

Despite the increasing success in preparing these types of materials, bimetallic core-shell high faceted nanocrystals with sizes less than 15 nm in size have proved challenging to synthesize. Indeed, electrocatalytic activity is optimized on nanoparticles in the sub 10 nm size range [98, 99]. Tilley and co-workers have recently achieved this preparing highly faceted, icosahedral Au–Pd core–shell nanocrystals, far smaller than any previously produced [100]. The control of the size and shape was achieved by the extremely slow growth of the palladium shell on the gold core, thus enabling precise, layer by layer, control of the shell thickness. Initially, highly faceted, monodisperse icosahedral gold nanocrystal seeds (7.1 ± 0.4 nm in size) were synthesized and then added to a precursor solution containing palladium acetylacetonate and hexadecylamine in toluene. Slow growth was achieved by reducing the reaction mixture under hydrogen at 60 °C. The measured palladium shell thicknesses were plotted against reaction time (Fig. 8.14), and the rate of palladium shell growth calculated as one atomic layer of palladium deposited every 13 min.

8.5 Other Pt and Pd Morphologies with High-Index Facets

Up till now we have discussed primarily single nanocrystal surface structure control. Apart from surface structure also the overall nanocrystal morphology is important in determining electrocatalytic properties. It is indeed a challenge to synthesis NCs with both well-defined surface structures and also morphology. In reality, we seek to control simultaneously; crystallinity, morphology, surface structure, particle size, and chemical composition, all of which contribute to the electrocatalytic performance of the resulting structures. Metal nanowires and nanorods are a class of NCs which have shown increasing promise for both the ORR and SOM oxidation reactions. Here we present a few recent examples of the preparation and characterization of Pt and Pd-based materials, which are of particle interest in fuel cell applications.

8.5.1 Pd, Au, and Pt Nanowire Arrays

Highly ordered Pt and Pd-based nanowire arrays (NWAs) with high surface to volume ratios have been shown to be highly electrocatalytically active especially for alcohol electrooxidation [101–103]. The nanowire array architecture can improve electrocatalytic performance through greater noble metal utilization efficiency and can also act as a template for deposition and also are excellent current collectors. Recently, the synergistic effect of Au–Pd bimetallic surfaces in Au-covered Pd and Pt nanowires for ethanol electrooxidation in alkaline media has been reported [104, 105]. Cherevko and co-workers prepared highly ordered Pt, Pd, and Au nanowire arrays using a home-made AAO (Anodic Aluminum Oxide) template by electrodeposition (Fig. 8.15) [105]. After which the AAO template was removed. Decoration of Pt and Pd on the as prepared Au nanowire array was achieved using chemical reduction with ascorbic acid. The ethanol electrooxidation activities of the decorated materials was several times greater than the Pd or Pt only arrays although the synergistic effect of the two metals was not determined but could be due to an up-shift in the energy of the d-states.

8.5.2 Bimetallic Platinum and Palladium-Based Nanowires

1D noble metal nanowires (NWs) have been shown recently to display structure-dependant enhancements in electrocatalytic activity for both the ORR and methanol oxidation reactions (MOR) [106–112]. Koenigsmann et al. have shown that Pt nanostructures with ultrathin one-dimensional (1-D) nanowire morphologies can be prepared that maintain elongated single crystalline segments with smooth crystal planes [106]. These structures possess proportionally *less* surface low coordinated atoms (LCAs) as compared with conventional nanoparticles. Despite

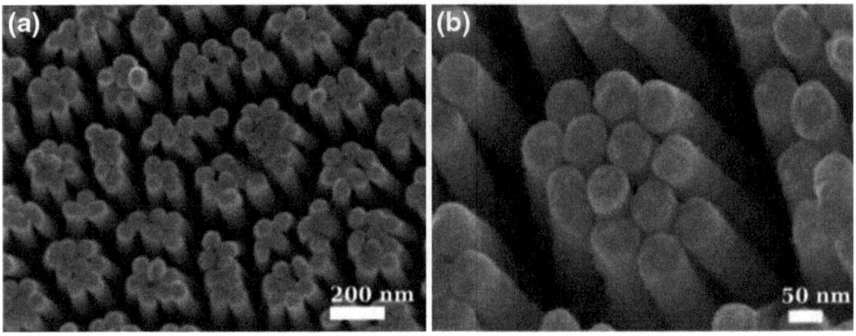

Fig. 8.15 SEM images of (**a**) Au and (**b**) $Pt_{30\ min}$/Au NWA electrodes. Reprinted from Ref. [104] with permission from Elsevier

having less LCAs, activities for methanol and ethanol electrooxidation are greatly enhanced [102, 108]. For example PtRu nanowire networks have been prepared and shown to have high activity for methanol electrooxidation [113]. The group of Adzic have prepared a range of ultrathin nanowire catalysts with Pt [108], PdAu [114], PdPt [115] compositions. In a recent report these researchers used a novel procedure utilizing a polycarbonate membrane to grow the nanowires without the need of surfactants [114]. In this way a series of bimetallic $Pd_{1-x}Au_x$ and $Pd_{1-x}Pt_x$ nanowires were prepared (Fig. 8.16). Two specific materials, Pd_9Au and Pd_4Pt, exhibited ORR activities much larger than analogous commercial Pt nanoparticles. The average wire diameters were in the order of 50 nm with lengths of up to 6 μm. HR-TEM analysis suggested that the long axis of the NWs is oriented along the (111) crystallographic direction. Core-shell Pt $\sim$ Pd_9Au nanowires were prepared by the Cu UPD/galvanic displacement deposition of a Pt monolayer. The enhanced activity of this material was accompanied by greatly improved stability over the 30,000 cycles of a durability test (cycling 0.6 to 1.0 V in 0.1 M $HClO_4$). HR-TEM showed no perceptible change in the structural integrity and the texture of the Pt $\sim$ Pd_9Au NWs after the stability test.

The morphology effect was also investigated by this group with respect to ethanol and CO oxidation in acidic solution [116]. Pt nanowires were found to have a two times higher catalytic activity with respect to nanoparticles despite having less LCAs. The enhance activity was likely to be due to a combination of terrace-step surface sites specific for EOR and efficient removal of chemisorbed CO from the nanowires.

8.5.3 Multiple Twinned Pt Nanorods

The electrochemical square wave potential method described in Sect. 8.4.2 has also been applied to form fivefold twinned Pt nanorods [24]. The growth of nanorods instead of NCs was achieved by exposing the glassy carbon electrode

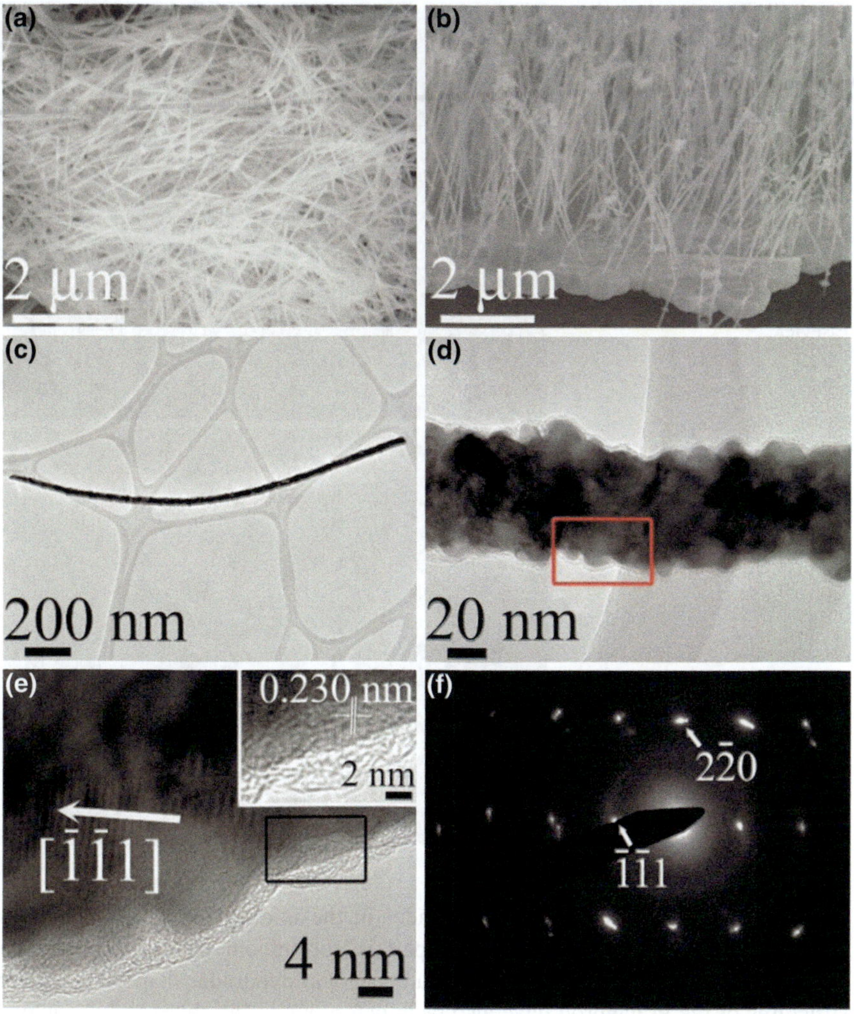

Fig. 8.16 Representative SEM images of the isolated Pd_9Au NWs (**a**) and of a free-standing NW array (**b**). TEM image of a single Pd_9Au NW (**c**) with a high-magnification image (**d**) highlighting the central region of the wire. The *red box* denotes where a high-resolution image (**e**) was obtained. *Inset* to panel **e** shows a selected area denoted by the *black box*, highlighting well resolved 111 lattice planes. Selected area electron diffraction pattern (**f**) corresponding to the images in (**d**) and (**e**) is shown. Reprinted with permission from Ref. [113]. Copyright (2010) American Chemical Society

loaded with Pt nanospheres to air for 3–5 h prior to treatment by the square wave potential. This air treatment made the GC surface inert so that new Pt nucleus would not be formed on the GC, instead they would form on the surface of the Pt nanospheres, thus growing into nanorods during the square wave potential treatment. The average length of the Pt nanorods is about 1 μm. The diameter is not

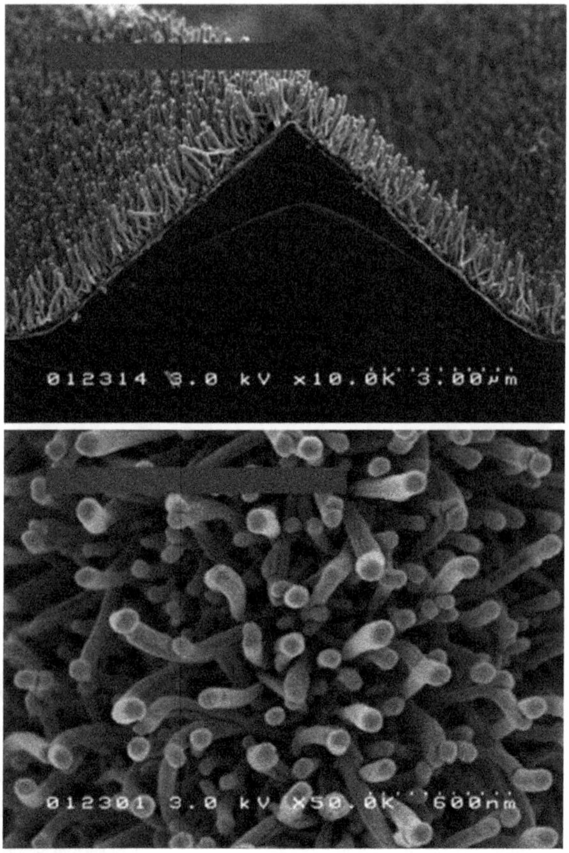

Fig. 8.17 Scanning electron micrographs of typical NSTF catalysts fabricated on a microstructured catalyst transfer substrate, (top) magnification of ×10,000, and (*bottom*) original magnification of ×50,000. The *dotted scale-bar* is shown in each micrograph. Reprinted from Ref. [116] with permission from Elsevier

uniform along the nanorod length, broader in the middle, gradually tapering to both ends. The surface of the middle part is not smooth with zigzag arranged facets. The ends are asymmetrical with a decagonal pyramidal shape. The surface structures of the Pt nanorod have also been determined. The sharp end is enclosed by {410} facets, the obtuse end by {320}, {210} or {730} facets, the middle part mainly by the zigzag-arranged {520} facets. The densities of stepped atoms on each facet are also not the same and range from 4 to 7 (10^{14} cm^{-2}).

8.5.4 Nanostructured Thin Film (NSTF) Catalysts

Pt-based NSTF ORR catalysts are to date the only practical catalyst that comes close to meeting the US DOE targets for automotive applications (see Chap. 4 for a detailed discussion) [117]. Produced by 3 M, the NSTF catalyst is strikingly different with respect to traditional carbon supported Pt nanoparticle catalysts

(Fig. 8.17) [118]. Once incorporated into an MEA, it contains neither carbon nor additional ionomer in the electrode layers that are 20–30 times thinner than conventional dispersed Pt/C-based MEAs. Figure 8.17 shows SEM images of the NSTF catalyst-coated whiskers on a transfer substrate, which has been fabricated by an all-dry continuous process. After which the NSTF coating is transferred onto the surfaces of a PEM to form a catalyst-coated membrane. The support structure consists of a thin monolayer of an oriented array of crystalline organic whiskers (<1 μm tall, 30–55 nm wide) which are non conducting and corrosion resistant. The Pt thin film is then applied by magnetron sputter coating. The advantages of this material are; (i) thin catalyst layer means better access to catalyst surface and lower volume, (ii) support corrosion resistance, (iii) high performance with ultralow PGM loadings, (iv) high-volume manufacturability, and (v) a three orders of magnitude lower release of membrane damaging free radicals produced during incomplete ORR.

References

1. C. Coutanceau, P. Urchaga, S. Brimaud, S. Baranton, Colloidal syntheses of shape- and size-controlled Pt nanoparticles for electrocatalysis. Electrocatalysis **3**, 75 (2012)
2. T. Olson, P. Hamill, A time-dependent approach to the kinetics of homogeneous nucleation. J. Chem. Phys. **104**, 210 (1996)
3. R. Van Hardeveld, F. Hartog, in *Advances in Catalysis,* vol. 22, ed. by D.D. Eley, H. Pines, P.B. Weisz (Academic, New York, 1972)
4. G.A. Somorjai, Surface science. Science **201**, 489 (1978)
5. F. Tao, M. Salmeron, In situ studies of chemistry and structure of materials in reactive environments. Science **331**, 171 (2011)
6. D.L. Feldheim, The new face of catalysis. Science **316**, 699 (2007)
7. Z.L. Wang, Transmission electron microscopy of shape-controlled nanocrystals and their assemblies. J. Phys. Chem. B **104**, 1153 (2000)
8. Y.N. Wen, H.M. Zhang, Surface energy calculation of the fcc metals by using the MAEAM. Solid State Commun. **144**, 163 (2007)
9. N.P. Lebedeva, M.T.M. Koper, J.M. Feliu, R.A. van Santen, Role of crystalline defects in electrocatalysis: mechanism and kinetics of CO adlayer oxidation on stepped platinum electrodes. J. Phys. Chem. B **106**, 12938 (2002)
10. S.G. Sun, A.C. Chen, T.S. Huang, J.B. Li, Z.W. Tian, Electrocatalytic properties of Pt(111), Pt(332), Pt(331) and Pt(110) single-crystal electrodes towards ethylene-glycol oxidation in sulfuric-acid-solutions. J. Electroanal. Chem. **340**, 213 (1992)
11. R.A. Van Santen, Complementary structure sensitive and insensitive catalytic relationships. Accounts Chem. Res. **42**, 57 (2009)
12. N. Tian, Z.Y. Zhou, S.G. Sun, Platinum metal catalysts of high-index surfaces: from single-crystal planes to electrochemically shape-controlled nanoparticles. J. Phys. Chem. C **112**, 19801 (2008)
13. M. Chen, B.H. Wu, J. Yang, N.F. Zheng, Small adsorbate-assisted shape control of Pd and Pt nanocrystals. Adv. Mater. **24**, 862 (2012)
14. J.F. Nicholas, *An Atlas of Models of Crystalline Surfaces* (Gordon & Breach, New York, 1965)
15. Z.Y. Zhou, N. Tian, J.T. Li, I. Broadwell, S.G. Sun, Nanomaterials of high surface energy with exceptional properties in catalysis and energy storage. Chem. Soc. Rev. **40**, 4167 (2011)

16. T. Ming et al., Growth of tetrahexahedral gold nanocrystals with high-index facets. J. Am. Chem. Soc. **131**, 16350 (2009)
17. M.L. Personick, M.R. Langille, J. Zhang, C.A. Mirkin, Shape control of gold nanoparticles by silver underpotential deposition. Nano Lett. **11**, 3394 (2011)
18. N. Tian, Z.Y. Zhou, S.G. Sun, Y. Ding, Z.L. Wang, Synthesis of tetrahexahedral platinum nanocrystals with high-index facets and high electro-oxidation activity. Science **316**, 732 (2007)
19. N. Tian, Z.Y. Zhou, N.F. Yu, L.Y. Wang, S.G. Sun, direct electrodeposition of tetrahexahedral Pd nanocrystals with high-index facets and high catalytic activity for ethanol electrooxidation. J. Am. Chem. Soc. **132**, 7580 (2010)
20. J.A. Zhang et al., Concave cubic gold nanocrystals with high-index facets. J. Am. Chem. Soc. **132**, 14012 (2010)
21. T.T. Tran, X.M. Lu, Synergistic effect of Ag and Pd ions on shape-selective growth of polyhedral Au nanocrystals with high-index facets. J. Phys. Chem. C **115**, 3638 (2011)
22. Y.Y. Ma et al., Synthesis of trisoctahedral gold nanocrystals with exposed high-index facets by a facile chemical method. Angew. Chem. Int. Edit. **47**, 8901 (2008)
23. X.Q. Huang, Z.P. Zhao, J.M. Fan, Y.M. Tan, N.F. Zheng, Amine-assisted synthesis of concave polyhedral platinum nanocrystals having {411} high-index facets. J. Am. Chem. Soc. **133**, 4718 (2011)
24. Z.Y. Zhou, N. Tian, Z.Z. Huang, D.J. Chen, S.G. Sun, Nanoparticle catalysts with high energy surfaces and enhanced activity synthesized by electrochemical method. Faraday Discuss. **140**, 81 (2008)
25. R. Narayanan, M.A. El-Sayed, Shape-dependent catalytic activity of platinum nanoparticles in colloidal solution. Nano Lett. **4**, 1343 (2004)
26. F.J. Vidal-Iglesias et al., Shape-dependent electrocatalysis: ammonia oxidation on platinum nanoparticles with preferential (100) surfaces. Electrochem. Commun. **6**, 1080 (2004)
27. E. Herrero, K. Franaszczuk, A. Wieckowski, Electrochemistry of methanol at low-index crystal planes of platinum—an integrated voltammetric and chronoamperometric study. J. Phys. Chem-US **98**, 5074 (1994)
28. J.W. Shin, C. Korzeniewski, Infrared spectroscopic detection of Co formed at step and terrace sites on a corrugated electrode surface plane during methanol oxidation. J. Phys. Chem-US **99**, 3419 (1995)
29. T.H.M. Housmans, M.T.M. Koper, Methanol oxidation on stepped Pt[n(111) x (110)] electrodes: a chronoamperometric study. J. Phys. Chem. B **107**, 8557 (2003)
30. T.H.M. Housmans, A.H. Wonders, M.T.M. Koper, Structure sensitivity of methanol electrooxidation pathways on platinum: An on-line electrochemical mass spectrometry study. J. Phys. Chem. B **110**, 10021 (2006)
31. S.C.S. Lai, N.P. Lebedeva, T.H.M. Housmans, M.T.M. Koper, Mechanisms of carbon monoxide and methanol oxidation at single-crystal electrodes. Top. Catal. **46**, 320 (2007)
32. A.V. Tripkovic, K.D. Popovic, Oxidation of methanol on platinum single crystal stepped electrodes from [110] zone in acid solution. Electrochim. Acta **41**, 2385 (1996)
33. D. Cao, G.Q. Lu, A. Wieckowski, S.A. Wasileski, M. Neurock, Mechanisms of methanol decomposition on platinum: a combined experimental and ab initio approach. J. Phys. Chem. B **109**, 11622 (2005)
34. S.G. Sun, Y.Y. Yang, Studies of kinetics of HCOOH oxidation on Pt(100), Pt(110), Pt(111), Pt(510) and Pt(911) single crystal electrodes. J. Electroanal. Chem. **467**, 121 (1999)
35. C.J. Sun Shigang, The electrocatalytic properties of platinum single crystal (210),(310) and (610) Stepped Surfaces in Formic Acid Oxidation. Chem. J. Chin. Univ. **11**, 998 (1990)
36. R.R. Adzic, A.V. Tripkovic, V.B. Vessovic, J. Electroanal. Chem. **204**, 329 (1986)
37. N. Hoshi, M. Nakamura, K. Kida, Structural effects on the oxidation of formic acid on the high index planes of palladium. Electrochem. Commun. **9**, 279 (2007)
38. N. Hoshi, K. Kida, M. Nakamura, M. Nakada, K. Osada, Structural effects of electrochemical oxidation of formic acid on single crystal electrodes of palladium. J. Phys. Chem. B **110**, 12480 (2006)

39. J.X. Wang, I.K. Robinson, B.M. Ocko, R.R. Adzic, Adsorbate-geometry specific subsurface relaxation in the CO/Pt(111) system. J. Phys. Chem. B **109**, 24 (2005)
40. M. Wakisaka, T. Ohkanda, T. Yoneyama, H. Uchida, M. Watanabe, Structures of a CO adlayer on a Pt(100) electrode in $HClO_4$ solution studied by in situ STM. Chem. Commun. **21**, 2710 (2005)
41. N.P. Lebedeva, M.T.M. Koper, E. Herrero, J.M. Feliu, R.A. van Santen, CO oxidation on stepped Pt[n(111) x (111)] electrodes. J. Electroanal. Chem. **487**, 37 (2000)
42. G. Garcia, M.T.M. Koper, Stripping voltammetry of carbon monoxide oxidation on stepped platinum single-crystal electrodes in alkaline solution. Phys. Chem. Chem. Phys. **10**, 3802 (2008)
43. G. Garcia, M.T.M. Koper, Carbon monoxide oxidation on Pt single crystal electrodes: understanding the catalysis for low temperature fuel cells. ChemPhysChem **12**, 2064 (2011)
44. K. Mikita, M. Nakamura, N. Hoshi, Situ infrared reflection absorption spectroscopy of carbon monoxide adsorbed on Pt(S)-[n(100)x(110)] electrodes. Langmuir **23**, 9092 (2007)
45. N.M. Markovic, H.A. Gasteiger, P.N. Ross, oxygen reduction on platinum low-index single-crystal surfaces in sulfuric-acid-solution—rotating ring-Pt(Hkl) disk studies. J. Phys. Chem-Us **99**, 3411 (1995)
46. M.D. Macia, J.M. Campina, E. Herrero, J.M. Feliu, On the kinetics of oxygen reduction on platinum stepped surfaces in acidic media. J. Electroanal. Chem. **564**, 141 (2004)
47. A. Kuzume, E. Herrero, J.M. Feliu, Oxygen reduction on stepped platinum surfaces in acidic media. J. Electroanal. Chem. **599**, 333 (2007)
48. A. Brouzgou, A. Podias, P. Tsiakaras, PEMFCs and AEMFCs directly fed with ethanol: a current status comparative review. J. Appl. Electrochem. **43**, 119 (2013)
49. C. Lamy, S. Rousseau, E.M. Belgsir, C. Coutanceau, J.M. Leger, Recent progress in the direct ethanol fuel cell: development of new platinum-tin electrocatalysts. Electrochim. Acta **49**, 3901 (2004)
50. S.Q. Song, P. Tsiakaras, Recent progress in direct ethanol proton exchange membrane fuel cells (DE-PEMFCs). Appl. Catal. B-Environ. **63**, 187 (2006)
51. S.Q. Song, V. Maragou, P. Tsiakaras, How far are direct alcohol fuel cells from our energy future? J. Fuel Cell Sci. Tech. **4**, 203 (2007)
52. A.S. Arico, P. Creti, P.L. Antonucci, V. Antonucci, Comparison of ethanol and methanol oxidation in a liquid-feed solid polymer electrolyte fuel cell at high temperature. Electrochem. Solid-State **1**, 66 (1998)
53. H. Luo et al., Conventional IVF and ICSI for sibling oocytes in couples with moderate oligoasthenoteratozoospermia: a prospective randomized study. Hum. Reprod. **22**, I163 (2007)
54. D.D. James, P.G. Pickup, Measurement of carbon dioxide yields for ethanol oxidation by operation of a direct ethanol fuel cell in crossover mode. Electrochim. Acta **78**, 274 (2012)
55. D. J. Tarnowski, C. Korzeniewski, Effects of surface step density on the electrochemical oxidation of ethanol to acetic acid. *J Phys Chem B* **101**, 253 (Jan 9, 1997)
56. S.G. Sun, Y. Lin, Kinetics of isopropanol oxidation on Pt(111), Pt(110), Pt(100), Pt(610) and Pt(211) single crystal electrodes—studies of in situ time-resolved FTIR spectroscopy. Electrochim. Acta **44**, 1153 (1998)
57. S.E. Habas, H. Lee, V. Radmilovic, G.A. Somorjai, P. Yang, Shaping binary metal nanocrystals through epitaxial seeded growth. Nat. Mater. **6**, 692 (2007)
58. A.R. Siekkinen, J.M. McLellan, J.Y. Chen, Y.N. Xia, Rapid synthesis of small silver nanocubes by mediating polyol reduction with a trace amount of sodium sulfide or sodium hydrosulfide. Chem. Phys. Lett. **432**, 491 (2006)
59. A.X. Yin, X.Q. Min, Y.W. Zhang, C.H. Yan, Shape-selective synthesis and facet-dependent enhanced electrocatalytic activity and durability of monodisperse sub-10 nm Pt-Pd tetrahedrons and cubes. J. Am. Chem. Soc. **133**, 3816 (2011)
60. A.X. Yin et al., Multiply twinned Pt-Pd nanoicosahedrons as highly active electrocatalysts for methanol oxidation. Chem. Commun. **48**, 543 (2012)

61. B.H. Wu, N.F. Zheng, G. Fu, Small molecules control the formation of Pt nanocrystals: a key role of carbon monoxide in the synthesis of Pt nanocubes. Chem. Commun. **47**, 1039 (2011)
62. G.X. Chen, Y.M. Tan, B.H. Wu, G. Fu, N.F. Zheng, Carbon monoxide-controlled synthesis of surface-clean Pt nanocubes with high electrocatalytic activity. Chem. Commun. **48**, 2758 (2012)
63. X.Q. Huang et al., Freestanding palladium nanosheets with plasmonic and catalytic properties. Nat. Nanotechnol. **6**, 28 (2011)
64. B.H. Wu, H.Q. Huang, J. Yang, N.F. Zheng, G. Fu, Selective hydrogenation of alpha, beta-unsaturated aldehydes catalyzed by amine-capped platinum-cobalt nanocrystals. Angew. Chem. Int. Edit. **51**, 3440 (2012)
65. J.B. Wu, A. Gross, H. Yang, Shape and composition-controlled platinum alloy nanocrystals using carbon monoxide as reducing agent. Nano Lett. **11**, 798 (2011)
66. J.B. Wu et al., Icosahedral platinum alloy nanocrystals with enhanced electrocatalytic activities. J. Am. Chem. Soc. **134**, 11880 (2012)
67. L.T. Qu et al., Controlled removal of individual carbon nanotubes from vertically aligned arrays for advanced nanoelectrodes. J. Mater. Chem. **20**, 3595 (2010)
68. M.P. Soriaga et al., Electrochemistry of the I-on-Pd single-crystal interface—studies by UHV-EC and in-situ STM. Surf. Sci. **335**, 273 (1995)
69. A. Carrasquillo, J.J. Jeng, R.J. Barriga, W.F. Temesghen, M.P. Soriaga, Electrode-surface coordination chemistry: Ligand substitution and competitive coordination of halides at well-defined Pd(100) and Pd(111) single crystals. Inorg. Chim. Acta **255**, 249 (1997)
70. Y.J. Xiong et al., Synthesis and mechanistic study of palladium nanobars and nanorods. J. Am. Chem. Soc. **129**, 3665 (2007)
71. X.Q. Huang, H.H. Zhang, C.Y. Guo, Z.Y. Zhou, N.F. Zheng, Simplifying the creation of hollow metallic nanostructures: one-pot synthesis of hollow palladium/platinum single-crystalline nanocubes. Angew. Chem. Int. Edit. **48**, 4808 (2009)
72. Y.W. Zhang et al., Highly selective synthesis of cataytically active monodisperse rhodium nanocubes. J. Am. Chem. Soc. **130**, 5868 (2008)
73. X.Q. Huang, S.H. Tang, H.H. Zhang, Z.Y. Zhou, N.F. Zheng, Controlled formation of concave tetrahedral/trigonal bipyramidal palladium nanocrystals. J. Am. Chem. Soc. **131**, 13916 (2009)
74. Z.Q. Niu, Q. Peng, M. Gong, H.P. Rong, Y.D. Li, Oleylamine-mediated shape evolution of palladium nanocrystals. Angew. Chem. Int. Edit. **50**, 6315 (2011)
75. M.S. Jin, H. Zhang, Z.X. Xie, Y.N. Xia, Palladium nanocrystals enclosed by {100} and {111} facets in controlled proportions and their catalytic activities for formic acid oxidation. Energ. Environ. Sci. **5**, 6352 (2012)
76. G. Ertl, J. Tornau, Z. Phys. Chem. **104**, 301 (1977)
77. Y. Ding et al., Facets and surface relaxation of tetrahexahedral platinum nanocrystals. Appl. Phys. Lett. **91**, (2007)
78. Z.Y. Zhou et al., High-index faceted platinum nanocrystals supported on carbon black as highly efficient catalysts for ethanol electrooxidation. Angew. Chem. Int. Edit. **49**, 411 (2010)
79. Z.Y. Zhou et al., Shape transformation from Pt nanocubes to tetrahexahedra with size near 10 nm. Electrochem. Commun. **22**, 61 (2012)
80. Y.X. Chen et al., Electrochemical milling and faceting: size reduction and catalytic activation of palladium nanoparticles. Angew. Chem. Int. Edit. **51**, 8500 (2012)
81. V. Bambagioni et al., Ethanol oxidation on electrocatalysts obtained by spontaneous deposition of palladium onto nickel-zinc materials. ChemSusChem **2**, 99 (2009)
82. L. Wang, M. Bevilacqua, Y-X. Chen, J. Filippi, M. Innocenti, A. Lavacchi, A. Marchionni, H. Miller, F. Vizza, Enhanced electro-oxidation of alcohols at electrochemically treated polycrystalline palladium surface. J. Power Sources **242**, 872 (2013)

83. Y. Zhang et al., Impaired stimulatory effect of ETB receptor on D-3 receptor in immortalized renal proximal tubule cells of spontaneously hypertensive rats. Kidney Blood Press. Res. **34**, 75 (2011)
84. F. Lu et al., Truncated ditetragonal gold prisms as nanofacet activators of catalytic platinum. J. Am. Chem. Soc. **133**, 18074 (2011)
85. C.W. Yang et al., Fabrication of Au-Pd core-shell heterostructures with systematic shape evolution using octahedral nanocrystal cores and their catalytic activity. J. Am. Chem. Soc. **133**, 19993 (2011)
86. B.I. Abelev et al., System size dependence of associated yields in hadron-triggered jets STAR collaboration. Phys. Lett. B **683**, 123 (2010)
87. B.I. Abelev et al., Center of mass energy and system-size dependence of photon production at forward rapidity at RHIC. Nucl. Phys. A **832**, 134 (2010)
88. F. Wang et al., Heteroepitaxial growth of high-index-faceted palladium nanoshells and their catalytic performance. J. Am. Chem. Soc. **133**, 1106 (2011)
89. Y.W. Lee et al., kinetically controlled growth of polyhedral bimetallic alloy nanocrystals exclusively bound by high-index facets: AuPd Hexoctahedra. Small **9**, 660 (2013)
90. M.A. Mahmoud, C.E. Tabor, M.A. El-Sayed, Y. Ding, Z.L. Wang, A new catalytically active colloidal platinum nanocatalyst: the multiarmed nanostar single crystal. J. Am. Chem. Soc. **130**, 4590 (2008)
91. M.J. Jiang et al., Epitaxial overgrowth of platinum on palladium nanocrystals. Nanoscale **2**, 2406 (2010)
92. H. Kobayashi et al., Seed-mediated synthesis of Pd-Rh bimetallic nanodendrites. Chem. Phys. Lett. **494**, 249 (2010)
93. L. Wang, Y. Yamauchi, Autoprogrammed synthesis of triple-layered Au@Pd@Pt core-shell nanoparticles consisting of a Au@Pd bimetallic core and nanoporous Pt shell. J. Am. Chem. Soc. **132**, 13636 (2010)
94. L. Wang, Y. Yamauchi, Strategic synthesis of trimetallic Au@Pd@Pt core-shell nanoparticles from Poly(vinylpyrrolidone)-based aqueous solution toward highly active electrocatalysts. Chem. Mater. **23**, 2457 (2011)
95. H. Zhang et al., Controlling the morphology of rhodium nanocrystals by manipulating the growth kinetics with a syringe pump. Nano Lett. **11**, 898 (2011)
96. B. Lim et al., Facile synthesis of bimetallic nanoplates consisting of Pd cores and Pt shells through seeded epitaxial growth. Nano Lett. **8**, 2535 (2008)
97. C.L. Lu, K.S. Prasad, H.L. Wu, J.A.A. Ho, M.H. Huang, Au nanocube-directed fabrication of Au-Pd core-shell nanocrystals with tetrahexahedral, concave octahedral, and octahedral structures and their electrocatalytic activity. J. Am. Chem. Soc. **132**, 14546 (2010)
98. A.T. Bell, The impact of nanoscience on heterogeneous catalysis. Science **299**, 1688 (2003)
99. M.H. Shao, A. Peles, K. Shoemaker, Electrocatalysis on platinum nanoparticles: particle size effect on oxygen reduction reaction activity. Nano Lett. **11**, 3714 (2011)
100. A.M. Henning et al., Gold-palladium core-shell nanocrystals with size and shape control optimized for catalytic performance. Angew. Chem. Int. Edit. **52**, 1477 (2013)
101. G.Y. Zhao, C.L. Xu, D.J. Guo, H. Li, H.L. Li, Template preparation of Pt-Ru and Pt nanowire array electrodes on a Ti/Si substrate for methanol electro-oxidation. J. Power Sources **162**, 492 (2006)
102. S.M. Choi, J.H. Kim, J.Y. Jung, E.Y. Yoon, W.B. Kim, Pt nanowires prepared via a polymer template method: its promise toward high Pt-loaded electrocatalysts for methanol oxidation. Electrochim. Acta **53**, 5804 (2008)
103. H. Wang, C.W. Xu, F.L. Cheng, S.P. Jiang, Pd nanowire arrays as electrocatalysts for ethanol electrooxidation. Electrochem. Commun. **9**, 1212 (2007)
104. F.L. Cheng et al., Synergistic effect of Pd-Au bimetallic surfaces in Au-covered Pd nanowires studied for ethanol oxidation. Electrochim. Acta **55**, 2295 (2010)
105. S. Cherevko, X.L. Xing, C.H. Chung, Pt and Pd decorated Au nanowires: extremely high activity of ethanol oxidation in alkaline media. Electrochim. Acta **56**, 5771 (2011)

106. C. Koenigsmann, W.P. Zhou, R.R. Adzic, E. Sutter, S.S. Wong, Size-dependent enhancement of electrocatalytic performance in relatively defect-free, processed ultrathin platinum nanowires. Nano Lett. **10**, 2806 (2010)
107. H.J. Zhou, W.P. Zhou, R.R. Adzic, S.S. Wong, Enhanced electrocatalytic performance of one-dimensional metal nanowires and arrays generated via an ambient, surfactantless synthesis. J. Phys. Chem. C **113**, 5460 (2009)
108. M. Ablikim et al., Observation of chi(cJ) decaying into the p(p)over-barK(+)K(-) final state. Phys. Rev. D **83**, (2011)
109. L. Xiao, L. Zhuang, Y. Liu, J.T. Lu, H.D. Abruna, Activating Pd by morphology tailoring for oxygen reduction. J. Am. Chem. Soc. **131**, 602 (2009)
110. H.W. Liang et al., A free-standing Pt-nanowire membrane as a highly stable electrocatalyst for the oxygen reduction reaction. Adv. Mater. **23**, 1467 (2011)
111. S.H. Sun, F. Jaouen, J.P. Dodelet, Controlled growth of Pt nanowires on carbon nanospheres and their enhanced performance as electrocatalysts in PEM fuel cells. Adv. Mater. **20**, 3900 (2008)
112. J.M. Patete et al., Viable methodologies for the synthesis of high-quality nanostructures. Green Chem. **13**, 482 (2011)
113. B. Li et al., Highly active Pt-Ru nanowire network catalysts for the methanol oxidation reaction. Catal. Commun. **18**, 51 (2012)
114. C. Koenigsmann, E. Sutter, T.A. Chiesa, R.R. Adzic, S.S. Wong, Highly enhanced electrocatalytic oxygen reduction performance observed in bimetallic palladium-based nanowires prepared under ambient, surfactantless conditions. Nano Lett. **12**, 2013 (2012)
115. C. Koenigsmann et al., Enhanced electrocatalytic performance of processed, ultrathin, supported Pd-Pt core-shell nanowire catalysts for the oxygen reduction reaction. J. Am. Chem. Soc. **133**, 9783 (2011)
116. W.P. Zhou et al., Morphology-dependent activity of Pt nanocatalysts for ethanol oxidation in acidic media: nanowires versus nanoparticles. Electrochim. Acta **56**, 9824 (2011)
117. M.K. Debe, Electrocatalyst approaches and challenges for automotive fuel cells. Nature **486**, 43 (2012)
118. M.K. Debe, A.K. Schmoeckel, G.D. Vernstrorn, R. Atanasoski, High voltage stability of nanostructured thin film catalysts for PEM fuel cells. J. Power Sources **161**, 1002 (2006)

Chapter 9
Monolayer Decorated Core Shell and Hollow Nanoparticles

9.1 Key Concepts

A general consensus now exists in the field of nanostructured electrocatalysts, that bimetallic catalysts are essential if they are to be employed successfully in electrolytic devices, primarily fuel cells and electrolyzers. Only in PEMFC anodes that operate on ultrapure H_2 can monometallic low loading Pt materials be practically used. In the case of fuel cells which are fed with alcohols or hydrogen containing impurities such as in the case of reformate gas (H_2 and CO), bimetallic catalysts need to be used. The most common objective is to reduce the affinity of the catalyst surface for poisoning species such as CO. On the cathode side there also exists the problem of atmospheric pollutants, e.g., SO_2, H_2S, NO, NO_2, and NH_3 that can be found in the air feed to the cathode in a PEMFC stack. The most common example of a bimetallic catalyst is the Pt–Ru alloy, although an incredibly large range of other combinations as well as ternary and quaternary catalyst formulations have been studied.

In addition the durability of existing platinum catalysts in working fuel cells is unsatisfactory [1]. There are various mechanisms that lead to the decomposition of platinum catalysts, which is equivalent in real terms to a loss of electrochemically active surface area of the platinum nanoparticles. For example, platinum nanoparticles can dissolve and redeposit on other Pt nanoparticles (this is known as Ostwald ripening and agglomeration, see Fig. 9.1) [2]. Platinum can deposit on the membrane by the reduction of Pt^{2+} by H_2 diffusing from the anode (hydrogen crossover) thus forming a band. These phenomena occur mostly during potential cycling that happens for example during stop and go driving of fuel cell cars [1].

The principle objective of research in this area involves primarily the cathode catalyst of a fuel cell, where O_2 is reduced (ORR) and where Pt is the most commonly used metal [3, 4]. The ORR reduction kinetics on Pt are sluggish, so the amount of Pt and as a consequence both the cathode catalyst cost and overall fuel cell stack costs are high [5]. The cost of the Pt-based catalyst is estimated to be around 22 % of the total system cost [1]. Thus, the high catalyst material cost of ~\$50 g_{Pt} means that the current drawn from every gram of Pt must be maximized

A. Lavacchi et al., *Nanotechnology in Electrocatalysis for Energy*,
Nanostructure Science and Technology 170, DOI: 10.1007/978-1-4899-8059-5_9,

© Springer Science+Business Media New York 2013

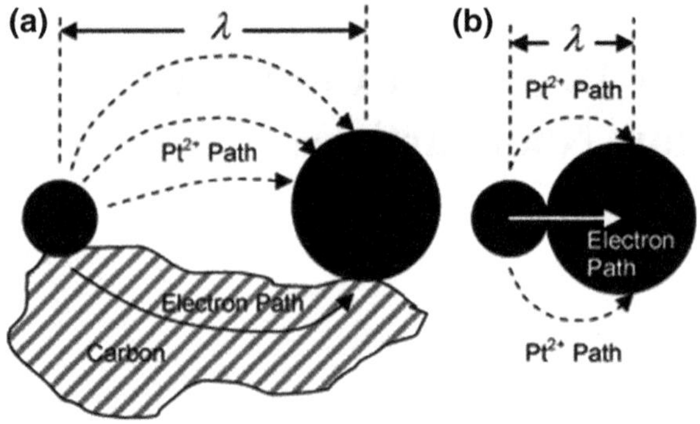

Fig. 9.1 Mechanism of particle growth by **a** Ostwald ripening and **b** agglomeration/sintering. Both occur by a coupled transport of Pt^{2+} (or Pt^{4+}) ions through ionomer/aqueous media and electron transport through the carbon support (or through direct particle to particle contact). Net Pt transport occurs from smaller particles to larger particles. Reprinted from reference [2] with permission of Elsevier

while simultaneously seeking to double current durability levels of around 2,000–3,000 h to 5,000 h [1]. Since it was discovered that alloying carbon supported platinum catalysts with 3D transition metals (e.g., Co, Cu, Ni, Cr) leads to a sharp increase in Pt mass activity, many researchers have endeavored to determine the origin of this activity increase at the same time seeking to improve the stability under working conditions. It is now clear that the Pt–M alloy is not stable under the harsh conditions of the PEMFC cathode where the non-noble metal is quickly dissolved. Researchers have generally followed two strategies; (i) they seek to create a monolayer Pt only shell on a non-PGM core or (ii) remove the non-PGM metal(s) to create a hollow Pt shell particle. In both cases the increase in ORR activity has been found to be closely related to the surface strain-induced d-band shifts created on the Pt shell that reduces the affinity of the Pt surface for chemisorbed O or OH that hinders the ORR. Mass activity improvements of up to a factor of five (based on total precious metal content) have been reported. Despite their monolayer coverage, these core–shell catalysts also exhibit reasonable durability under standard potential cycling conditions in liquid electrolytes. The challenge now for research in this area apart from continued improvements in activity and stability is the preparation of industrial scale quantities of these materials for future application in PEMFC stacks.

9.2 Core–Shell Nanoparticles

It is now well accepted that the alloying of nanostructured Pt catalysts with 3D transition metals such as Ni, Co, and Cr enhances considerably the ORR activity [6–13]. The enhanced activity has been attributed to a number of factors including electronic (ligand effects), geometric (Pt–Pt interatomic distance), particle size, and surface roughness effects [11, 14–16] and also a decrease in the surface Pt d-band occupancy [17]. It has also been shown that alloying Pt with 3D metals reduces the stability of chemisorbed O or OH groups thereby enhancing ORR activity [17], while calculations on small Pt clusters have confirmed the relationship between the ORR activity and the calculated O and OH binding strength [18]. A detailed relationship between the calculated oxygen-binding energy and the ORR activity was first introduced by Norskov et al. [19]. In a schematic kinetic analysis, the ORR activity was shown to be governed by the O–O dissociation rate, either *via* direct O_2 dissociation or via the formation of OHH, and by the proton/electron transfer rates to the adsorbed oxygen or hydroxide species for a given potential and pH.

Both rates depend differently on the oxygen-binding energy, leading to so-called Volcano behavior. The model developed by Nørskov et al. further suggests that the ORR activity of Pt can be improved by slightly reducing the oxygen-binding energy. Following this prediction, different approaches have been used to improve the ORR activity of Pt. Using the calculated oxygen-binding energy as a guide, Greeley et al. screened Pt-based alloys and reported that alloying Pt can improve the ORR activity by a factor up to 6–10 [20, 21].

Core–shell or core–shell-like nanostructures are a convenient way to build multifunctionality, higher activity, and stability in electrocatalysts based upon metal nanoparticles, which are generally very small (2 and 5 nm). In the case of core–shell catalysts, the core exists of one metal component or a multi-metal alloy that is surrounded by a skin of the active catalyst which can be a single element like Pt or consist of a mixed metal shell.

In principle, the synthesis of multimetallic core–shell nanoparticles or core–shell-like heterogeneous nanostructures is usually favored. The synthesis is possible as the heterogeneous nucleation of the second metal component on the existing nanoparticle seed or core has a lower critical energy barrier, that is, the overall excess free energy, than the homogenous nucleation. Depending on the overall excess energy, which is largely related to the surface and interfacial energy terms, and the strain energy because of lattice mismatch at the interface, three different major types of nanostructures form, namely, layer-by-layer, island-on-wetting layer, and island growth modes (Fig. 9.2) [22].

The recent advances in the aberration-corrected HAADF-STEM technique has allowed the imaging of spatial distributions of metal elements in previously unobtainable detail. Angstrom scale structural and compositional investigations can be undertaken on individual core shell nanoparticles. In combination with electron energy loss spectroscopy (EELS), the complex mapping of the fine structure of shell and core can be clearly elucidated. For example, Strasser and co-workers have carefully examined dealloyed Pt–Co nanoparticles with diameters of around 10–15 nm. The study revealed a complex three-dimensional fine structure including subsequent shells of Co depletion and enrichment (an example is shown in Fig. 9.3) [23].

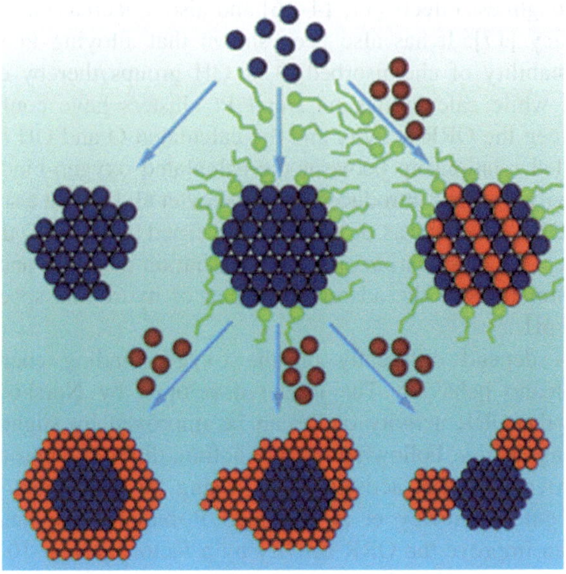

Fig. 9.2 Synthesis of multimetallic core–shell or core–shell-like nanoparticles from metal precursors (from reference [22]). Reprinted from reference [22] with permission from John Wiley and Sons

9.3 Synthesis of Platinum and Platinum Alloy Shells

The synthetic strategies that have been developed to build heterogeneous structures of bi or multi-metal core shell nanoparticles can be grouped into three main classes.

(1) Electrochemical methods which include both the underpotential deposition (UPD) and dealloying methods. When copper is used as the sacrificial layer, a Pt monolayer can be deposited on different metallic nanoparticles through the galvanic replacement reaction.
(2) In the second strategy, the electrochemical removal of more reactive metals from platinum-based alloys (dealloying) leads to a Pt-enriched shell.
(3) The third method generally involves first the preparation of a metal or alloy core nanoparticle of suitable dimensions. The desired shell is then obtained either by (i) annealing at high temperatures which results in a selective phase separation of Pt to the shell (ii) galvanic displacement of a non-noble metal from the surface of the core with Pt or (iii) the core is used as a seed onto which the shell is grown through various chemical deposition methods.

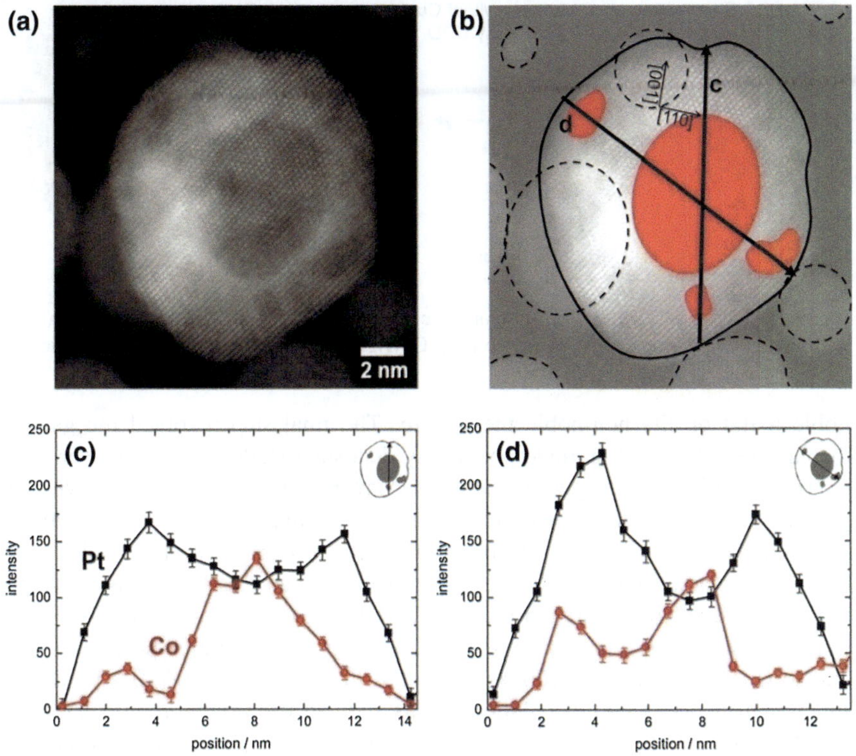

Fig. 9.3 **a** Ångstrom-scale resolution HAADF micrograph providing "Z contrast" conditions of a dealloyed Pt–Co bimetallic nanoparticle. **b** Co-enriched areas (*red areas*) are highlighted. **c** and **d** EELS intensity profiles across the particle for Pt (*black squares*) and Co (*red circles*), respectively. Reprinted from reference [23]. Copyright (2012) American Chemical Society

9.3.1 Underpotential Deposition (UPD) Replacement

Adzic and co-workers were the first to report a new class of multimetallic core shell ORR electrocatalysts based upon a monolayer of Pt deposited on metal or alloy carbon supported nanoparticles [15, 24, 25]. The Pt mass activities of these materials for the ORR were several times higher than commercial Pt/C catalysts. Various nanostructures of a carbon supported Pt monolayer on metal core have been prepared, e.g., Pd ($Pd@Pt_{ML}/C$), Pd_9Au_1 ($Pd_9Au_1@Pt_{ML}//C$), $AuNi_{10}$ ($AuNi_{10}@Pt_{ML}/C$), $AuCo_5$ ($AuCo_5@Pt_{ML}/C$) [26, 27]. These core–shell particles were prepared by UPD followed by replacement reactions (Fig. 9.4) [28]. The general procedure involves first the formation of mixed metal alloyed nanoparticles on a carbon support. Surface segregation of the noble metal component predicted by density functional theory (DFT) calculations, is achieved at high temperatures under flowing H_2 (600–850 °C). This forms a 1–2 monolayer of

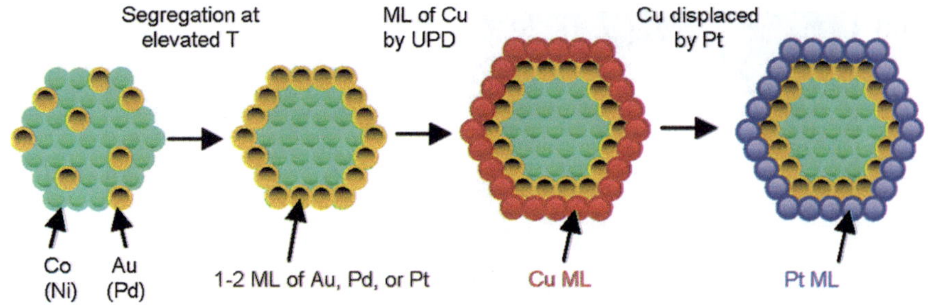

Fig. 9.4 Model of the synthesis of Pt monolayer catalysts on non-noble metal-noble metal core–shell particles. Reprinted from reference [28]. Cpyright (2005) American Chemical Society

noble metal on the non-noble metal core. The final step involved the galvanic displacement by Pt of an underpotentially deposited (UPD) Cu monolayer.

The resulting monolayer Pt catalysts showed remarkable mass activity and stability [27]. The enhancement in mass activity compared to commercial Pt/C for these catalysts ranges from 2.5 to 4 times with respect to the total noble metal content. Remarkably with only one single layer of Pt atoms, the Pt mass activity at 0.9 V (vs. reversible hydrogen electrode, RHE) was shown to retain 80 % of its initial value, that is, a drop from 0.30–0.24 A mg^{-1} after 60,000 potential cycles, and to 0.19 A mg^{-1} after 100,000 potential cycles between 0.7–0.9 V. Density functional theory (DFT) calculation results showed the structural changes within the Pd cores affected the stability of the Pt monolayer. Partial dissolution of Pd helps to improve the interaction between the Pt shells and Pd cores, thus resulting in better stability of the Pt surface atoms. By replacing Pd with a Pd_9Au_1 alloy as the core, the $Pd_9Au_1@Pt_{ML}/C$ catalysts had Pt mass activity of 0.20 A mg^{-1} after 200,000 potential cycles in an expanded testing range between 0.6 and 1.0 V. This final Pt mass activity is still higher than that of a freshly made Pt/C catalyst (ca. 0.13 A mg^{-1}) [27]. The authors assigned the improvement in activity to a combination of three factors [1, 29]; the underlying Pd induces a slight contraction of the atomic arrangement of the surface Pt causing the d-band center of the Pt monolayer to shift negatively. This causes a decrease in the Pt reactivity causing a weakening of the Pt–OH bond, reducing the coverage of OH or O that inhibit O_2 reduction (ii) formation of smooth services reducing the concentration of low coordination atoms that bind OH strongly and (iii) the so-called self healing mechanism of the Pt_{ML} catalysts whereby during Fuel Cell potential cycling partial dissolution of the Pd occurs, resulting in a small contraction of the Pt_{ML} shell, giving rise to a more stable structure with increased dissolution resistance and specific activity. HAADF images of isolated Pt monolayer shell on a Pd core nanoparticles are shown in Fig. 9.5.

More recently, the same authors have attempted to tune the catalytic activity of ML Pt on Ruthenium core–shell nanoparticles by varying the shell thickness [30]. The same UPD-based synthesis method described above was used to prepare 1, 2, and 3 ML Pt on Ru nanoparticles. The Pt mass specific activity showed the

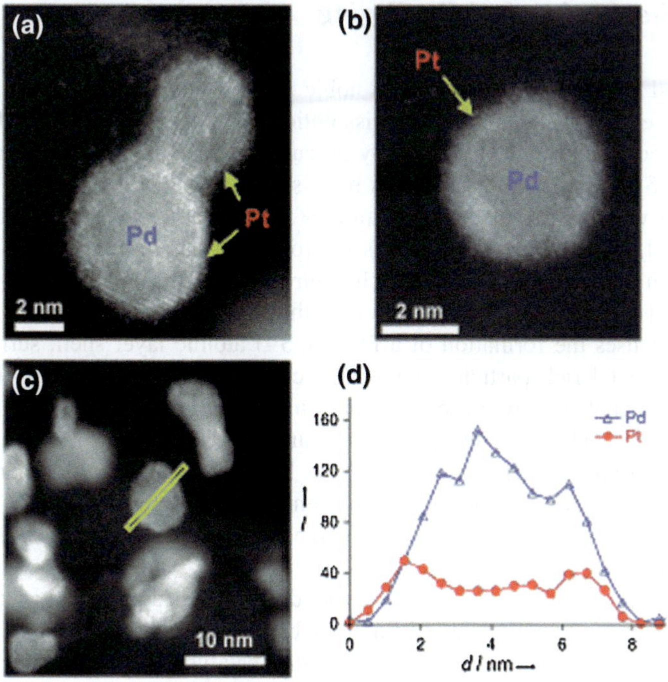

Fig. 9.5 **a–c** HAADF images of the sample of Pt monolayer shell on a Pd core nanoparticle, the $Pd@Pt_{ML}/C$ electrocatalyst. **d** Distribution of components in a $Pd@Pt_{ML}/C$ nanoparticle in **c** obtained by a line-scan analysis using EDS. Reprinted from reference [27] with permission from John Wiley and Sons

following trend; $Ru@Pt_{2ML}/C > Ru@Pt_{3ML}/C > Ru@Pt_{1ML}/C$ (Pt/C) > Pt/C ($Ru@Pt_{1ML}/C$). DFT calculations confirmed that two layers of Pt on Ru was optimal because O and OH hydrogenation are more facile than Pt; yet O–O bond scission kinetics are still fast.

Zhang and co-workers have recently confirmed the results of Adzic's group with similar Pdcore–Ptshell catalysts using a much simpler preparation method [31]. They made use of a commercial 20 wt % Pd/C (BASF) catalyst as substrate and Pt was deposited onto the Pd nanoparticles by a chemical reduction method. The thus obtained Pd@Pt/C catalysts were shown to have a Pdcore–Ptshell structure. The Pd@Pt/C catalysts were prepared in one step without further treatment and the nanoparticles had uniform dispersion on the carbon black support. In the half cell and single cell testing, the Pd@Pt/C catalysts exhibited better ORR activity and electrochemical durability than commercial Pt/C catalysts. Upon single cell cycling Pd dissolution was also observed and this led to improved stability and performance. The improved durability is believed to be associated with the dissolution of Pd and the corresponding structure transformation from core–shell structure to a Pt–Pd alloy with a Pt-rich surface.

9.3.2 Electrochemical Dealloying

Dealloyed Pt–M alloyed nanoporous noble metal catalysts form during the partial selective electrochemical surface dissolution of a less noble metal M from a homogeneous noble metal poor alloy precursor [32–34]. Pioneering work by the group of Strasser on these materials have shown that they exhibit improved catalytic activity for the electro-reduction of molecular oxygen at the cathode of PEMFCs [35–39]. The origin of this improvement has been shown to be due to changes in the geometric strain which controls the surface reactivity of bimetallic particles. [12, 32, 40]. The dissolution of the transition metal from the surface of the core causes the formation of a Pt-rich, 3–5 atomic layer shell, surrounding a transition metal rich particle core (single core–shell structure) [41]. This causes a lattice mismatch between the transition metal rich cores and Pt shells and as a consequence modifies the electronic structure of Pt leading to significantly altered catalytic activity. Experiments carried out by Strasser et al. with a series of dealloyed Pt nanoparticles confirms the presence of a compressive surface strain that leads to a weakened chemisorption of oxygen, which in turn improves the catalytic rate of the ORR [32, 42].

The Pt alloy precursors were prepared by a process which involved first impregnation, freeze drying, and finally subsequent reductive annealing [40, 43]. In the first step a commercial 30 wt % Pt/C catalyst was impregnated with an aqueous solution of the metal nitrate(s), followed by freeze drying in liquid N_2. The final step involved annealing in a tube furnace at 650–900 °C for 7 h in 4 vol. % H_2/96 vol. % Ar flow. Electrochemical dealloying was achieved by depositing this Pt precursor on the working electrode of a 3 electrode electrochemical cell. The electrochemical dealloying process consisted of three CV scans (0.06–1 V RHE) at 100 mV/s, followed by 200 fast scans at 500 mV/s, and finally three scans at 100 mV/s. The electrochemical dealloying of the Pt alloy precursors was also achieved in situ in a Fuel Cell MEA using cathode potential cycling (MEA conditions, $T_{\text{cell}} = 80$ °C, 100 % RH (H_2/N_2), 101.3 $\text{kPa}_{\text{(abs)}}$). [36, 44].

A wide range of binary, e.g., PtM_3 (M = Cu, Co, Ni) and ternary, e.g., $PtNi_3M$ (M = Cu, Co, Fe, Cr) dealloyed catalysts have been prepared and investigated by Strasser's group [39, 44, 45].

In order to understand better the relationship between the activity enhancement and the structural composition of these core–shell particles, these researchers carried out a combined HAADF STEM and EELS study of individual Pt–Co dealloyed core–shell nanoparticles with an average particle size of 10–15 nm [23]. The study revealed an unexpected compositional fine structure composed of spherical 1–2 nm wide Co depletion and enrichment zones in the shell. The authors rationalized the formation mechanism of the enrichment features through a combination of rapid selective Co surface electrochemical dissolution and vacancy injection, followed by an inverse Kirkendall effect of outward vacancy annihilation associated with opposite very short range Co segregation (Fig. 9.6) [23]. In more detail the authors describe the phenomenon represented in the figure as

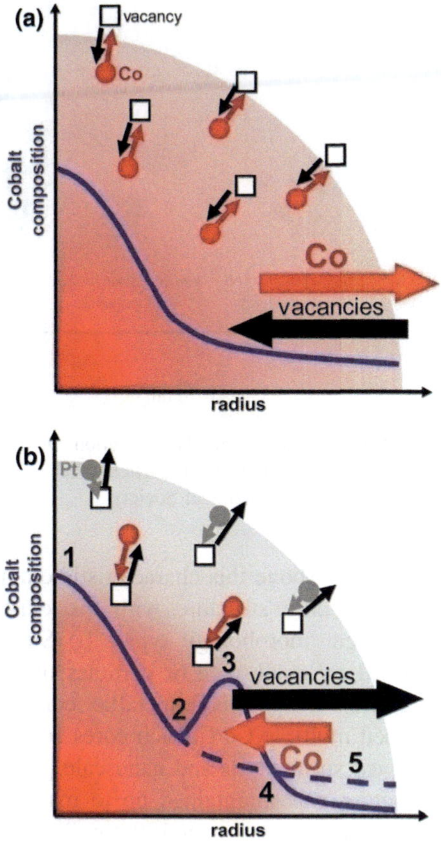

Fig. 9.6 **a** and **b** Schematic representation of the formation mechanism of the multishell dealloyed Pt-Co nanoparticles. A quarter of a spherical cross-section through a dealloyed Pt–Co particle with a Co-rich core (*red color* near origin) and Pt-rich shell (*gray color*) is shown as the background in both figures. The *solid blue line* represents the relative Co composition through the particle (reference [23]). Reprinted from reference [23] with permission from the American Chemical Society

follows: in Fig. 9.6a the blue solid line shows the early stage Co composition as a consequence of rapid initial surface dealloying, Co bulk downhill diffusion (red arrow), and associated with vacancy injection into the particle (black arrow). In Fig. 9.6b Pt surface diffusion acts as a vacancy sink and causes vacancy back migration toward the particle surface (inverse Kirkendall effect) followed by subsequent vacancy ejection/annihilation at the surface. The migration of vacancies is coupled to an opposite uphill diffusion of the faster diffusing Co atoms (red arrow) (inverse Kirkendall effect) inducing a sharp Co drop-off at position 4. The resulting Co compositional profile (solid blue curve) is consistent with the experimentally observed compositional multilayer structure involving adjacent Co maxima and minima.

In a more complex study, the same authors have studied the dependence of morphology and intraparticle composition on the initial particle size of dealloyed Pt–Co and Pt–Cu alloy nanoparticle catalysts [46]. This study showed the existence of a characteristic particle diameter (10–15 nm), which defines particle morphology. Smaller particles were found to usually have a simple core–shell

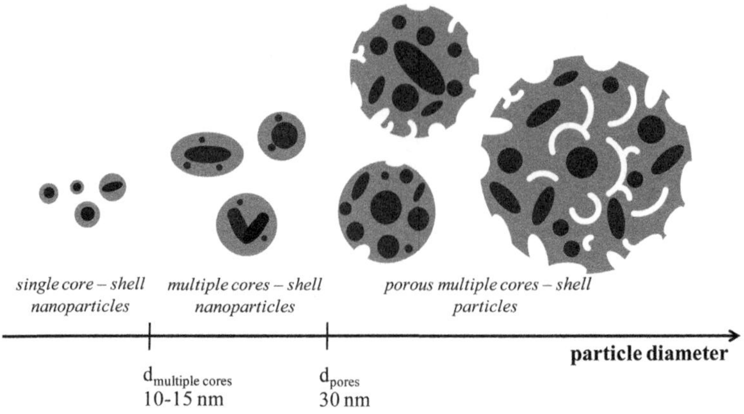

Fig. 9.7 Illustration of the evolution in size-dependent morphology and composition of dealloyed Pt–Cu and Pt–Co particle electrocatalysts. Reprinted from reference [46]. Copyright (2012) American Chemical Society

structure. Above this characteristic diameter, larger particles showed an irregular multiple core structure with the existence of three distinctly different size-dependent morphology types: (i) Single core–shell particles were exclusively formed by dealloying of particles of diameter below 10–15 nm (ii) Above this diameter nonporous bimetallic core–shell particles dominated with irregular shaped multiple Co/Cu rich cores and (iii) Particles with diameters above 30 nm showed surface pits and nanoscale pores next to multiple Co/Cu rich cores. This structure was maintained up to macroscopic bulk like dealloyed particles with diameters of more than 100 nm. This evolution in particle morphology is nicely summarized in Fig. 9.7.

In summary, of all the materials investigated the dealloyed $PtCu_3$ catalyst was shown to exceed the cathode catalyst activity targets in RDE experiments and real MEAs for automotive applications. The authors conclude that the significantly improved performance is due to the induced lattice strain in the Pt-rich shell. This strain significantly effects the chemisorption of oxygenated reactive intermediates and results in a fourfold to sixfold activity enhancement of the oxygen reduction activity. In a very recent report dealloyed Pd–Cu thin films also showed a twofold increase in ORR activity with respect to pure Pd films [47]. As with the analogous Pt–Cu films the improvement in activity was primarily due to the compressive strain in the surface layer. The thickness of the Pd shell was about 1.5 nm, larger than the Pt layer (1 nm). The thicker layer leads to a smaller compressive strain in the Pd layer and hence the improvement in ORR activity is less (two times) than Pt (2.4 times). Pd formed a thicker layer as more Cu is dissolved from the surface layer which results from faster diffusion of Pd than Pt to the surface during dealloying for the same solution.

9.3.3 Annealing and Stepwise Chemical Approaches

Recently, Volcano behavior of the ORR activity was nicely demonstrated for a series of model Cu/Pt(1 1 1) near-surface alloys with a varying subsurface Cu coverage. An eightfold variation in ORR activity was reported when the sub-monolayer concentration of Cu in Pt(1 1 1) was changed [48]. Saeys and co-workers looked to improve the activity of carbon supported PdPt-based ORR catalysts for direct methanol fuel cells, proposing to substitute the subsurface Pt atoms by 3D transition metal atoms and create core–shell structures. These authors first reported DFT calculations that show that the oxygen-binding energy for $Pd_3M@Pd_3Pt$ core–shell structures can be tuned [49], which is expected to enhance the ORR activity. A series of carbon supported PdM@PdPt (M = Pt, Ni, Co, Fe, and Cr) core–shell electrocatalysts were therefore prepared by a replacement reaction between PdM nanoparticles and an aqueous solution of $PtCl_4^{2-}$ and tested for their ORR activity. Optimal activity and high methanol tolerance were observed for the sample PdFe@PdPt/C [49].

Amongst the range of bimetallic Pt-based electrocatalysts that have been developed recently, considerable attention has been shown to Pt–Co alloy catalysts because of their relatively high activity and stability for ORR in acidic environments [9, 50, 51]. Abruna and co-workers have recently reported a new class of Pt–Co nanocatalysts composed of ordered Pt_3Co intermetallic cores with a 2–3 atomic layer thick platinum shell ($Pt_3Co@Pt$) [52]. The Pt shell is strained to the ordered intermetallic core, which has a 0.8 % smaller lattice constant than the disordered alloy. The simple two step synthetic procedure involved first in the preparation of carbon supported Pt_3Co nanoparticles by an impregnation-reduction method. The nanostructure of the carbon supported Pt_3Co nanoparticles (Pt_3Co/C) was controlled by a post-treatment of the as-prepared Pt_3Co/C, at different temperatures under a flowing H_2/N_2 mixed gas atmosphere. Treatment at 400 °C produced a Pt_3Co/C catalyst (Pt_3Co/C-400) with the same structure of traditional disordered Pt_3Co/C catalysts. The authors then used various techniques such as ADF-STEM to show that the chemical nanostructure of the material treated at 700 °C (Pt_3Co/C-700) showed a Pt-rich surface indicating that the Pt_3Co core is surrounded by a pure Pt shell with a shell thickness of about 0.5 nm, which is approximately 2–3 atomic layers (Fig. 9.8).

The durability of these catalysts was evaluated by potential cycling between +0.05 and +1.00 V for 5,000 cycles. The ECSA of Pt_3Co/C-400 reached a maximum value after 500 cycles, then decreased rapidly with further cycling. After 5,000 cycles, the ECSA had decayed by about 30 % of its maximum value. However, for the Pt_3Co/C-700 catalyst, the ECSA increased by about 20 % after 2,000 cycles, and then stabilized. The ordered core–shell structure was shown to have remained intact after the potential cycling tests. The same authors have used the same synthetic method to prepare carbon supported $Pd_3Co@Pd/C$ nanoparticles [53] which were shown to have an activity comparable to Pt/C but at the same time exhibited a much higher tolerance to methanol crossover. The electronic

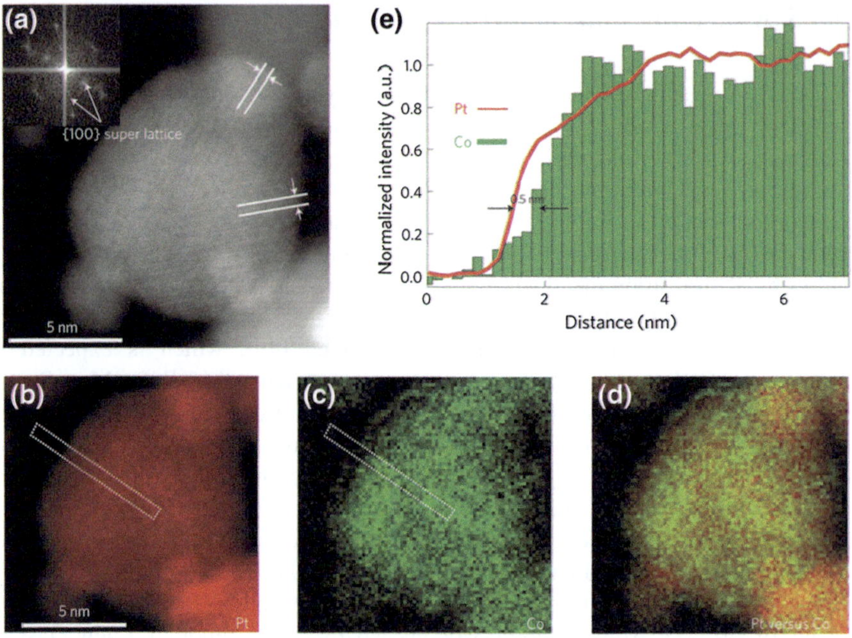

Fig. 9.8 **a** ADF-STEM image of a Pt_3Co/C-700 nanoparticle, with two parallel lines along with arrow marks indicating {100} lattice spacing. **b–d** 2D EELS maps of Pt (**b**), Co (**c**) and the composite Pt versus Co map (**d**). **e** Line profiles extracted from the boxed area in (**b**) and (**c**) across the facet showing that the Pt shell is ~0.5 nm thick. Reprinted from reference [52] with permission from Nature Publishing Group

properties of Pd were modulated by controlling the Co content in the alloy core. The electrocatalytic activity toward the ORR was greatly enhanced by the deposition of a further monolayer equivalent of Pt on the surface.

A unique approach has been developed by Sun and co-workers using a 5 nm Pd core and a FePt shell of which the thickness can be controlled from 1 to 3 nm [54]. The same authors have also reported a facile synthesis of sub 10 nm Pd/Au NPs having a gold shell (thickness 1–2 nm) [55]. These core–shell particles were used as seeds for coating with a further FePt shell to give quaternary Pd@Au@FePt NPs. The Pd@Au NPs were much more active for the ORR than Au NPs of a similar size and it was shown that activities were dependant on the thickness of the gold shell, with the thinnest (1 nm) shell exhibiting the highest activity [56].

The Pd@Au NPs were synthesized by controlled coating of Au over Pd NPs. In the synthesis, 5 nm Pd NPs were first prepared by the reduction of $[Pd(acac)_2]$ (acac = aceylacetonate) in the presence of oleylamine (OAm) and borane–tert-butylamine at 75 °C. These Pd NPs were then used as seeds, and gold coating on palladium was achieved by reducing $HAuCl_4$ with OAm at 80 °C in the presence of Pd NPs in 1-octadecene. These as-synthesized Pd@Au NPs were also used as seeds for FePt coating to give quaternary Pd@Au@FePt NPs, with Au and FePt forming two distinct metallic layers over the Pd NP core. The 2 nm FePt coating over the 5/1 nm Pd@Au NPs was achieved by controlled reduction of [Pt(acac)$_2$] and thermal decomposition of iron pentacarbonyl ([Fe(CO)$_5$] in the presence of OAm, oleic acid and Pd@Au NPs in 1-octadecene solvent.

Researchers at the state university of new York at Birmingham have investigated the preparation of the following core/shell nanoparticles (Au@Pt, Pt@Au, Fe_3O_4@Au@Pt) and have attempted to correlate the nanostructure with the electrocatalytic properties for both the methanol oxidation reaction (MOR) and the oxygen reduction reaction (ORR) [57]. The general approach to the synthesis of M1@M2 (M1 or M2: e.g., Au or Pt) nanoparticles was as follows: Briefly, for the synthesis of Au@Pt nanoparticles, Au nanoparticle cores encapsulated with tetraoctylammonium monolayer shells synthesized by a two-phase method were used as seeds for subsequent Pt coating by reaction with platinum acetylacetonate (Pt(acac)$_2$) in the presence of oleylamine using 1,2-hexadecanediol as a reducing agent. The thickness of the Pt or Au shells was adjusted by controlling the reaction temperature. The mass activity of the Fe_3O_4@Au@Pt sample was higher with respect to the other core shell materials for both ORR and MOR. The origin of the synergistic effect of the nano scale oxide core on the Au and Pt surface sites was however not determined.

Chen and co-workers have prepared various carbon supported Ni@Pt core–shell nanoparticles by the deposition of Pt onto Ni particles with an average particle size of 5 nm [58]. The thickness of the Pt shell was tuned by varying the amount of Pt in the synthesis. The activity and stability exhibited a volcano type of dependence on Pt coverage. A Pt/Ni atomic ratio of 3:10, which roughly corresponds to a monolayer of Pt on Ni nanoparticles was found to be the most optimized composition.

Surfaces with different strains can be achieved by depositing a pseudomorphic metal monolayer (e.g., Pt or Pd) onto substrates with different lattice constants, and from the preferential dissolution of the more reactive component from the Pt-based bimetallic alloy [59–61]. Simulation results reveal that heterogeneous Pt catalysts with a similar lattice parameter between the substrate and the Pt layer can achieve the highest catalytic activity [9, 60, 62].

Two recent examples have been provided by Ying and co-workers [63, 64]. These researchers prepared two types of bimetallic core particles AgPd and AuCu onto which a thin Pt shell was applied by reduction of Pt(acac)$_2$. Both examples show how the core composition can be tuned to reduce the tensile strain effect of the core atoms on the Pt layers. In one example Pt deposited onto a pure Au core which has a larger lattice parameter than Pt induced a tensile strain effect on the Pt shell, thus decreasing the electrocatalytic activity of Pt for ORR [64]. The addition of Cu lead to a dramatic increase in activity which was attributed to the compressive strain effect exerted by the AuCu core on the Pt shell. In more detail the authors explained the phenomenon as follows: The enhancement of ORR activity could be explained by the core–shell structure of the AuCu@Pt nanoparticles leading to lattice strain effect in catalysis. For the AuCu@Pt nanoparticles, the AuCu core has a slightly smaller lattice spacing ($d_{110} = 0.269$ nm) than the Pt shell ($d_{110} = 0.277$ nm), hence compressing the Pt atoms laterally in the shell as compared to their bulk form, thereby lowering their d-band centers. The down-shifts of the d-band centers and the associated general reduction in the adsorption strengths could inhibit oxygen adsorption, but facilitate the rate-limiting step of ORR, i.e., the removal of surface OH groups on the Pt shell in AuCu@Pt, thus improving the ORR activity. In contrast, the catalytic activity of the core–shell Au@Pt nanoparticles toward ORR was found to be much lower than that of the core–shell AuCu@Pt nanoparticles because Au has a much larger lattice parameter than Pt (0.408 vs. 0.392 nm). Consequently, tensile surface strain was induced when Pt was deposited on the Au core, causing up-shifts of the d-band center of Pt and increasing the adsorption strength, thereby hampering the removal of the OH groups and decreasing the ORR activity.

9.4 Non-Platinum Metal Shells

Non-platinum core shell particles are becoming of more interest in applications such as formic acid or direct alcohol fuel cells especially in alkaline conditions. Chen and co-workers have prepared Au@Pd core–shell nanostructures using a novel three-phase transfer procedure [65]. In the first step Au NPs with uniform size and morphology were prepared using a standard procedure [66]. The thin Pd shell was formed as follows; Pd^{2+} ions in the aqueous phase were transported by a phase transfer (sodium oleate) agent in the intermediate phase (EG) to the organic phase (oleic acid), where they are reduced to Pd atoms and homogeneously deposited on the surfaces of the Au NPs. The Au@Pd core–shell NPs were obtained with uniform and thin Pd shells (<1 nm). They showed remarkably high activity and stability for the electrooxidation of methanol in alkaline conditions. It was revealed that the upshift of the d-band center induced mainly by lattice expansion, accelerates the removal of the poisoning intermediates during the MOR and suppresses Pd oxidation.

The intrinsic challenges associated with the synthesis of multimetallic core–shell nanostructures make it very difficult to control the nanostructures when sizes are less than 10 nm. A recent report by Ozoemena and co-workers has shown for the first time the combination of top-down nanostructuring with precious metal decoration for core–shell particle formation [67]. A precursor FeCo@Fe/C material was prepared by annealing under H_2/Ar at 300–500 °C. These large sized (>210 nm) core–shell particles had Co concentrated at the core while the shell (approx. 7.2 nm) was predominantly Fe. A rapid solvothermal microwave reaction in the presence of a palladium salt and ethylene glycol lead to the downsizing of the core particles and decoration of Pd (particle sizes 3–7 nm). This material showed a much higher ECSA (314.80 $cm^2 mg^{-1}$) with respect to a Pd/C (29.47 $cm^2 mg^{-1}$) sample prepared under the same conditions.

9.5 Hollow Nanoparticles

As already discussed in the previous section greatly improved catalytic performance and stability of Pt for ORR has been obtained using bimetallic alloyed Pt–M/C nanostructured catalysts (M is a 3D transition metal like Co, Ni, or Fe). The increase in activity has been observed both in terms of mass activity (per g of Pt) and specific activity (per real cm^2 of Pt) [6, 7, 68–71]. Stamankovic and co-workers further improved the ORR activity by enriching the surface of the Pt_3M nanoparticles in Pt, by annealing at high temperatures (>800 °C) thus creating a Pt-skin structure [9, 14]. Alternatively, Pt skeleton structures are formed when the Pt surface enrichment arises from the dissolution of the 3D transition metal after the alloyed material comes in contact with an acidic environment [6]. In both cases the Pt atoms located on the outmost surface layer chemisorb oxygenated and spectator species more weakly than both pure Pt and the alloyed Pt_3M nanoparticles [6, 7]. Thermodynamic calculations (for the Pt_3Co particles) show there is a strong driving force for the segregation of Co atoms to the surface. Rossinot and co-workers showed through both theoretical calculations and experimental results that Co atoms contained in Pt_3Co/C nanoparticles segregate to the surface under long-term operation in a PEMFC and as a consequence the structure of the fresh Pt_3Co/C nanoparticles evolves toward core/shell particles having a Pt-enriched shell and a depleted Co core [72]. This is known as the Kirkendall effect which is a non reciprocal atomic diffusion process occurring through vacancy exchange in a diffusion couple [73–75]. The unbalanced atomic fluxes of the counter-flowing atoms with different diffusivities will be balanced by a compensating vacancy wind that blows in a direction opposite to the net mass flow [76]. In the example of Pt_3Co/C the leaching of Co from the surface together with the large coverage of oxygenated species during PEMFC operation means the diffusion of Co is much faster than Pt. Hence a flux of vacancies diffuses in the opposite direction to counterbalance the net Co flux. These vacancies aggregate into nanoscale voids and ultimately form a cavity located in the center of the initial nanoparticle. In a careful study Dubau et al. have shown that during PEMFC operation at high cathode potentials hollow spherical Pt nanoparticles are formed, which have thick Pt shells and Co content less than 5 % and that the intrinsic catalytic activity is better than that of the original alloyed electrocatalyst [72]. The improvement in performance is believed to be due to both (i) geometric effects like the large particles size of the hollow Pt nanoparticles >8 nm, which decreases the contribution of low coordination sites binding oxygenated species very strongly and (ii) the hollow-induced lattice contraction [72].

Since this discovery, several research groups have endeavored to reproduce the enhanced activity and stability of hollow core supported Pt monolayer catalysts through synthetic means. Various strategies have been developed that include (i) the kirkendall effect [77, 78] (ii) transmetallation [79, 80] (iii) galvanic replacement [81–83] and (iv) the electrochemical removal of core atoms.

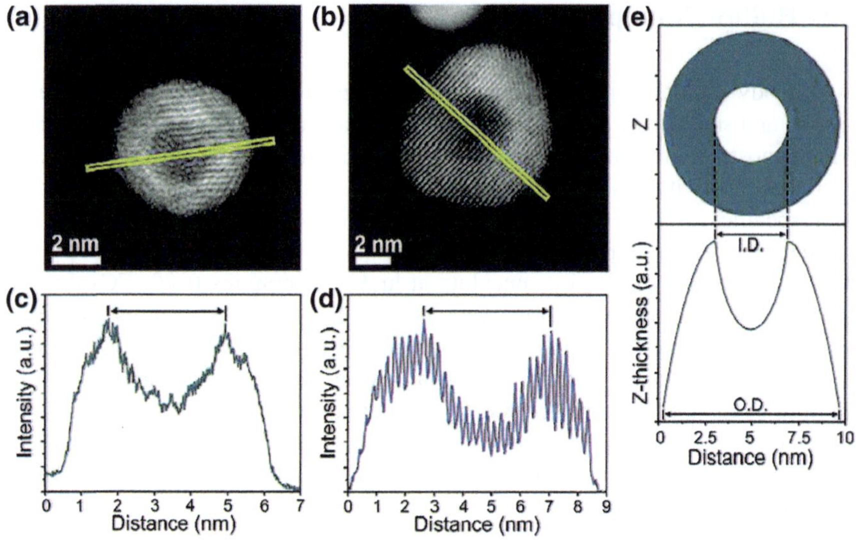

Fig. 9.9 **a** and **b** High-resolution STEM images of Pt hollow particles. **c** and **d** Line-scans of the intensity profile nearly parallel (**c**) and perpendicular (**d**) to the direction of the lattice plane. **e** Calculated z-thickness versus x-distance at the $y = 0$ position for a hollow sphere. Reprinted from reference [77]. Copyright (2011) American Chemical Society

The group of Adzic has recently developed two methods for the formation of hollow core supported Pt monolayer catalysts. The first strategy involved the electrodeposition of Ni nanoparticles on a carbon powder support [77]. A Pt shell was then deposited onto these Ni particles under an inert atmosphere by galvanic displacement. The Ni core was then completely removed by potential cycling in acidic solution. High-resolution STEM images confirmed the formation of compact, nonporous Pt hollow nanospheres (6.5 nm) (see Fig. 9.9). Favorable lattice contraction was achieved on the hollow nanoparticles resulting in increased ORR activity. In addition hollow nanoparticles have generally a larger particle size with respect to solid particles with the same mass, as a consequence increasing the amount of high coordination surface sites per mass of Platinum, essential for ORR activity and stability. Maintaining high ORR active area per mass of Pt with larger particle sizes 3–12 nm greatly reduces the problem of Pt dissolution inherent in small solid particles, especially below 5 nm.

Recently same researchers have prepared Pt monolayer shell on compact hollow Pd–Au core nanoparticles [78]. The synthetic procedure involved in the formation of Ni nanoparticles with average diameter of 9 nm by electrodeposition. Partial galvanic displacement of Ni atoms by Pd and Au ions yielded noble metal shells on nickel particles, which formed Pd–Au hollow particles upon dissolution of the remaining Ni in acidic solution at room temperature. The total metal mass activities for ORR were found to be 2.2 times that of a Pt monolayer on Pd core catalyst. The enhancement was scribed to the combination of smooth surface

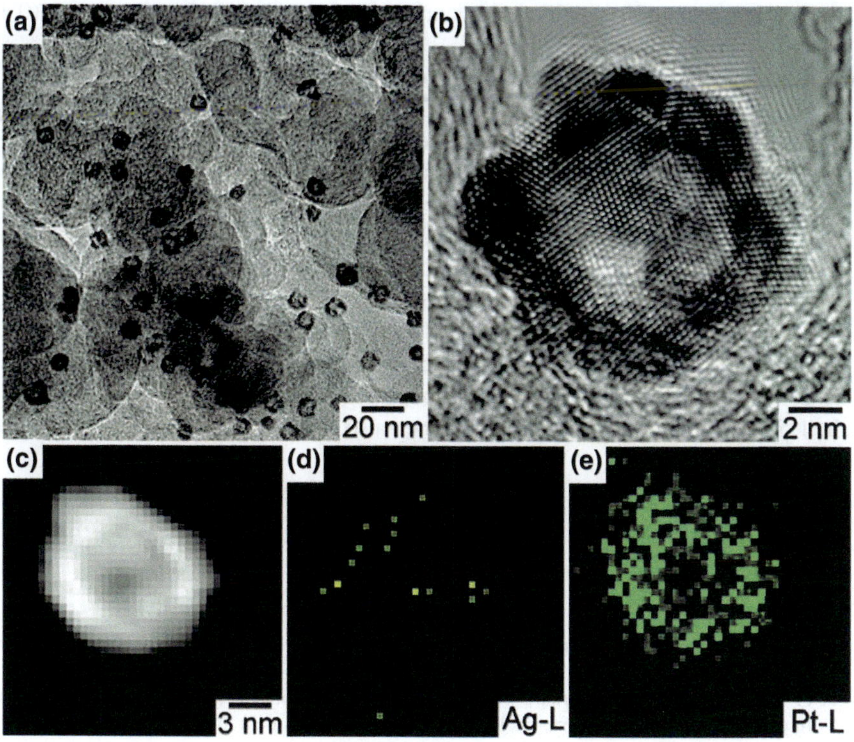

Fig. 9.10 Representative. **a** TEM **b** HR-TEM, and **c** STEM images and **d–e** Elemental maps of carbon supported Pt hollow structures made from carbon supported Pt on Ag nanoparticles. Reprinted from reference [88]. Copyright (2010) American Chemical Society

morphology and hollow-induced lattice contraction. Smooth surface reduces the site-blocking oxygen containing species and the compressive strain, induced by the lattice mismatch weakens oxygen binding causing the observed enhancement for the ORR on core–shell catalysts [32, 84, 85].

A facile one-pot preparation of carbon supported PtNi alloy hollow nanoparticles has been reported by Sung Jong Bae and co-workers [86]. The procedure involved the addition of a reducing agent ($NaBH_4$) to a mixture of the metal precursors $Pt(NH_3)_4Cl_2$ and $NiCl_2.6H_2O$ and Vulcan XC-72 in water. Aqueous HCL was continuously added to the reaction solution reaching 3 M prior to filtration. The formation of PtNi shell and hollow core was achieved using a platinum precursor with a relatively low standard reduction potential that slowed down the Pt reduction thus allowing the Ni metal cluster to form before the Pt galvanic reaction could occur. Alloy hollow nanoparticles with an average diameter of 8.3–8.5 nm (shell thickness 2.2–2.5 nm) were uniformly distributed over the carbon support. The measured ORR mass activity of 0.5 A per mg of Pt exceeds the activity target of fuel cell catalysts originally proposed by Gasteiger et al. (0.44 A per mg Pt @ 0.9 V RHE) [87].

A group of researchers from the University of Rochester New York have used a sacrificial Ag core to form hollow Pt nanoparticles [88]. The platinum hollow nanoparticles were obtained by selectively removing the silver cores from the Pt on Ag nanoparticles. First, platinum was deposited onto the surface of as prepared Ag nanoparticles. The Ag core was selectively removed under acidic conditions by electrochemical potential cycling (0–1.3 V RHE). The Ag nanoparticles had an average size of 9.5 nm and the 3 nm Pt nanoparticles nucleated and grew on the Ag surface. After removal of the Ag core the Pt hollow structure remained intact (see Fig. 9.10) and the carbon supported material exhibited a mass activity 3.22 A per g Pt (0.9 V RHE) twice that of a commercial carbon supported Pt catalyst. Other researchers have shown that hollow PtAg (0.15) nanoparticles (21 nm with shell 5 nm) prepared in a similar manner also show improved catalytic performance [83].

References

1. S.S. Kocha, *Electrochemical Degradation: Electrocatalyst and Support Durability* (Academic Press, Boston , 2012), p. 89
2. P. Parthasarathy, A.V. Virkar, Effect of stress on dissolution/precipitation of platinum: Implications concerning core–shell catalysts and cathode degradation in proton exchange membrane fuel cells. J. Power Sources **196**, 9204 (2011)
3. M. Watanabe, D.A. Tryk, M. Wakisaka, H. Yano, H. Uchida, Overview of recent developments in oxygen reduction electrocatalysis. Electrochim. Acta. **84**, 187 (2012)
4. J. Lee, B. Jeong, J.D. Ocon, Oxygen electrocatalysis in chemical energy conversion and storage technologies. Curr. Appl. Phys. **13**, 309 (2013)
5. A. Brouzgou, S.Q. Song, P. Tsiakaras, Low and non-platinum electrocatalysts for PEMFCs: current status, challenges and prospects. Appl. Catal. B-Environ. **127**, 371 (2012)
6. T. Toda, H. Igarashi, H. Uchida, M. Watanabe, Enhancement of the electroreduction of oxygen on Pt alloys with Fe, Ni, and Co. J. Electrochem. Soc. **146**, 3750 (1999)
7. U.A. Paulus et al., Oxygen reduction on high surface area Pt-based alloy catalysts in comparison to well defined smooth bulk alloy electrodes. Electrochim. Acta. **47**, 3787 (2002)
8. U.A. Paulus, T.J. Schmidt, H.A. Gasteiger, R.J. Behm, Oxygen reduction on a high-surface area Pt/Vulcan carbon catalyst: a thin-film rotating ring-disk electrode study. J. Electroanal. Chem. **495**, 134 (2001)
9. V.R. Stamenkovic et al., Trends in electrocatalysis on extended and nanoscale Pt-bimetallic alloy surfaces. Nat. Mater. **6**, 241 (2007)
10. S. Koh, N. Hahn, C.F. Yu, P. Strasser, Effects of composition and annealing conditions on catalytic activities of dealloyed Pt-Cu nanoparticle electrocatalysts for PEMFC. J. Electrochem. Soc. **155**, B1281 (2008)
11. S. Chen, H.A. Gasteiger, K. Hayakawa, T. Tada, Y. Shao-Horn, Platinum-alloy cathode catalyst degradation in proton exchange membrane fuel cells: nanometer-scale compositional and morphological changes. J. Electrochem. Soc. **157**, A82 (2010)
12. S. Koh, P. Strasser, Electrocatalysis on bimetallic surfaces: Modifying catalytic reactivity for oxygen reduction by voltammetric surface dealloying. J. Am. Chem. Soc. **129**, 12624 (2007)
13. Z.C. Liu, S. Koh, C.F. Yu, P. Strasser, Synthesis, dealloying, and ORR electrocatalysis of PDDA-stabilized Cu-rich Pt alloy nanoparticles. J. Electrochem. Soc. **154**, B1192 (2007)

14. V.R. Stamenkovic, B.S. Mun, K.J.J. Mayrhofer, P.N. Ross, N.M. Markovic, Effect of surface composition on electronic structure, stability, and electrocatalytic properties of Pt-transition metal alloys: Pt-skin versus Pt-skeleton surfaces. J. Am. Chem. Soc. **128**, 8813 (2006)
15. R.R. Adzic et al., Platinum monolayer fuel cell electrocatalysts. Top. Catal. **46**, 249 (2007)
16. M. Watanabe, K. Tsurumi, T. Mizukami, T. Nakamura, P. Stonehart, Activity and stability of ordered and disordered Co-Pt Alloys for phosphoric-acid fuel-cells. J. Electrochem. Soc. **141**, 2659 (1994)
17. S. Mukerjee, S. Srinivasan, M.P. Soriaga, J. Mcbreen, Role of Structural and Electronic-Properties of Pt and Pt Alloys on electrocatalysis of oxygen reduction—an in-situ xanes and exafs investigation. J. Electrochem. Soc. **142**, 1409 (1995)
18. A.B. Anderson, O-2 reduction and CO oxidation at the Pt-electrolyte interface. The role of H2O and OH adsorption bond strengths. Electrochim. Acta. **47**, 3759 (2002)
19. J.K. Norskov, T. Bligaard, J. Rossmeisl, C.H. Christensen, Towards the computational design of solid catalysts. Nat. Chem. **1**, 37 (2009)
20. J. Greeley et al., Alloys of platinum and early transition metals as oxygen reduction electrocatalysts. Nat. Chem. **1**, 552 (2009)
21. I.E.L. Stephens, A.S. Bondarenko, U. Gronbjerg, J. Rossmeisl, I. Chorkendorff, Understanding the electrocatalysis of oxygen reduction on platinum and its alloys. Energ. Environ. Sci. **5**, 6744 (2012)
22. H. Yang, Platinum-based electrocatalysts with core–shell nanostructures. Angew. Chem. Int. Edit. **50**, 2674 (2011)
23. M. Heggen, M. Oezaslan, L. Houben, P. Strasser, Formation and analysis of core–shell fine structures in Pt bimetallic nanoparticle fuel cell electrocatalysts. J. Phys. Chem. C **116**, 19073 (2012)
24. J. Zhang et al., Platinum monolayer electrocatalysts for O-2 reduction: Pt monolayer on Pd (111) and on carbon-supported Pd nanoparticles. J. Phys. Chem. B **108**, 10955 (2004)
25. J. Zhang, M.B. Vukmirovic, K. Sasaki, F. Uribe, R.R. Adzic, Platinum monolayer electro catalysts for oxygen reduction: effect of substrates, and long-term stability. J. Serb. Chem. Soc. **70**, 513 (2005)
26. K. Sasaki et al., Recent advances in platinum monolayer electrocatalysts for oxygen reduction reaction: scale-up synthesis, structure and activity of Pt shells on Pd cores. Electrochim. Acta. **55**, 2645 (2010)
27. K. Sasaki et al., Core-protected platinum monolayer shell high-stability electrocatalysts for fuel-cell cathodes. Angew. Chem. Int. Edit. **49**, 8602 (2010)
28. J. Zhang et al., Platinum monolayer on nonnoble metal-noble metal core–shell nanoparticle electrocatalysts for O-2 reduction. J. Phys. Chem. B **109**, 22701 (2005)
29. R. Adzic, Platinum monolayer electrocatalysts: tunable activity, stability, and self-healing properties. Electrocatalysis **3**, 163 (2012)
30. L.J. Yang et al., Tuning the catalytic activity of Ru@Pt core–shell nanoparticles for the oxygen reduction reaction by varying the shell thickness. J Phys Chem C **117**, 1748 (2013)
31. G. Zhang, Z. Shao, W. Lu, F. Xie, H. Xiao, X. Qin, B. Yi, Core–shell Pt modified Pd/C as an active and durable electrocatalyst for the oxygen reduction reaction in PEMFCs. Appl. Catal. B **132–133**, 183 (2013)
32. P. Strasser et al., Lattice-strain control of the activity in dealloyed core–shell fuel cell catalysts. Nat. Chem. **2**, 454 (2010)
33. J. Snyder, I. McCue, K. Livi, J. Erlebacher, Structure/processing/properties relationships in nanoporous nanoparticles as applied to catalysis of the cathodic oxygen reduction reaction. J. Am. Chem. Soc. **134**, 8633 (2012)
34. R.Z. Yang, P. Strasser, M.F. Toney, Dealloying of Cu3Pt (111) studied by surface x-ray scattering. J. Phys. Chem. C **115**, 9074 (2011)
35. F. Hasche, M. Oezaslan, P. Strasser, Activity, structure and degradation of dealloyed PtNi3 nanoparticle electrocatalyst for the oxygen reduction reaction in PEMFC. J. Electrochem. Soc. **159**, B25 (2012)

36. R. Srivastava, P. Mani, P. Strasser, In situ voltammetric de-alloying of fuel cell catalyst electrode layer: a combined scanning electron microscope/electron probe micro-analysis study. J. Power Sources **190**, 40 (2009)
37. R. Srivastava, P. Mani, N. Hahn, P. Strasser, Efficient oxygen reduction fuel cell electrocatalysis on voltammetrically dealloyed Pt-Cu-Co nanoparticles. Angew. Chem. Int. Edit. **46**, 8988 (2007)
38. M. Oezaslan, F. Hasche, P. Strasser, PtCu3, PtCu and Pt3Cu alloy nanoparticle electrocatalysts for oxygen reduction reaction in alkaline and acidic media. J. Electrochem. Soc. **159**, B444 (2012)
39. M. Oezaslan, F. Hasche, P. Strasser, Oxygen electroreduction on PtCo3, PtCo and Pt3Co alloy nanoparticles for alkaline and acidic PEM fuel cells. J. Electrochem. Soc. **159**, B394 (2012)
40. P. Strasser, Dealloyed core–shell fuel cell electrocatalysts. Rev. Chem. Eng. **25**, 255 (2009)
41. C.F. Yu, S. Koh, J.E. Leisch, M.F. Toney, P. Strasser, Size and composition distribution dynamics of alloy nanoparticle electrocatalysts probed by anomalous small angle X-ray scattering (ASAXS). Faraday Discuss. **140**, 283 (2008)
42. T. Anniyev et al., Complementarity between high-energy photoelectron and L-edge spectroscopy for probing the electronic structure of 5d transition metal catalysts. Phys. Chem. Chem. Phys. **12**, 5694 (2010)
43. Z.C. Liu, C.F. Yu, I.A. Rusakova, D.X. Huang, P. Strasser, Synthesis of Pt(3)Co alloy nanocatalyst via reverse Micelle for oxygen reduction reaction in PEMFCs. Top. Catal. **49**, 241 (2008)
44. P. Mani, R. Srivastava, P. Strasser, Dealloyed binary PtM3 (M = Cu, Co, Ni) and ternary PtNi3 M (M = Cu, Co, Fe, Cr) electrocatalysts for the oxygen reduction reaction: Performance in polymer electrolyte membrane fuel cells. J. Power Sources **196**, 666 (2011)
45. M. Oezaslan, P. Strasser, Activity of dealloyed PtCo3 and PtCu3 nanoparticle electrocatalyst for oxygen reduction reaction in polymer electrolyte membrane fuel cell. J. Power Sources **196**, 5240 (2011)
46. M. Oezaslan, M. Heggen, P. Strasser, Size-dependent morphology of dealloyed bimetallic catalysts: linking the nano to the macro scale. J. Am. Chem. Soc. **134**, 514 (2012)
47. R.Z. Yang, W.Y. Bian, P. Strasser, M.F. Toney, Dealloyed PdCu3 thin film electrocatalysts for oxygen reduction reaction. J. Power Sources **222**, 169 (2013)
48. I.E.L. Stephens et al., Tuning the activity of Pt(111) for oxygen electroreduction by subsurface alloying. J. Am. Chem. Soc. **133**, 5485 (2011)
49. Q.T. Trinh, J.H. Yang, J.Y. Lee, M. Saeys, Computational and experimental study of the Volcano behavior of the oxygen reduction activity of PdM@PdPt/C (M = Pt, Ni, Co, Fe, and Cr) core–shell electrocatalysts. J. Catal. **291**, 26 (2012)
50. E. Antolini, J.R.C. Salgado, E.R. Gonzalez, The stability of Pt–M (M = first row transition metal) alloy catalysts and its effect on the activity in low temperature fuel cells - A literature review and tests on a Pt–Co catalyst. J. Power Sources **160**, 957 (2006)
51. X.G. Li, H.R. Colon-Mercado, G. Wu, J.W. Lee, B.N. Popov, Development of method for synthesis of Pt-Co cathode catalysts for PEM fuel cells. Electrochem Solid St **10**, B201 (2007)
52. D.L. Wang et al., Structurally ordered intermetallic platinum-cobalt core–shell nanoparticles with enhanced activity and stability as oxygen reduction electrocatalysts. Nat. Mater. **12**, 81 (2013)
53. D.L. Wang et al., Facile synthesis of carbon-supported Pd-Co core–shell nanoparticles as oxygen reduction electrocatalysts and their enhanced activity and stability with monolayer Pt decoration. Chem. Mater. **24**, 2274 (2012)
54. V. Mazumder, M.F. Chi, K.L. More, S.H. Sun, Core/shell Pd/FePt nanoparticles as an active and durable catalyst for the oxygen reduction reaction. J. Am. Chem. Soc. **132**, 7848 (2010)
55. V. Mazumder, M.F. Chi, K.L. More, S.H. Sun, Synthesis and characterization of multimetallic Pd/Au and Pd/Au/FePt core/shell nanoparticles. Angew. Chem. Int. Edit. **49**, 9368 (2010)

56. Y. Lee, A. Loew, S.H. Sun, Surface- and structure-dependent catalytic activity of Au nanoparticles for oxygen reduction reaction. Chem. Mater. **22**, 755 (2010)
57. J. Luo et al., Core/shell nanoparticles as electrocatalysts for fuel cell reactions. Adv. Mater. **20**, 4342 (2008)
58. Y.M. Chen, Z.X. Liang, F. Yang, Y.W. Liu, S.L. Chen, Ni-Pt core–shell nanoparticles as oxygen reduction electrocatalysts: effect of Pt shell coverage. J. Phys. Chem. C **115**, 24073 (2011)
59. L.A. Kibler, A.M. El-Aziz, R. Hoyer, D.M. Kolb, Tuning reaction rates by lateral strain in a palladium monolayer. Angew. Chem. Int. Edit. **44**, 2080 (2005)
60. J.L. Zhang, M.B. Vukmirovic, Y. Xu, M. Mavrikakis, R.R. Adzic, Controlling the catalytic activity of platinum-monolayer electrocatalysts for oxygen reduction with different substrates. Angew. Chem. Int. Edit. **44**, 2132 (2005)
61. P. Strasser, S. Koha, J. Greeley, Voltammetric surface dealloying of Pt bimetallic nanoparticles: an experimental and DFT computational analysis. Phys. Chem. Chem. Phys. **10**, 3670 (2008)
62. Z. Peng, H. Yang, Synthesis and oxygen reduction electrocatalytic property of Pt-on-Pd bimetallic hetero nanostructures. J. Am. Chem. Soc. **131**, 7542 (2009)
63. J.H. Yang, J. Yang, J.Y. Ying, Morphology and lateral strain control of Pt nanoparticles via core–shell Construction using alloy AgPd core toward oxygen reduction reaction. ACS Nano. **6**, 9373 (2012)
64. J.H. Yang, X.J. Chen, X.F. Yang, J.Y. Ying, Stabilization and compressive strain effect of AuCu core on Pt shell for oxygen reduction reaction. Energ. Environ. Sci. **5**, 8976 (2012)
65. Q. Tan et al., Highly efficient and stable nonplatinum anode catalyst with Au@Pd core–shell nanostructures for methanol electrooxidation. J. Catal. **295**, 217 (2012)
66. J.T. Zhang, Y. Tang, L. Weng, M. Ouyang, Versatile strategy for precisely tailored core@shell nanostructures with single shell layer accuracy: the case of metallic shell. Nano. Lett. **9**, 4061 (2009)
67. O.O. Fashedemi, , B. Julies, K.I. Ozoemena, Synthesis of Pd-coated FeCo@Fe/C core–shell nanoparticles: microwave-induced 'top-down' nanostructuring and decoration. Chem. Commun. **49**, 2034 (2013)
68. F. Maillard, M. Martin, F. Gloaguen, J.M. Leger, Oxygen electroreduction on carbon-supported platinum catalysts. Particle-size effect on the tolerance to methanol competition. Electrochim. Acta. **47**, 3431 (2002)
69. S. Chen et al., Origin of oxygen reduction reaction activity on "Pt3Co" Nanoparticles: atomically resolved chemical compositions and structures. J. Phys. Chem. C **113**, 1109 (2009)
70. S. Chen et al., Enhanced activity for oxygen reduction reaction on "Pt(3)CO" nanoparticles: Direct evidence of percolated and sandwich-segregation structures. J. Am. Chem. Soc. **130**, 13818 (2008)
71. L. Dubau, F. Maillard, M. Chatenet, J. Andre, E. Rossinot, Nanoscale compositional changes and modification of the surface reactivity of Pt3Co/C nanoparticles during proton-exchange membrane fuel cell operation. Electrochim. Acta. **56**, 776 (2010)
72. L. Dubau et al., Further insights into the durability of Pt3Co/C electrocatalysts: Formation of "hollow" Pt nanoparticles induced by the Kirkendall effect. Electrochim. Acta. **56**, 10658 (2011)
73. H.J. Fan, U. Gosele, M. Zacharias, Formation of nanotubes and hollow nanoparticles based on Kirkendall and diffusion processes: a review. Small **3**, 1660 (2007)
74. Y.D. Yin et al., Formation of hollow nanocrystals through the nanoscale Kirkendall Effect. Science **304**, 711 (2004)
75. M. Ibanez et al., Means and limits of control of the shell parameters in hollow nanoparticles obtained by the Kirkendall effect. Chem. Mater. **23**, 3095 (2011)
76. M.E. Glicksman, *Diffusion in Solids: Field theory, Solid State principles, and Applications* (Wiley, Ney York, 2000)

77. J.X. Wang et al., Kirkendall effect and lattice contraction in nanocatalysts: a new strategy to enhance sustainable activity. J. Am. Chem. Soc. **133**, 13551 (2011)
78. Y. Zhang et al., Hollow core supported Pt monolayer catalysts for oxygen reduction. Catal. Today **202**, 50 (2013)
79. P.R. Selvakannan, M. Sastry, Hollow gold and platinum nanoparticles by a transmetallation reaction in an organic solution. Chem. Commun. **13**, 1684 (2005)
80. M. Sastry, A. Swami, S. Mandal, P.R. Selvakannan, New approaches to the synthesis of anisotropic, core–shell and hollow metal nanostructures. J. Mater. Chem. **15**, 3161 (2005)
81. S.E. Skrabalak, L. Au, X.D. Li, Y.N. Xia, Facile synthesis of Ag nanocubes and Au nanocages. Nat. Protoc. **2**, 2182 (2007)
82. S.J. Guo, S.J. Dong, E. Wang, A general method for the rapid synthesis of hollow metallic or bimetallic nanoelectrocatalysts with urchinlike morphology. Chem-Eur. J. **14**, 4689 (2008)
83. M.R. Kim, D.K. Lee, D.J. Jang, Facile fabrication of hollow Pt/Ag nanocomposites having enhanced catalytic properties. Appl. Catal. B-Environ. **103**, 253 (2011)
84. V.R. Stamenkovic et al., Improved oxygen reduction activity on Pt3Ni (111) via increased surface site availability. Science **315**, 493 (2007)
85. P. Mani, R. Srivastava, P. Strasser, Dealloyed Pt-Cu core–shell nanoparticle electrocatalysts for use in PEM fuel cell cathodes. J. Phys. Chem. C **112**, 2770 (2008)
86. S.J. Bae et al., Facile preparation of carbon-supported PtNi hollow nanoparticles with high electrochemical performance. J. Mater. Chem. **22**, 8820 (2012)
87. H. A. Gasteiger, S.S. Kocha, B. Sompalli, F.T. Wagner, Activity benchmarks and requirements for Pt, Pt-alloy, and non-Pt oxygen reduction catalysts for PEMFCs. Appl. Catal. B-Environ. **56**, 9 (2005)
88. Z.M. Peng, J.B. Wu, H. Yang, synthesis and oxygen reduction electrocatalytic property of platinum hollow and platinum-on-silver nanoparticles. Chem. Mater. **22**, 1098 (2010)

Chapter 10
Molecular Complexes in Electrocatalysis for Energy Production and Storage

10.1 Key Concepts

The employment of metal complexes as electrocatalysts represents a potentially very important development in the field of energy production and storage. From a practical perspective, a molecular metal complex, soluble in different solvents or easily dispersible on very small surfaces as well as bound to electrodes, but capable of promoting an electrochemical reaction, has the potential advantage to overcome the drawbacks that present the technologies based on metal nanoparticles. Their role is to coordinate a substrate (e.g., H_2, O_2, CO_2, CO, alcohols, or sugars) through the metal center and favor transformation reactions that occur through the gain or loss of electrons. In this case, the electrode plays a less important role acting as both electron reservoir and for favoring electron transfer. In addition to the metal center, the ligands play a fundamental role in determining the chemical and electrochemical characteristics of the system. In this way they can accelerate or modify a chemical reaction which may otherwise be impossible. In some cases the oxidation state of the central metal atom is changed or it may remain the same while the oxidation state of the ligand atoms change. Every single metal atom is involved in the electrochemical reaction in contrast to what occurs on metal electrode surfaces or on metal nanoparticles.

Indeed, the established methods of organometallic synthesis, or coordination chemistry, leading to well-defined molecular metal complexes offer enormous advantages in the rational design and optimization of electrocatalysts, including a reduced metal loading due to the fact that all metal sites are active, unlike catalysts based on metal nanoparticles. The possibilities and range of applications of molecular complexes employed in electrocatalysis is very large in view of the fact that these molecular metal complexes can be easily embedded in a huge variety of nanosized conductive supports of relevance in electrocatalysis and electrophotocatalysis such as functionalized fullerenes, carbon nanotubes, nanofibers, and other nanosized matrices, and for example titania nanotubes. The tuning of well-defined molecular architectures and combination with a matching support material, will in the future, mimicking natural processes, allow for further development including scale up of devices for energy production and storage.

A. Lavacchi et al., *Nanotechnology in Electrocatalysis for Energy*,
Nanostructure Science and Technology 170, DOI: 10.1007/978-1-4899-8059-5_10,
© Springer Science+Business Media New York

10.2 Rhodium Molecular Catalysts for Organometallic Fuel Cells (OMFCs)

The combination of the production of renewable energy (with no CO_2 emissions) together with the production of industrially relevant feedstocks (such as aldehydes, ketones, and carboxylic acids) has been recently achieved using direct alcohol fuel cells or DAFCs. In this way the free energy of an alcohol is converted into electrical energy and the alcohol itself is transformed into an oxidation product or products which are different from CO_2. Two types of DAFCs have been developed for this purpose: (1) traditional devices, where the anode and cathode are separated by an anion-exchange membrane and are coated with an electrocatalyst, generally a nanostructured noble metal, supported on a conductive material [1–6] and (2) enzymatic biofuel cells (EBFC) utilizing oxidation enzymes such as dehydrogenases in conjunction with an electron-transfer mediator (Fig. 10.1c) [7, 8].

Figure 10.1a shows a simplified working scheme of a DAFC where alcohols (in this case ethanol) are selectively converted into carboxylates ($RCOO^-$) where the electrolyte is an anion-exchange membrane (eqns). On the anode, ethanol is oxidized releasing four electrons which are then utilized to reduce one oxygen molecule to four OH^- groups at the cathode.

Anode $C_2H_5OH{+}5OH^- \rightarrow CH_3COO^-{+}4H_2O{+}4e^-$
Cathode $O_2{+}2H_2O{+}4e^- \rightarrow 4OH^-$
Overall $C_2H_5OH{+}O_2{+}OH^- \rightarrow CH_3COO^-{+}2H_2O$

Efficient devices of this type have been recently developed for a variety of renewable alcohols and polyalcohols such as ethylene glycol, glycerol, and 1,2-propandiol [4, 5]. Typical power density curves for air-breathing cells fueled with aqueous solutions of ethanol, ethylene glycol, or glycerol are shown in Fig. 10.1b.

Recently a third type of fuel cell has been developed, known as an organometallic fuel cell (OMFC), operating in alkaline media where the anode catalyst is a molecular metal complex [9]. In this device, the metal complex evolves through fast chemical equilibrium to form specific catalysts for ethanol dehydrogenation, aldehyde dehydrogenation, and H^+/electron transfer. The idea of using an organometallic complex as an anode was developed after that H. Grutzmacher et al. [10–12] noted a study in which the organometallic complex [Rh(trop$_2$N)(PPh$_3$)] [1] (Scheme 10.1) was shown to catalyzes the oxidation of primary alcohols to give carboxylic acid derivatives in the presence of hydrogen acceptors such as RR'C=O, olefins, or palladium nanoparticles in either homogeneous or heterogeneous phase.

The amido complex 1 is simply accessed by deprotonation of the precursor complex [Rh(OTf)- (trop2NH)(PPh3)] [2] under basic conditions (OTf = trifluoro methanesulfonate).

Subsequently, it has been shown by Grutzmacher, Bianchini, Vizza et al., that the [Rh(OTf)(trop$_2$NH)(PPh$_3$)] complex [2] can be deposited intact onto Vulcan XC-72, a conductive carbon support that is often utilized for the preparation of

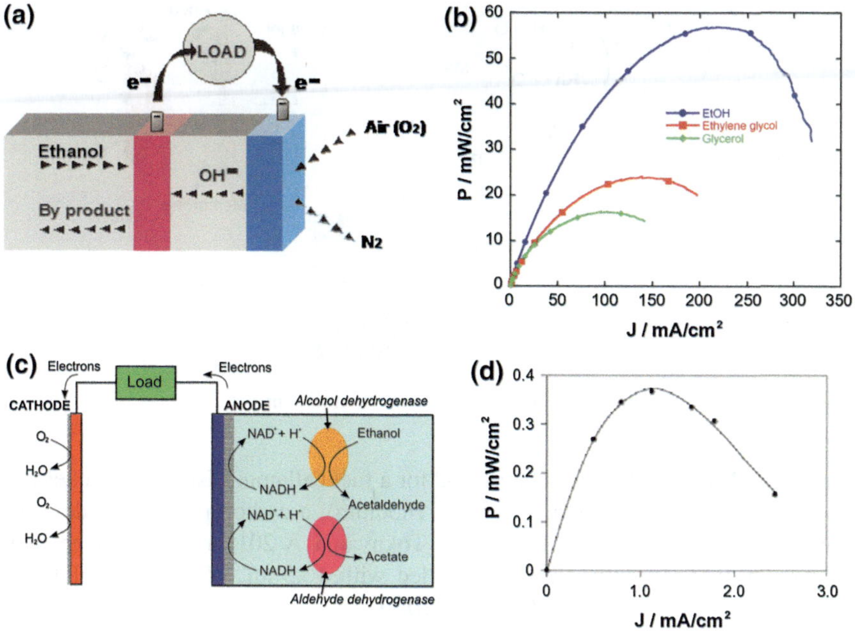

Fig. 10.1 **a** Working scheme of a DAFC in alkaline environment. **b** Power density curves of air-breathing DAFCs fuelled with ethanol (10 wt%), ethylene glycol (5 wt%), or glycerol (5 wt%) in 2 M KOH at 22 °C. *Anode* Pd-(CeO_2)/C on Ni mesh; *cathode* Fe-Co/C on carbon paper; membrane Tokuyama A201. **c** Ethanol/O_2 enzymatic biofuel cell, where alcohol dehydrogenase and aldehyde dehydrogenase catalyze a stepwise oxidation of ethanol to acetaldehyde and then to acetate, passing electrons to the anode via the mediator NAD^+/NADH. **d** Power density of the cell operating in buffered solution at pH 7.15, containing 1 mM ethanol and 1 mM NAD

Scheme10.1 Acceptor-assisted catalytic oxidation of primary alcohols with water to give carboxylates. Reproduced from Ref. [9] with permission from John Wiley and Sons

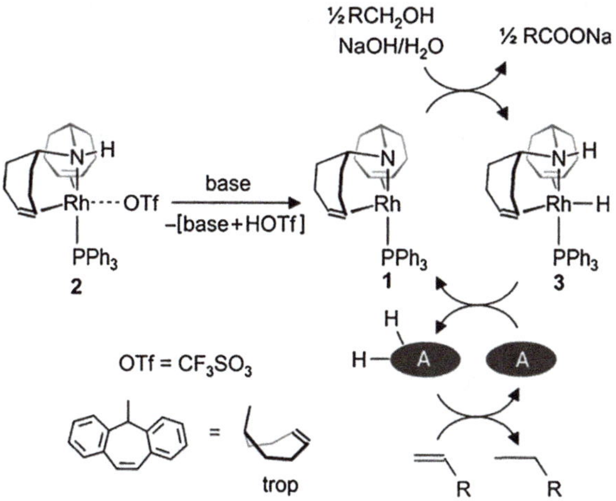

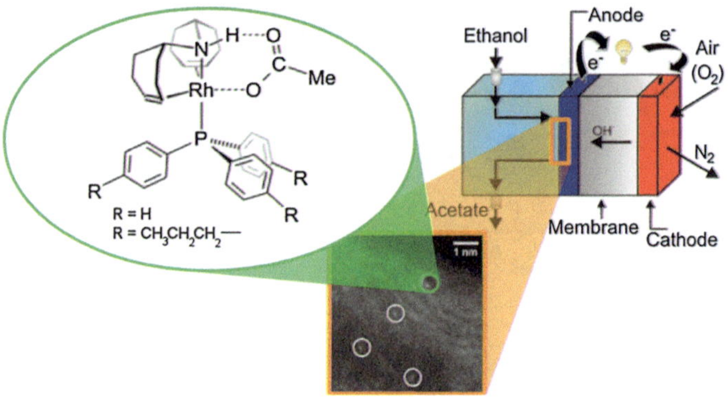

Fig. 10.2 Working scheme of a OMFC in alkaline environment

electrocatalysts. A MEA was fabricated for a fuel cell comprising of a nickel foam anode coated with 2@C (ca. 1 mg cm^2 rhodium), a carbon-paper cathode coated with a Fe–Co/C electrocatalyst and a Tokuyama A-201 anion-exchange membrane. The anode compartment was filled with 10.5 mL of a water solution of ethanol (10 wt%) and 2 M KOH (Fig. 10.2).

Figure 10.3 (blue trace) shows the polarization and power density curves of this passive cell recorded at 22 °C. A maximum power density of 7 mW cm^{-2} was obtained, which is far higher than that of any biofuel cell, yet slightly lower than that observed with a DAFC equipped with a palladium-based anode as described above. The power density supplied by the OMFC increases remarkably by increasing the working temperature of the MEA in an active cell under control of the oxygen and fuel fluxes. Indeed, 24 $mWcm^{-2}$ was obtained at 60 °C with a fuel flow of 4 mL min^{-1} and an oxygen flow 0.2 L min^{-1} (Fig. 10.3, red trace) Such a value is still lower than that obtainable with the best palladium-based anode electrocatalysts reported to date (Pd-(CeO_2)/C) [13], yet it falls in the upper range of power densities produced by the vast majority of DAFCs containing nanosized noble metal electrocatalysts.

The passive OMFC was subjected to galvanostatic experiments at low current intensities. The cell ran for 44.3 h, producing selectively 14.4 mmol of potassium acetate, which corresponds to 48 % conversion of the starting ethanol. Experimental evidence (i.e., NMR, XRPD, CV, etc.) led the authors to propose the mechanism of ethanol oxidation shown in Scheme 10.2.

On the electrode surface, the precursor 2@C is rapidly converted into the hydroxo complex 5eq@C, which is in a rapid equilibrium with the amide 1@C and water. The amide [Rh(trop2N)(PPh3)] dehydrogenates ethanol to acetaldehyde; the aldehyde reacts further with OH^- to form the acetate ion and the hydride 3@C. The latter complex is oxidized at the electrode releasing two H^+ (neutralized to give water under the basic conditions) and two electrons with regeneration of the amide 1@C. There is some resemblance with the enzymatic biofuel cell, but the main characteristic of this system is that one molecular rhodium complex is

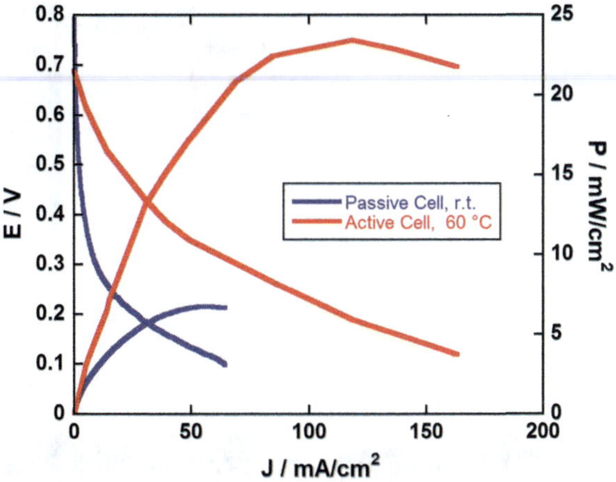

Fig. 10.3 Polarization and power density curves of OMFCs fuelled with 10 wt% ethanol in 2 M KOH (*anode* 1@C on Ni mesh; *cathode* anode, Fe–Co/C on carbon paper; Tokuyama A006 membrane). (*Blue trace*) air-breathing OMFC at 22 °C; (*red trace*, passive cell) active OMFC at 60 °C (*fuel flow* 4 mL min^{-1}; oxygen flow 0.2 L min^{-1}, active cell). Reprinted from Ref. [9] with permission of John Wiley and Sons

capable of evolving through fast chemical equilibria in the course of the catalytic cycle to form a specific catalyst for alcohol dehydrogenation (the amide 1@C), a specific catalyst for aldehyde dehydrogenation (the hydroxo complex 5@C), and a specific catalyst for the H^+/electron transfer (the hydride 3@C).

The success of the OMFC is based on a series of different properties all related to the molecular complex architecture and its interaction with the carbon conductive support. It is well known that the support morphological features, such as surface area and pore distribution, together with the structural presence of different functional surface groups, can strongly affect the complex deposition morphology on carbon (constituting the complete composition of anode catalyst) during the wet impregnation step. Furthermore, a crystalline or amorphous phase of the supported molecular complex directly affects all the catalytic reaction steps occurring on the anode surface.

It has been observed that the stability of the acetate complex 4@C (Scheme 10.2) is responsible for the decline in performance of the OMFC. In fact, at about 50 % conversion of the ethanol in acetate, the displacement of the acetate complex 4@C by OH^- to regenerate 5@C becomes too slow and current flow stops.

Recent studies have addressed this negative point by a rational modulation of the catalyst morphology. An improvement has been obtained by changing the carbon conductive support from Vulcan XC-72 (labeled Cv) to Ketjenblack ED600 (labeled Ck), characterized by surface areas of 254 and 1,400 m^2/g and pore volumes of about 170 and 500 mL/100 g, respectively. Under the same experimental conditions, two OMFC devices using the two different anode catalyst compositions, i.e., [Rh(trop$_2$N)(PPh$_3$)OTf]@Cv versus [Rh(trop$_2$N)(PPh$_3$)OTf]@Ck, have shown a fundamental difference: the OMFC based on [Rh(trop$_2$N)(PPh$_3$)OTf]@Ck greatly improved the efficiency of the cell in galvanostatic experiments [14].

A second strategy developed to improve the catalyst has involved a decrease in the precious metal Rh loading on the anode electrode. Maintaining the same

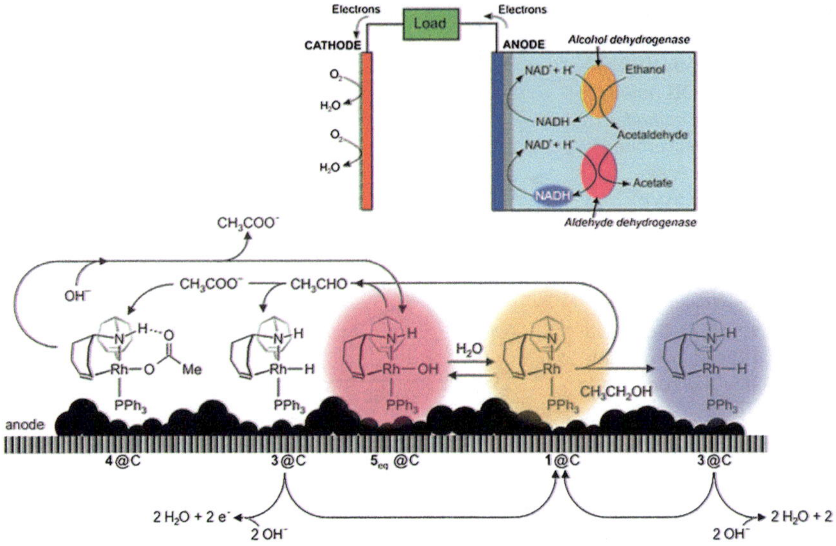

Scheme 10.2 Proposed mechanism for the reactions occurring on the surface of the OMFC anode coated with the molecular complex catalyst. The function of an enzymatic biofuel cells (EBFC; *top*) is included for comparison. Similar colors relate to similar functions (*pink* aldehyde dehydrogenation, *orange* alcohol dehydrogenation, *violet* hydrogen/electron transfer). Reprinted from Ref. [9] with permission from John Wiley and Sons

electrode composition (i.e., [Rh(trop$_2$N)(PPh$_3$)OTf]@Ck) and reducing only the total Rh content in the range from ca. 1.0 to 0.1 mg/cm^2. The same cell recyclability was obtained, without affecting the performance of the cell in terms of current density and conversion of ethanol to acetate.

In order to obtain a reduction in Rh loading without affecting the performance, this strategy also required a rational modification of the molecular architecture of the complex. PPh$_3$ ligand modification to P(n-butyl Ph)$_3$ lead to two different Rhodium-based complexes: [Rh(trop$_2$N)P(n-butylPh)$_3$OTf] and [RhOAc(trop$_2$N)P(n-butyl Ph)$_3$] (Scheme 10.3), both then used in OMFC devices. When combined to the carbon Ketjenblack, both of them showed a reduced crystalline form with respect to [Rh(trop$_2$N)(PPh$_3$)OTf] and only [RhOAc(trop$_2$N)P(n-butylPh)$_3$] lead to the extreme amorphous form. From an electrochemical viewpoint this aspect resulted in a consistent efficiency gain: notably, the [RhOAc(trop$_2$N)P(n-butylPh)$_3$]@Ck—based OMFC has shown a comparable efficiency to that of the [Rh(trop$_2$N)(PPh$_3$) OTf]@Ck—based device, at the same time with 1/10 of Rh loading.

In conclusion, changing the carbon black support from Vulcan XC-72 (Cv) to Ketjenblack EC 600JD (Ck) and/or the axial phosphane to produce noncrystalline complexes has been found to remarkably change the electrochemical properties of the organorhodium catalysts, especially in terms of specific activity and durability. An in-depth study has shown that either Ck or P(4-*n*-butylPh)$_3$ favor the formation of an amorphous Rh-acetato phase on the electrode, leading to a much more efficient and recyclable catalyst as compared to a crystalline Rh-acetate complex which is formed on Cv with PPh_3 as ligand. The ameliorating effect of the amorphous phase has been ascribed to its higher number of surface complex molecules as compared to the crystalline phase. A specific activity as high as 10,000 A g_{Rh}^{-1} has been found in half cell, which is the highest value ever reported for ethanol electrooxidation.

From a practical viewpoint, a molecular metal complex, soluble in different solvents and hence easily dispersible on very small surfaces, but capable of delivering high power densities upon oxidation of alcohols and sugars, will pave the way to the further miniaturization of fuel cells for biological applications as well as biosensors [7, 8, 14, 15].

In conclusion, well-established methods of organometallic synthesis, leading to well-defined molecular metal complexes, offer enormous advantages in the rational design and optimization of fuel cell catalysts, including a reduced metal loading due to the fact that all metal sites are active. The possibilities and range of applications of OMFC technology are very large in view of the fact that these molecular metal complexes can be easily embedded in a huge variety of nanosized conductive supports of great relevance in drug delivery, (electro)catalysis, and photocatalysis (e.g., functionalized fullerenes, carbon nanotubes, nanofibers, and other nanosized matrices, for example titania nanotubes).

10.3 Bimetallic Ni–Ru Molecular Complexes as Electrocatalysts for PEMFCs

Despite the fact that molecular hydrogen is today considered one of the most promising energy vectors, the development of cheap and efficient hydrogen-based fuel devices has not yet been realized. Currently, PEMFCs (polymer electrolyte membrane fuel cells) perform by the oxidation of H_2 (anode reaction) and the

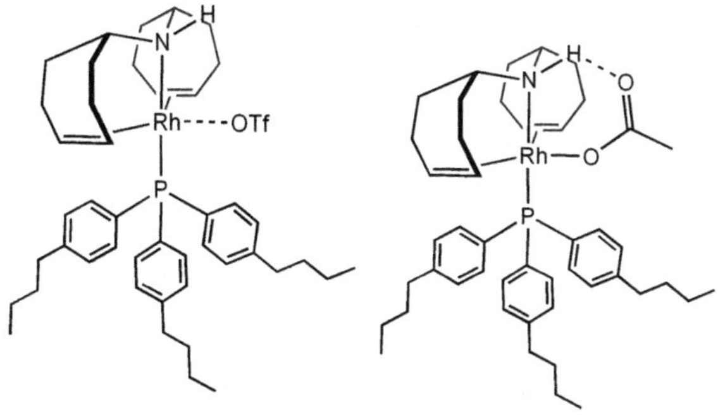

Scheme 10.3 [Rh(trop$_2$N)P(n-butylPh)$_3$OTf]Rh(trop2N)P(n-butylPh)$_3$OTf] (*left*) and [RhOAc (trop$_2$N)P(n-butylPh)$_3$] (*right*)

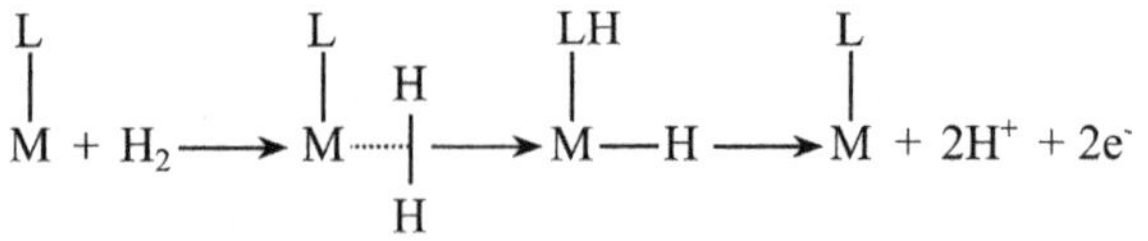

Scheme 10.4 Splitting heterolitic and oxidation of H_2

reduction of oxygen (cathode reaction), both on platinum-based electrodes. The development of fuel cells in general, e.g., PEMFC, DMFC (direct methanol fuel cell), and DFC (direct fuel cell), which utilize platinum is dramatically limited by the low natural abundance of this metal and hence its high cost.

A second limitation in the use of platinum-based catalysts in PEMFC but most importantly in direct alcohol fuel cells (DAFC) is the fact that platinum-based cathodes are sensitive to fuel crossover which causes significant cathode polarization. In addition platinum-based anodes are easily deactivated by very small quantities of carbon monoxide (CO) that is an intermediate in the oxidation of alcohols and is also contained in hydrogen gas obtained by reforming.

At high potentials platinum oxidatively decomposes water (between 0.6 and 0.8 V vs. RHE) thus limiting the capacity to oxidize adsorbed CO which results in large anodic over potentials.

Since their discovery, only a slight improvement in efficiency has been achieved in over a 100 years and thus a new model for fuel cells catalysis is required to achieve a fuel cell-economy, based on metal-mediated conversion of H_2 in energy [16–23].

Metal-mediated conversion of H_2 into a pair of protons and electrons has been studied extensively in relation to hydrogen metabolism, electrocatalysts for fuel cells and photocatalytic water splitting.

Currently, the accepted mechanism of the catalytic reaction of hydrogen oxidation is shown in Scheme 10.4 [24–29].

Hydrogenase enzymes, for example, following a similar reaction mechanism reported in Scheme 10.5 are known to catalyze the reversible interconversion between hydrogen and protons with remarkably high reaction rates (1,500–9,000 s^{-1} at pH 7 and 37 °C in water). Unfortunately, hydrogenases suffer from considerable oxygen sensitivity and, consequently, their production in the active form in industrial quantities would be problematic by preventing a real application.

Because catalysis hydrogenase enzymes requires only iron and nickel metal centers, these active sites are interesting sources of inspiration for the design of synthetic molecular-based catalysts as alternatives to platinum. Among a wide number of studies concerning the synthesis and electrochemical characterization of molecular electrocatalysts as bioinspired models of NiFe hydrogenases [7, 30–32], a new system developed by Gao et al. [33, 34] is noteworthy in terms of the final catalyst application and characterization in a complete fuel cell. The authors showed that a bimetallic Ni–Ru organometallic complex, soluble in water, was capable of catalyzing the oxidation of H_2 to two protons and two electrons in homogeneous

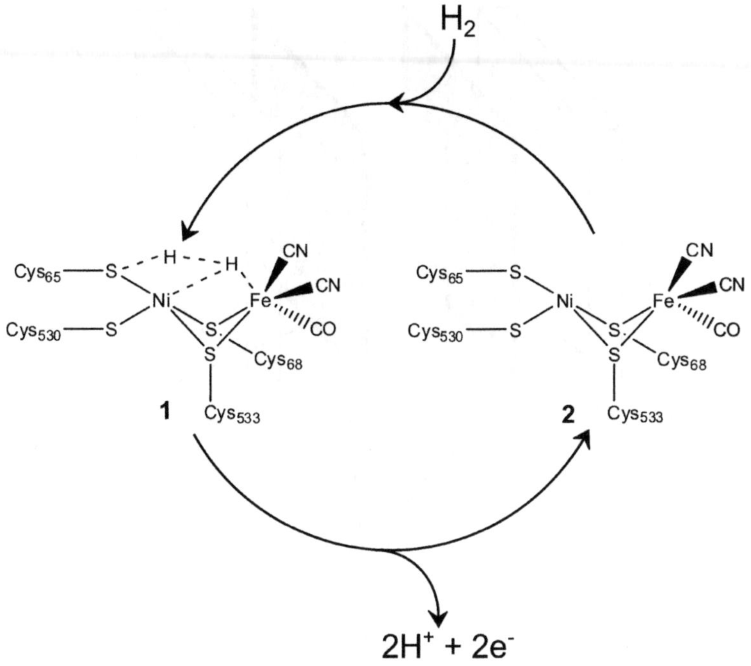

Scheme 10.5 Hydrogen oxidation by Ni–Fe hydrogenase

aqueous phase. The organometallic catalyst precursor $[Ni^{II}LRu^{II}(H_2O)(H_6–C_6Me_6)](NO_3)_2]$,[1], (L=N,N′-dimethyl-N,N′-bis(2-mercaptoethyl)-1,3-propanediamine), the stable hydride intermediate $[Ni^{II}(H_2O)L(\mu\text{-}H)Ru^{II}(\eta^6–C_6Me_6)](NO_3)$ $([2](NO_3))$ [35, 36] and two H_2 molecules are involved in the oxidative catalytic cycle (Fig. 10.4). The proposed mechanism for the oxidative reaction involves two molecules of H_2 and a Ni–Fe center. Furthermore, the authors were able to demonstrate the ability of the unusual Ni–Ru hydride complex **2** to promote not only H_2 oxidation but also O_2 reduction to H_2O; The ability to promote both types of redox processes has not been observed for any kind of molecular catalyst to date.

The organometallic catalysts were water soluble and therefore dissolve in the gas conduits in the cell. In solid-phase experiments, water-insoluble catalysts were immobilized on electrodes by exchanging the NO_3^- counter anion for $CF_3SO_3^-$. This change resulted in better performance with respect to the solution-phase experiments (OCV of 0.29 V at 60 °C, a current density of 11 μW cm-2 at a cell voltage of 0.17 V, maximum power density of 11 μW cm^{-2} (Fig. 10.5, 2 used for both anode and cathode).

Overall, this type of fuel cell underperforms conventional Pt-based devices. Research efforts should be focused on the development of highly efficient, low-cost catalysts that combine the flexibility of molecular design with the knowledge of the mechanism reaction that occurs at both the anode and cathode electrodes.

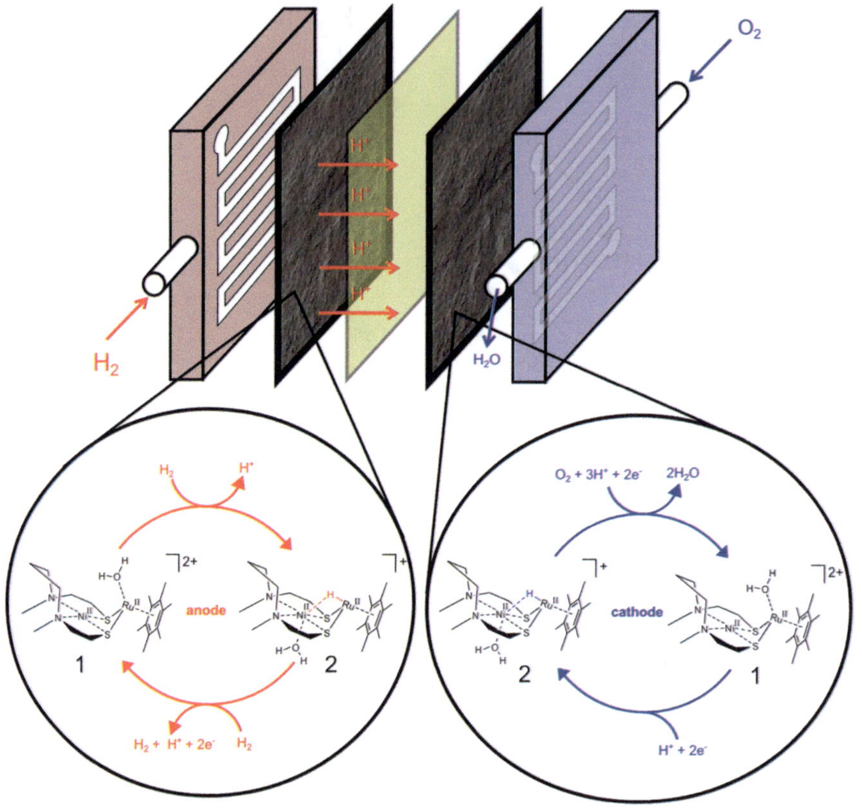

Fig. 10.4 Working scheme of an OMFC fed with H_2/O_2. Water-saturated H_2 and O_2 gases flow through conduits in the anode and cathode electrodes. Protons cross the polymer electrolyte and electrons flow through an electrical circuit from the anode to the cathode. Adapted from Ref. [33] with permission from John Wiley and Sons

Fig. 10.5 Polarization and power density curves for a H_2–O_2 fuel cell at 60 °C. Reproduced from Ref. [33] with permission of John Wiley and Sons

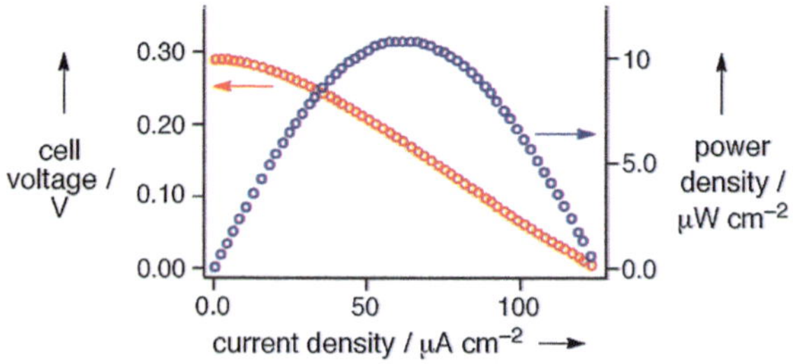

10.4 Fe and Ni Molecular Catalysts for Hydrogen Production by Electrocatalysis

Generally, homogeneous catalysis, is described in terms of purely metal-centered elementary steps (e.g., olefin insertion, H_2 oxidative addition and reductive elimination, CH activation, alkane reductive elimination, carbonilation, etc.). Hence, auxiliary ligands such as arylphosphines or amines are represented in generalized mechanistic schemes as 2-electron donors (L_n) and chemically inert. Notwithstanding this, several metal complexes as well as metallo-enzymes, which have first and secondary coordinating spheres around the metal containing functional groups which can simultaneously activate substrates. Belonging to this class of enzymes are the Fe–Fe hydrogenase *desulfovibrio desulfuricans* that catalyze the formation of hydrogen (H_2) from water with an overpotential as low as 100 mV, with the di–iron subcluster as active site (Fig. 10.6a) [37–40].

Theoretical and experimental studies suggest that a crucial role in delivering or removing protons to and from the correct atom(s) of the active site of the enzyme is played by the dithiolate bridge which contains an amine functional group which acts as a bronsted base, deprotonating a dihydrogen ligand in the H–H bond heterolysis step [37–41].

One of the objectives of many research groups consists in the development of effective catalysts capable of mimicking enzymes such as the *desulfovibrio desulfuricans* that are involved in the production of energy through electron-transfer process. As previously introduced in Sect. 10.3, hydrogenase enzymes efficiently catalyze not only the oxidation but also the production of H_2 using earth abundant metals. Detailed information about the production process has been obtained from protein films by voltammetry experiments, but enzymes are difficult to obtain in sufficient amounts to adapt for a commercial use. For this reason, in recent years synthetic complexes based on Ni, Co, Fe, and Mo have been developed for H_2 production [42–44].

A synthetic bioinspired Ni complex, containing 1,5-diaza-3,7-diphosphinocyclooctane ligands was developed by V. Artero, S. Palacin et al. [45], as an approach to try to mimic the catalytic rate of H_2 production of these biological materials. The complex was linked to carbon nanotubes and then coated with Nafion on a glassy carbon electrode. This high surface area cathode material showed high catalytic activity under the strongly acidic conditions of proton-exchange membrane electrolysers. Hydrogen evolved from aqueous sulfuric acid solution with very low overvoltages (20 mV) and the catalyst exhibited exceptional stability with more than 100,000 turnovers (Fig. 10.6b).

Daniel L. Dubois et al. [46–48], have studied the activity of mononuclear nickel(II) complexes with nitrogen atoms in the second coordination sphere, demonstrating that such nitrogen atoms function as proton relays and can remarkably accelerate the rates of intra and intermolecular proton transfers. Among the various molecular structures designed by Dubois et al. [49] are 1,5-diaza-3,7-diphosphacyclooctane ligands with different substituents on the phenyl

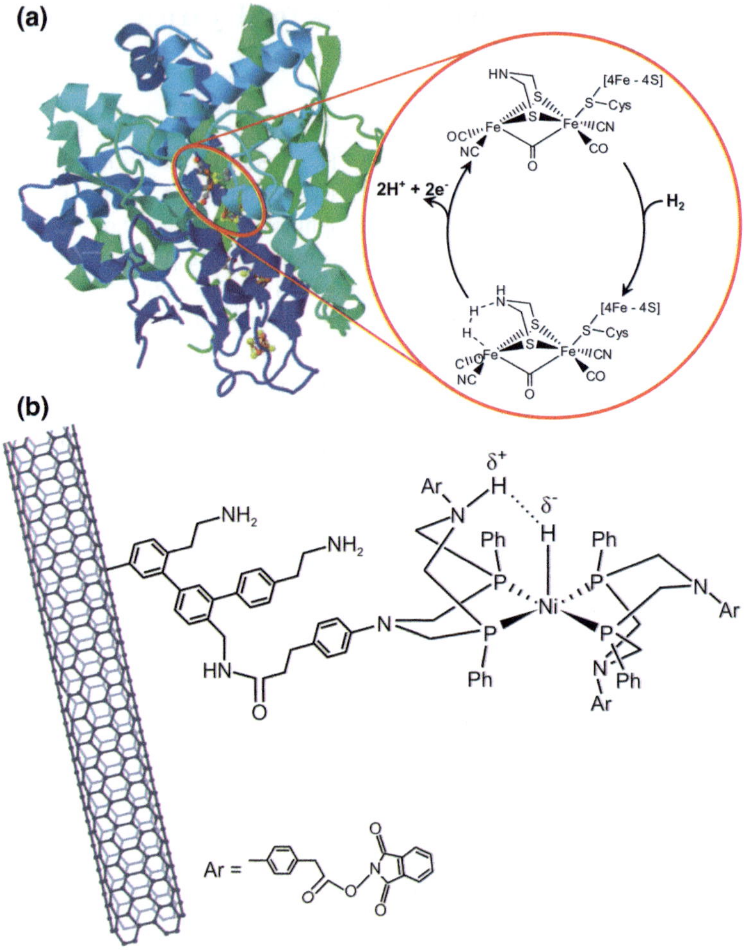

Fig. 10.6 **a** Structural model for [Fe,Fe]-H_2 ase *Desulfovibrio desulfuricans* with proposed mechanism for the H_2 heterolysis step; **b** bioinspired nickel electrocatalyst for H_2 oxidation/production

groups attached to either nitrogen or phosphorus atoms 1 (Scheme 10.6), as well as an 1-aza-3,6-diphosphacycloheptane ligand 2 (Scheme 10.6) [50]. The reactions of such ligands with $[Ni(CH_3CN)_6](BF_4)_2$ gave square-planar Ni(II) complexes (Scheme 10.6) that catalyze the production of hydrogen, using protonated dimethylformamide [(DMF)H]OTf as proton source (reaction 1), at overpotentials and turnover frequencies that depend strongly on the type and molecular structure of the ligand. The authors reported that the Ni(II) complex 3 (Scheme 10.6) catalyzed the reaction of hydrogen production with turnover frequencies of 33,000 s^{-1} in dry acetonitrile and 106,000 s^{-1} in the presence of 1.2 M of water at a potential of -1.13 V which corresponds to an overpotential of approximately 625 mV. Such turnover frequencies are remarkably higher than that exhibited by

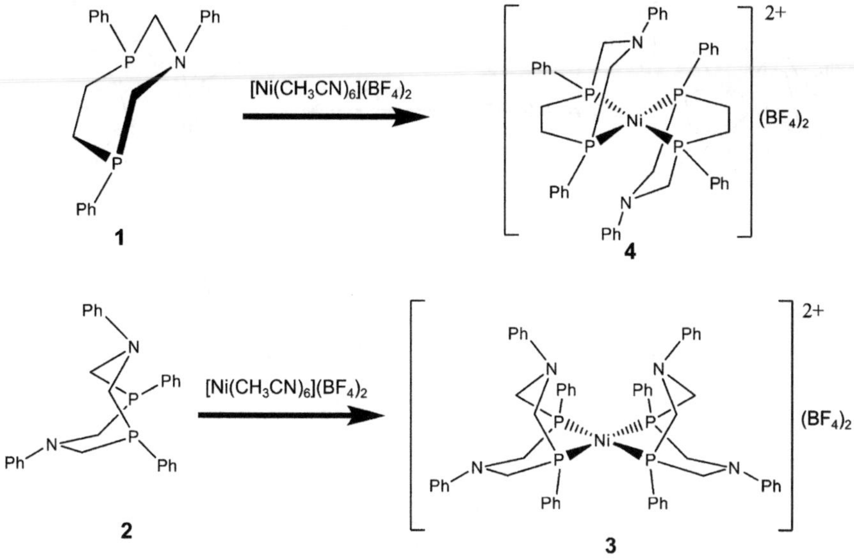

Scheme 10.6 Adapted from Ref. [50] with permission of John Wiley and Sons

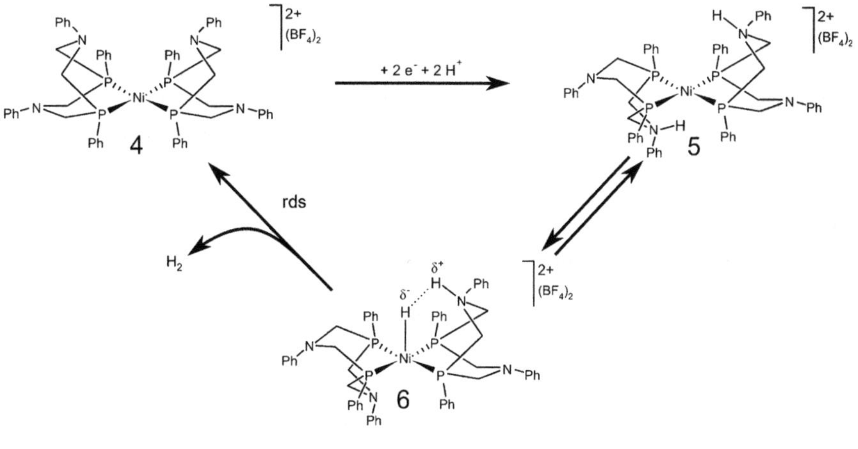

Scheme 10.7 Proposed mechanism for H_2 production by complex 4. Adapted from Ref. [50] with permission from John Wiley and Sons

the natural [FeFe] enzyme (9,000 s^{-1}) [51], as well as higher than those obtainable with Ni(II) complexes stabilized by the ligands of type 1 (350–1,850 s^{-1}) [44, 48].

On the basis of theoretical and kinetic studies with the diaza-complex 4 (Scheme 10.7), a mechanism has been proposed for which the rate-determining step (rds) for H_2 production is independent of acid concentration and involves one or more steps in the elimination of H_2 from the diprotonated Ni(0) complex 6

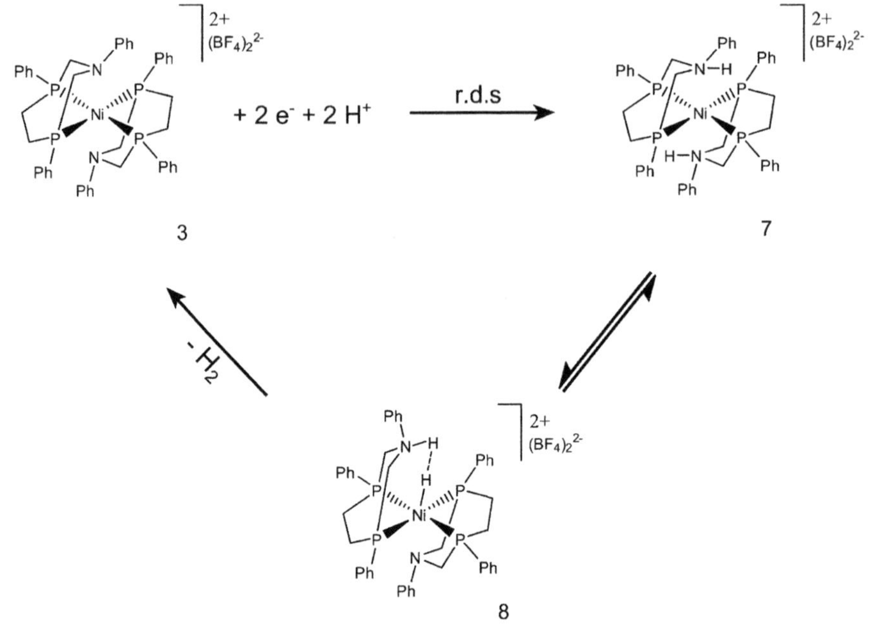

Scheme 10.8 Adapted from Ref. [50] with permission from John Wiley and Sons

(Scheme 10.7) [50]. Varying the electron-withdrawing character of the phenyl substituents on the nitrogen atoms of the ligand 1 (Scheme 10.6) has allowed the authors to establish important correlations between basicity of the nitrogen ligands and turnover frequencies/potential of the Ni(II)/Ni(I) couple at which catalysis occurs. A similar mechanism, yet with a different rds, has been proposed for the production of hydrogen catalyzed by complex **3** (Scheme 10.8).

The high turnover frequencies of H_2 production, greater than 100,000 s^{-1} showed by the systems, are remarkably higher than that exhibited by the natural [FeFe] enzymes. Unfortunately, the catalytic system presents poor stability under real electrolysis conditions.

Despite the growing number of molecular electrocatalysts that have up till now been reported, because of the poor stability of the catalytic systems, the application in real hydrogen production devices has yet to be reported in the scientific literature.

10.5 Molecular Catalysts for Electrochemical and Photoelectrochemical Reduction of CO_2

The electrochemical reduction of CO_2 may prove to be an important research in the development of alternate sources of fuels and chemicals.

In general, the direct electrochemical reduction on most metal electrodes requires highly negative potentials, up to −2.2 V versus SCE and results in a variety of products which distribution depends on the electrode materials as well as of the reactions condition, including CO_2 concentration, current density, and solvents.

This has prompted several research groups to attempt to reduce the thermodynamic barrier of CO_2 reduction, using suitable catalytic systems. Molecular catalysts are capable of lowering the overpotential by stabilizing the intermediate transition state between the linear CO_2 molecules and the final products. By varying the nature of the metal center and the ligand structures, the molecular complexes should lead to faster kinetics and long-term stability. In particular, a molecular catalyst should allow: (a) high faradic efficiency; (b) low overpotential, defined as the difference between the applied electrode potential and thermodynamic potential for conversion; (c) high TON (moles products/moles catalyst) or a high TOF (catalytic turnovers per unit time) and a high catalytic selectivity in products; (d) high catalytic selectivity, defined as moles of desiderate products/ moles other products.

In other words, efficient electrocatalysts should have formal potentials $E^0(Cat^{n+/0})$ well matched with $E^{0(products/substrates)}$ and good rate constants k_{cat} for the chemical reduction of substrates to products at a determined potential. Moreover, the rate constant for reduction of the electrocatalyst at the electrode must be high for an applied potential, near $E^0(Cat^{n+/0})$ [52].

Interesting approaches employ either modified electrodes (heterogeneous catalysis) or solutions (homogeneous catalysis) [53]. An electrocatalytic system can, for example, involve a metal complex dissolved in solution which is in contact with a chemically inert cathode (Fig. 10.7a). CO_2 is coordinated by the metal complex in solution and electrons are transferred to CO_2 from the electrode. Another method involves molecular complex particles embedded in a conductive material containing an appropriate electron relay which are then deposited onto a cathode surface (Fig. 10.7b). The use of modified electrodes containing surface-bound catalysts is advantageous for many reasons. The effective volume-concentration of electroactive material can reach high levels compared with a homogeneous solution. At high effective concentrations, the distance between adjacent metal complexes is sufficiently short to enhance cooperative effects. In addition, the use of surface immobilized electrocatalysts allows for the easy removal of the catalyst from the reaction vessel.

Since 1990 a number of studies have reported that molecular complexes such as metal-phthalocyanines (MPc), macrocyclic complexes, bipyridine complexes as well as transition metal phosphine complexes, can be used as catalysts for the CO_2 electroreduction process.

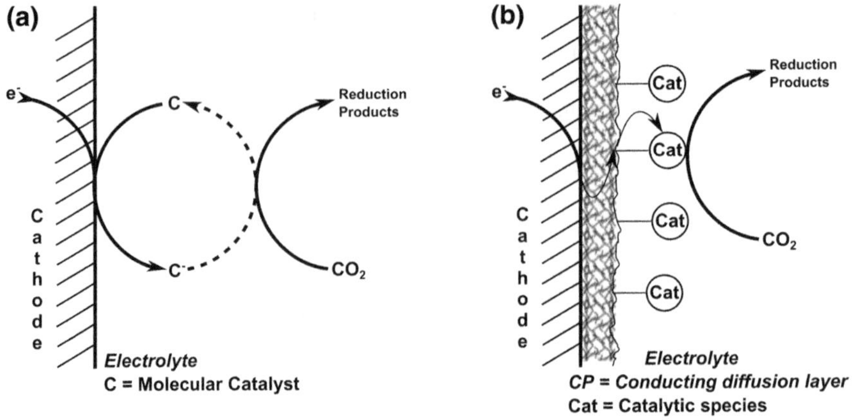

Fig. 10.7 **a** Molecular electrocatalysts in solution; **b** modified cathode by surface deposition of molecular electrocatalysts

10.5.1 Macrocyclic Complexes

Several metallophthalocyanines (MPCs) have been reported to be active toward the electroreduction of CO_2 [53–57]. Modified electrodes have been employed and in some cases, MPCs deposited onto electrode surfaces have been used with aqueous solutions. Polymeric complexes either immobilized on carbon electrodes or incorporated in coated Nafion® membranes have also been employed. The mechanism of catalysis for these systems is not well understood. Meshitsuka et al. [58] reported the first electrocatalyst for CO_2 reduction using cobalt and nickel phthalocyanines but did not reported any TOF, TON or current efficiencies. The products obtained were oxalic acid and glycolic acid but no formic acid was observed. Other authors [59] with the same CoPC using 1 atm of CO_2, at pH 5 and an applied potential of -1.15 V (vs. SCE), obtained manly CO and H_2 with a remarkable TOF $>10^5$.

These results were interpreted in terms of the formation of a CO_2 adduct with the reduced MPC complex [60]. Despite the fact that different complexes with CO_2 or CO have been isolated and characterized [61, 62], at this stage reaction mechanisms are only highly speculative [63].

Using a modified graphite electrode coated with cobaltoctabutoxyphthalocyanine (CoPC(BuO)$_8$), or octacynophthalocyanine (CoPC(CN)$_8$), Abe et al. [54, 64] investigated electrocatalytic CO_2 reduction. CO was the major product: at pH 4.4, most selective CO_2 reduction was achieved at -1.30 V with a TON $\sim 1.1 \times 10^6$ h^{-1} and the CO/H_2 ratio $\sim$4.2 with (CoPc(BuO)$_8$). Instead, using (CoPc(CN)$_8$) selective CO_2 reduction was achieved at -1.20 V (Ag/AgCl) with a ratio of produced CO/H_2 around 10 at pH 9.3.

The first reported cobalt and nickel aza-macrocyclic complexes with high current efficiency (up to 98 %) but with low TON (from 2 to 9 h^{-1}) were reported in 1980. [65] Ni^{II}(cyclam) complexes were studied by Sauvage et al. [66–68]. The complexes are extremely stable, highly selective, and showed production of CO in aqueous conditions, with high faradaic efficiencies (up to 96 %) at −0.86 V (vs. SCE). The nickel macrocycle complexes were shown to be very sensitive to the pH and required an Hg electrode surface in order to recycle the catalyst. Later it was observed that the high electrocatalytic activity originated from $[Ni(cyclam)]^+$ strongly absorbed on a mercury electrode surface [69]. Detailed studies have shown that the selective form of the catalyst was not $[Ni(cyclam)]^+$ but its adsorbed form on mercury, which possesses the geometry and electron density that allows the coordination of CO_2. This evidence led to the conclusion that only the adsorbed species was catalytically active and the homogeneous catalyst activity was only of minor importance [70].

Metalloporphyrins are effective electrocatalysts for the electroreduction of CO_2 to CO in aqueous and non-aqueous media. Co^{II}- tetraphenylporphyrin (Co^{II}tpp) fixed on glassy carbon electrodes was shown to catalyze the reduction of CO_2 to CO at potentials 100 mV more positive than the water-soluble $cobalt^{II}$ porphyrins [71] (turnover number $>10^5$) [66, 72, 73].

Savéant et al., found that Fe^0 porphyrins were able to catalyze the reduction of CO_2 to CO with a TOF as high as 350 h^{-1} in the presence of a weak Brønsted acid, with a catalyst decay rate of 1 % per catalytic cycle. These systems were shown however to require electrode potentials too high for their practical application [74].

10.5.2 Metal Bipyridine Complexes

A number of bipyridine complexes of nickel [75], cobalt [76], rhenium [77–81], ruthenium [82, 83], rhodium [84], and iridium [71] have been investigated for the electroreduction of CO_2. Tanaka et al. [82] reported that $[Ru(bpy)_2)(CO)_2]^{2+}$ electrocatalitically reduces CO_2 to afford H_2, CO, and $HCOO^-$. The most plausible reaction mechanism is shown in Scheme 10.9. $[Ru(bpy)_2)(CO)_2]^{2+}$ undergoes an irreversible two-electron reduction at −0.95 V versus SCE to afford $[Ru(bpy)_2)(CO)_2]^0$ with 20 electrons which in turn may liberate CO to gives the pentacoordinated $[Ru(bpy)_2)(CO)]^0$ with 18 electrons. In presence of a large excess of CO_2, $[Ru(bpy)_2)(CO)(COO^-)]^+$ is formed which under acidic conditions (pH 6) gains another proton to lose water and regenerate $[Ru(bpy)_2)(CO)_2]^{2+}$. Thus, the reduction of CO_2 under acidic conditions produces only CO. However, in weakly alkaline media (pH 9.5) the catalyst $[Ru(bpy)_2)(CO)(COO^-)]^+$ or $[Ru(bpy)_2)(CO)C(O)OH)]^+$ may undergo a two-electron reduction with the participation of a proton to create $HCOO^-$ and regenerate the five coordinate Ru(0) complex $[Ru(bpy)_2)(CO)]^0$. The evolution of CO as the same time may come from $[Ru(bpy)_2)(CO)_2]^{2+}$ existing as minor component.

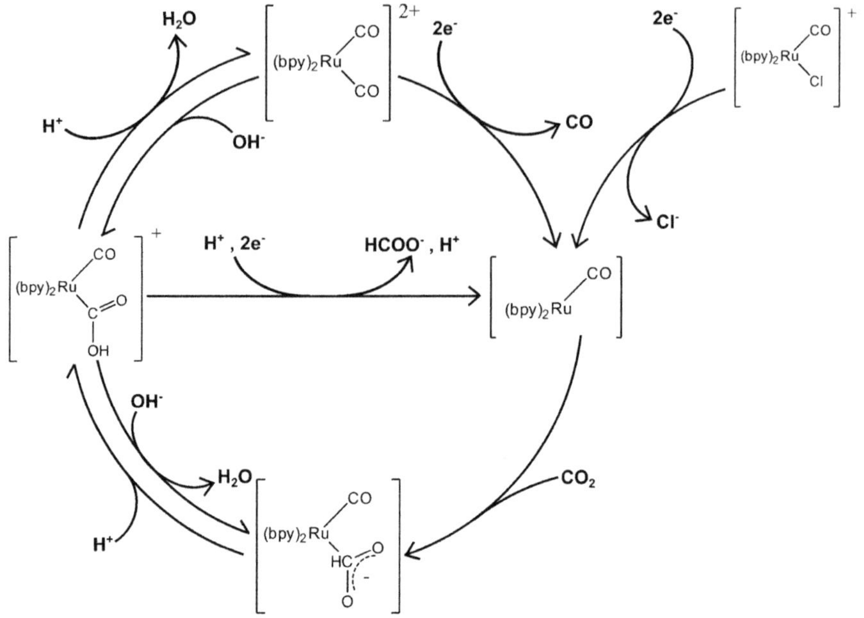

Scheme 10.9 Proposed mechanism of electroreduction of CO_2 for the $Ru[Ru(bpy)_2)(CO)_2]^{2+}$. Adapted from Ref. [83]. Copyright (2013) American Chemical Society

This system is limited by a low TOF and low selectivity but on the other hand is important to understand the mechanism of electroreduction of CO_2 through several keys intermediates.

Among the catalysts investigated, the [Re(bpy)CO_3Cl] complex is certainly one of the most robust and indeed since these initial studies much work have been performed on the electrocatalytic [79, 85] and photocatalytic [77, 86, 87] activity of similar molecular complexes.

Recently, Kubiak et al. [88, 89] explored complexes of the type [Re(bpy-R)(CO_3)(L)] (where bpy-R is 4,4′- disubstituted-2, 2′- bipyridine and L is a halide or a neutral ligand with OTf) and reported that [Re(bpy-*t*Bu)$(CO)_3$Cl] is a pre-catalyst for the electrochemical reduction of CO_2 to CO with a high turnover frequency (>200 s^{-1}) compared with the previous reports on [Re(bpy)$(CO)_3$Cl] [77, 78]. These high rates, however, are still much lower with respect to natural enzymes, which can operate near the thermodynamic potential and equilibrate the reduced and oxidized species [90].

Ni^{II} *N*-heterocyclic-carbene-pyridine complexes exhibit high selectivity for CO_2 electroreduction to CO over water reduction, but show low turnover frequencies of ca. 4 h^{-1}. This low catalytic activity is due to the formation of thermodynamically stable dimers that represent, as in many other cases, the death of the catalyst [91].

10.5.3 Metal Phosphine Complexes

A number of transition metal complexes of CO_2 with various phosphine or arsine ligands have been isolated and characterized [92–95].

[Rh(dppe)$_2$]Cl (dppe = 1,2-bis(diphenylphosphino) ethane) has been studied for CO_2 electroreduction in anhydrous CH_3CN, at −1.55 V (Ag wire). In this system the products upon reduction of CO_2 were found to be the formate anion with small percentages of cyanoacetate. Current efficiencies for the generation of the formate anion were approximately 42 % for short electrolysis runs and 22 % for longer runs. While mechanistic studies were not undertaken, it was concluded that CH_3CN was the proton source necessary for the formation of $HCOO^-$ [96].

Dubois et al. [97] have developed catalysts for CO_2 reduction based on iron, cobalt, and nickel complexes containing polyposphine ligands as well as weakly coordinating solvent molecules, for example, $[M(P)n(CH_3CN)_x]^{2+}$. The authors hypothesized that weakly bound solvent molecules would be important for catalysis. In fact, for many catalytic processes, vacant coordination sites are often required. In particular, two adjacent sites, one for coordination of a hydride ligand and the second for coordination of CO_2, could facilitate the C–H bond formation of formate. Only the iron complex showed catalytic activity toward CO_2 reduction, even though this occurred at high overpotentials and slow rates.

Despite this, these results led the DuBois group to study the electrocatalytic activity of $[Pd(triphosphine)(CH_3CN)]^{2+}$ complexes having structure **2** (Scheme 10.10). This catalyst is active for the electrochemical reduction of CO_2 to CO. At low acid concentrations, the catalytic rates are first order with respect to the catalyst, first order with respect to CO_2 concentration, and second order with respect to acid concentration. At higher acid concentrations, the catalytic rates become independent of the acid concentration, but remain first order with respect to catalyst and CO_2 [98–100]. In addition, it was found that monodentate phosphine ligands strongly inhibited the catalytic reaction, as did strongly coordinating solvents such as dimethylsulfoxide. These results supported the hypothesis on the requirement for dissociation of a weakly coordinating solvent molecule, during the catalytic cycle. It was also observed that electron-donating substituents on the triphosphine ligand increased catalytic rates. The more active catalysts of this class exhibited second-order rate constants between 5 and 300 M^{-1} s^{-1}. Based on these results, the catalytic mechanism was proposed and is shown in Scheme 10.10. The rate-determining step at low acid concentrations is the cleavage of the C–O bond in which species **8** is converted to **9** while the metallocarboxylic **7** is protonated to form the dihydroxycarbene **8** while the C–O bond cleavage reaction that transforms **9** to **2**, is the rate-determining step [101]. However, in solutions of low acid concentrations the rate of the catalytic cycle remains low.

Although the electrocatalytic nature of the process was clearly evident, the catalytic activity only lasted for short periods during electrolysis (around 0.5 h). The incorporation of bulky substituents on the terminal phosphorus atoms of the triphosphine ligands (which helps avoid the formation of bimetallic palladium

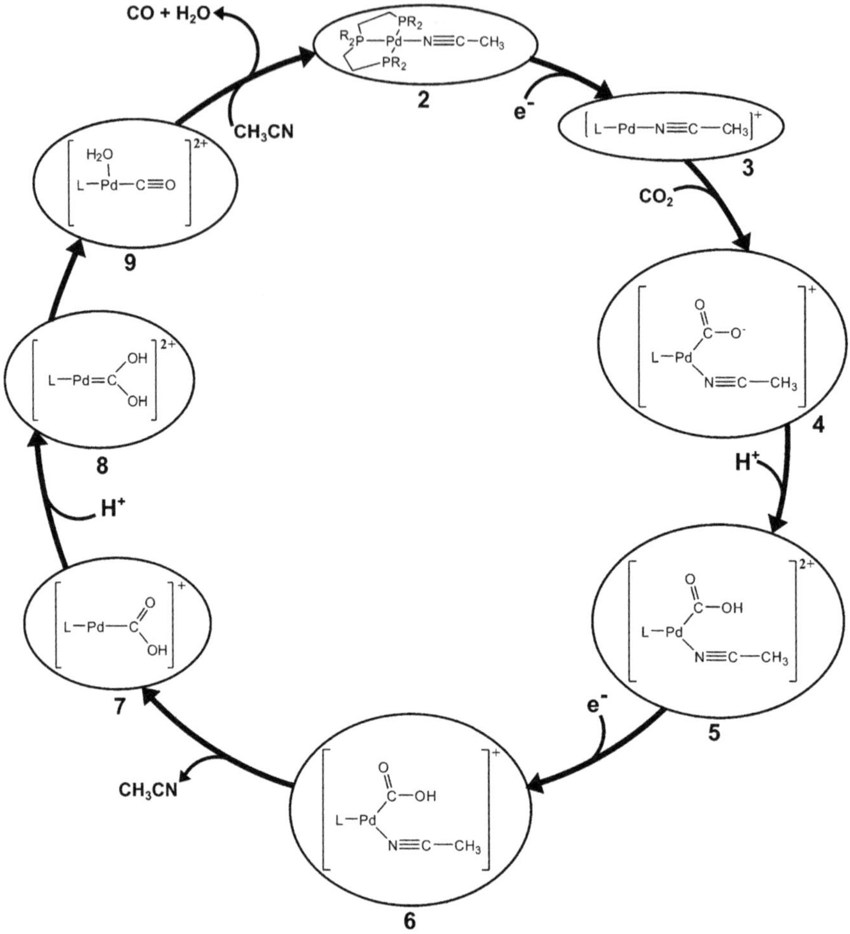

Scheme 10.10 Supposed catalytic cycle for CO_2 electroreduction. Adapted from Ref. [46]. Copyright (2013) American Chemical Society

complexes) resulted in an increased turnover number for these materials. In fact, $[Pd(etpC)(CH_3CN)]^{2+}$ (where etpC is $PhP(CH_2CH_2PCy_2)_2$) has a turnover number of 130 [102], and complex **11** (Scheme 10.11) has a turnover number greater than 200 with less than 20 % degradation and a rate constant of ~50 M^{-1} s^{-1} under normal operating conditions [102]. Complex 11 was shown to function as two independent catalytic sites with no evidence for cooperative interactions between the two metal centers, and the rates were typical of mononuclear [Pd(triphosphine)$(CH_3CN)]^{2+}$ catalysts.

The versatility of phosphine ligands allows for variation of both the electronic and steric effects of their complexes. The central donating atom of the tridentate ligand X can be varied from P to C, N, S as shown in Scheme 10.12.

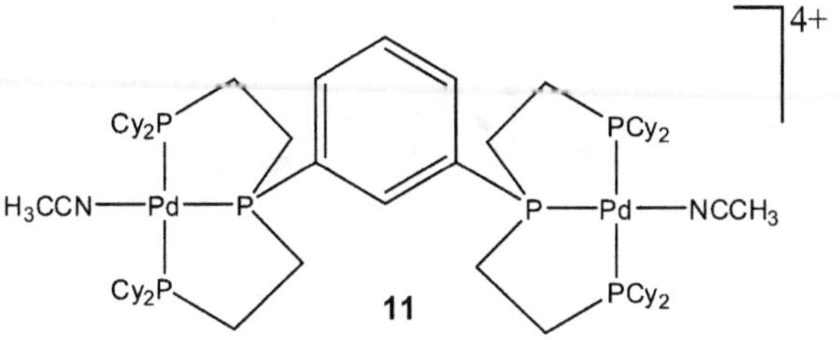

Scheme 10.11 1,3-bis (4-(1,1,7,7-tetracyclohexyl, 1,4,7-triphospha-epthyl))benzene

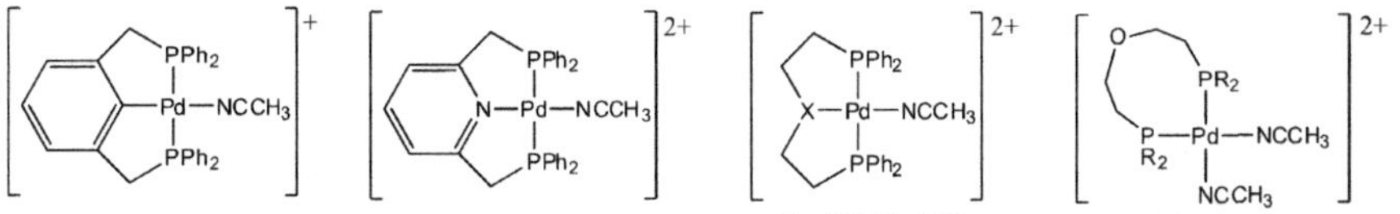

X = NH, S, AsPh

Scheme 10.12 Palladium tridentate phoshine complexes for CO_2 electroreduction

DuBois, from the study of these complexes, hypothesized that H_2 or CO production depends on the redox potential of Pd hydride formation; a less negative potential favors protonation of the coordinated CO_2 oxygen to form CO and H_2O.

10.5.4 Carbon Monoxide Dehydrogenases Enzymes

The most active known catalysts for the reversible reduction of CO_2 to CO are the carbon monoxide dehydrogenase enzymes of bacteria. One of these enzymes [NiFe] CO dehydrogenase of anaerobic bacteria *Carboxydothermus hydrogenoformans* catalyzes the rapid reduction of CO_2 to CO with turnover rates of 31,000 s^{-1} [103]. Structural studies have provided important insights into those features of the active site that contribute to high catalytic activity. The structure of the active site was resolved in three different reduced states which reveals a complex $NiFe_4S_4$ center (Scheme 10.13): state 1 at a potential of −320 mV, state **2** is the proposed transition state of the reaction in which CO binds to the Ni^{2+} ion and reacts with the Fe– bound OH group; state **3** at –600 mV+CO_2 is used as a model for the stabilization of the metal carboxylate state. State **4** at –600 mV.

Upon addition of CO_2 to the reduced state, CO_2 binds to both the Ni and Fe. This binding causes minimal geometry changes and occupies the fourth position around Ni completing the square-planar geometry. In the coordination of CO_2, nickel acts as the Lewis base, while the iron acts as the Lewis acid, and the partial negative charge on the oxygen is stabilized through hydrogen bonding provided by the protein surroundings. The positions of the Ni and Fe are held in place by the Fe_3S_4 framework and are essentially unchanged by the presence or absence of

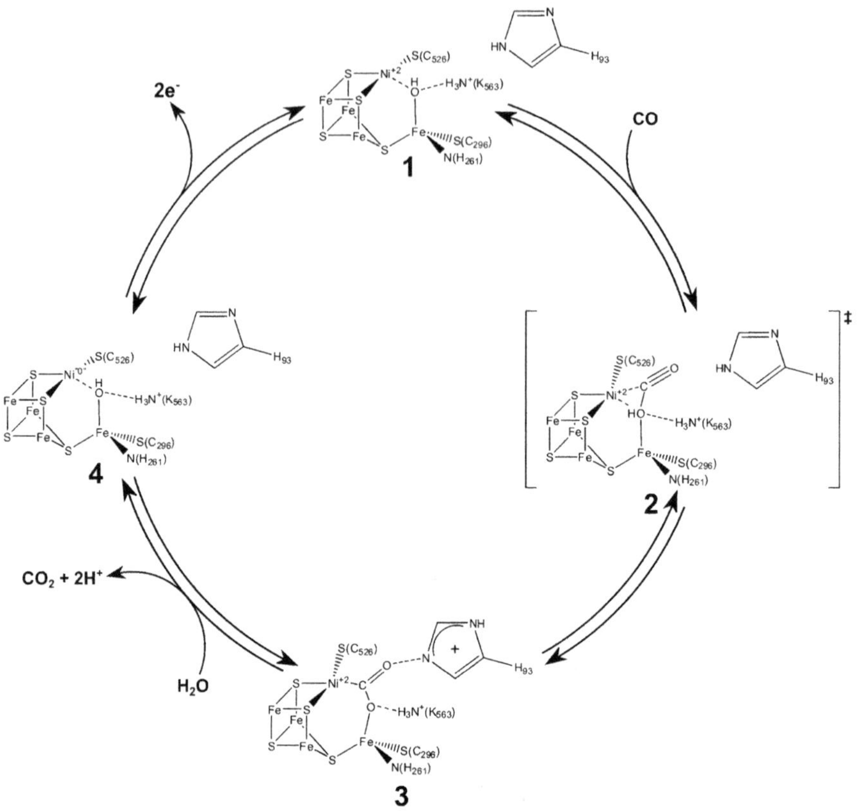

Scheme 10.13 Proposed reaction mechanism of reversible reduction of CO_2 to CO at anaerobic dehydrogenase enzymes

CO_2. The cluster also serves to act as an electronic buffer stabilizing the electronic charges on Fe and Ni during the catalytic cycle.

DuBois et al., reported very high catalytic rates ($k > 104\ M^{-1}\ s^{-1}$) for CO_2 reduction using the bimetallic palladium complex **3** and **4** reported in Scheme 10.14 in which two triphosphine units were bridged by a methylene group [104]. The complex shares some interesting structural features with the enzyme described above, where the active site of *Carboxydothermus hydrogenoformans* in the reduced form is depicted in structure 1, and the reduced form plus CO_2 is shown by structure 2. Complex 3 was designed to interact with CO_2 in much the same way as the enzyme, as shown in structure 4. Both the enzyme and the bimetallic palladium complex form seven-membered rings upon binding CO_2. It is these structural features that are thought to contribute to the remarkably high activity observed for 3 for the electrocatalytic reduction of CO_2 to CO.

However, the turnover number for **3** is too low, approximately 10. The structure of the CO dehydrogenase active site suggests that a more rigid ring system for separating the two metals may be important in preventing Ni–Fe bond formation.

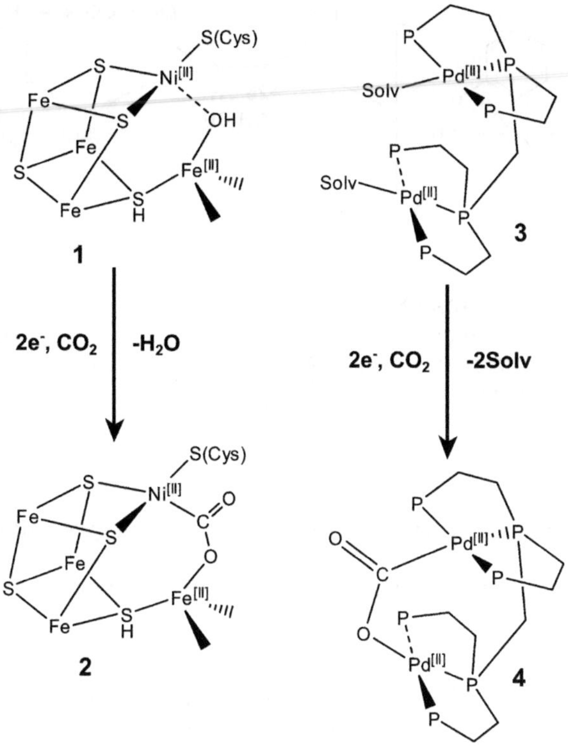

Scheme 10.14 Adapted from Ref. [46]. Copyright (2009) American Chemical Society

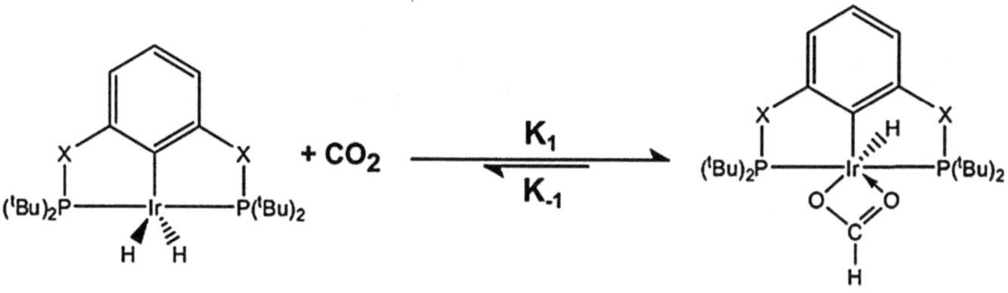

1 (X = O) (POCOP)
3 (X = CH2) (PCP)

2^{CO2} (X = O)
4^{CO2} (X = CH2)

Scheme 10.15 Adapted from Ref. [109]. Copyright (2012) American Chemical Society

So far we have shown that most of the examined electrocatalysts lead to the predominant formation of CO. Although CO formation is an important step, further reduction to formic acid, methanol, or higher hydrocarbons is desirable to obtain compounds with higher energy density. A few molecular electrocatalysts

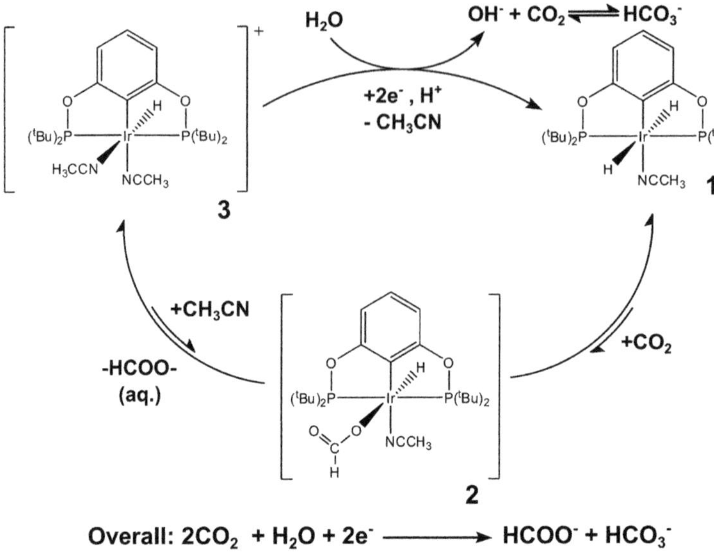

Scheme 10.16 Adapted from Ref. [109]. Copyright (2012) American Chemical Society

yield selectively formate/formic acid but these catalysts often lead to the nonselective formation of CO and H_2 [105]. In contrast, formate dehydrogenase [106] selectively reduces CO_2 to formate at the thermodynamic potential with a high turnover frequency of ca. 280 s^{-1}. Since formic acid could serve as a hydrogen storage material [107], or precursor to methanol as well as in direct formate fuel cells [108], more efficient electocatalysts for selective conversion of CO_2 to formic acid are desirable.

Mayer, Brookhart et al. [109] have reported a selective electrocatalytic reduction of CO_2 to formate with stable Iridium dihydride pincer complexes. As shown in Scheme 10.15 the treatment of either (POCOP)IrH_2 [1] or (PCP)IrH_2 [3] with CO_2 (1 atm) in THF at 25 °C rapidly yields the corresponding formate complexes 2CO_2 and 4CO_2.

Controlled-potential electrolysis at −1.45 V in 5 % H_2O/MeCN yielded formic acid as the predominant product upon acidic workup, with a turnover number of ca. 40 and a Faradaic efficiency of 85 %. No CO was formed and a small amount of H_2 (15 %) was found as a side product from background reduction of water. Scheme 10.16 shows an electrocatalytic mechanism consistent with the experimental observations. The dihydride-acetonitrile adduct 1 is in rapid equilibrium with **3** which likely forms via formate complex **2**. At water concentrations above 4 %, the dominant form of the catalyst in the electrocatalytic steady state is **3**. Two-electron, one-proton reduction of **3** yields dihydride 1. Water is the proton source, so hydroxide is generated and reacts with a second CO_2 to form bicarbonate, HCO_3^-.

10.5.5 Photoelectroreduction of CO_2

Among the various strategies currently being explored for the chemical recycling of carbon dioxide to useful fuels, photoelectrochemical systems are likely to play a fundamental role. The challenge is to convert the energy of sunlight into high energy density molecules such as liquid hydrocarbons. At present there are different systems used to reduce CO_2 using solar energy; for example, homogeneous photoreduction or electrochemical reduction by an electrolyzer powered by photovoltaic devices as well as photoelectrochemical reduction by a semiconductor photocathode. In this respect, a variety of different catalytic approaches, have been used to achieve light absorption, electron-hole separation, and electrochemical reduction of CO_2 to different products.

P-type semiconductor/liquid junctions, where the p-type semiconducting electrodes act as photocathodes for photoassisted CO_2 reduction, have been extensively studied [110–114].

Four different methods of photoassisted reduction of CO_2 have been reported: (a) direct heterogeneous CO_2 reduction by a semiconductor photocathode [112], (b) heterogeneous CO_2 reduction by metal particles on a semiconductor photocathode [115, 116] (c) homogeneous CO_2 reduction by a molecular catalyst through a semiconductor/molecular catalyst junction [117, 118], and (d) heterogeneous CO_2 reduction by a molecular catalyst attached to the semiconductor photocathode surface [119]. Each of these approaches has its own advantages and disadvantages.

Active photoreduction of CO_2 to CO was achieved with the [Re(bipy-But)(CO)$_3$Cl] complex (bipy-But = 4,4′-di- tert -butyl-2,2′-bipyridine) on p-type H–Si using a semiconductor-liquid rectifying junction [120, 121]. The selective photoreduction of CO_2 to CO was obtained on a planar p-type H–Si photocathode with a photovoltage of 570 (30 mV). The Faradaic efficiency was 97 $\pm$ 3 %, while the largest short-circuit quantum efficiency of ca. 60 % for CO production on p-type H–Si was observed for an illumination intensity of 95 mW cm^{-2}. A similar behavior was observed for polychromatic illumination with an open-circuit voltage higher than 650 mV, a short-circuit current density of 23 mA cm^{-2}, a fill factor of 56 $\pm$ 2 %, and an overall conversion efficiency of ca. 10 %. Both the H–Si and the rhenium catalyst showed excellent stability. Later the same authors studied not only the importance of the interactions of an electrocatalyst with the targeted catalytic substrate, but also the heterogeneous electron transfer from the electrode surface to the catalyst. In fact, it has been shown that the electrode surface makes a large difference in the catalytic current density by controlling the electron transfer from the electrode to the catalyst. The catalytic current densities have been increased threefold by modifying the p-Si surface with phenylethyl groups. The electrocatalytic system is very selective for the reduction of CO_2, with a rate that can be enhanced by the addition of a proton source.

Multijunction photoelectrolysis cells are presently used to overcome the high potential that is associated with water splitting [122–125]. Similar efforts for CO_2 photoreduction are at present lacking.

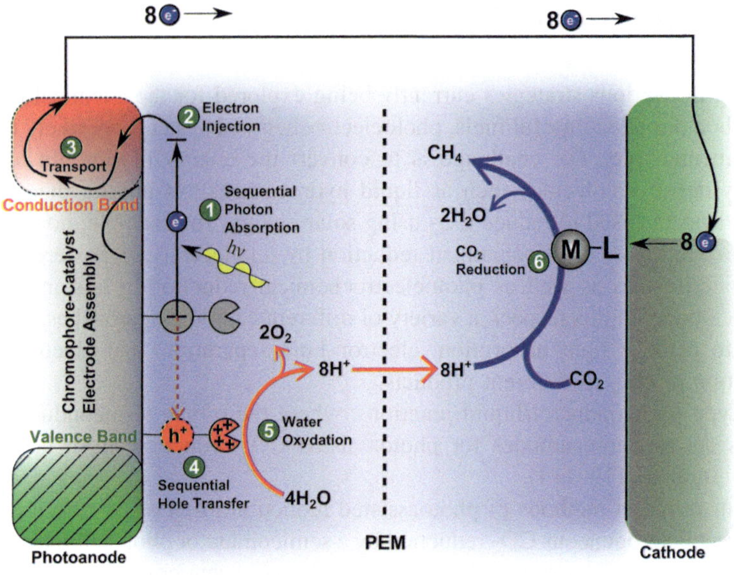

Fig. 10.8 Day sensized photoelectrosynthesis cell (DSPEC) for reduction of CO_2 to methane

An original hybrid photoelectrochemical approach has been proposed by Meyer et co-workers [126] (Fig. 10.8). The system consists of a dye-sensitized photolectrosynthesis cell (DSPEC) based on a photoanode for water oxidation [127–129]. DSPECs are closely related to dye-sensitized solar cells in that light absorption occurs at a light absorber (dye or chromophore), a molecule, cluster, or quantum dot-bound to the surface of a high band gap semiconductor, typically TiO_2. Light absorption and injection initiate a series of electron transfer events (from 1 to 6, Fig. 10.8) that lead to the reduction of CO_2 to fuel by nanostructured metal catalysts or molecular complexes.

The challenge currently facing researchers in this field for the realization of these types of cells is to adopt the same approach used for water-splitting/hydrogen-generation technology. The key will be the development of electrocatalyst materials that have low overpotentials for both CO_2 reduction and water oxidation, and the combination with effective proton-exchange membranes.

10.6 Molecular Complexes for Fuel Cell Cathodes

The oxygen reduction reaction (ORR) that transforms molecular oxygen into water is one of the most important reactions in energy conversion technologies such as fuel cells [130], microbial fuel cells [131], and metal-air batteries. Platinum and platinum alloys are generally considered the most active catalysts for ORR but

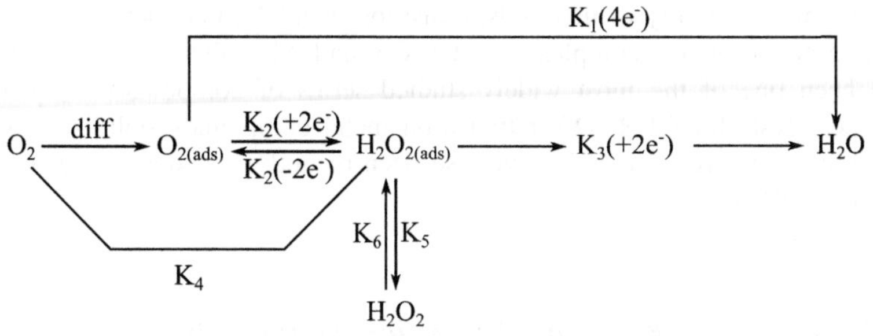

Scheme 10.17 Adapted from Ref. [134] with permission from Elsevier

serious concerns among which the poor availability in nature, its high cost and kinetic inefficiencies are a major barrier to the large-scale production of energy devices. The consequential quest for low cost alternatives to platinum has led to the study of nonprecious metal catalysts capable of promoting the direct four-electron reduction of O_2 to H_2O via a selective and complete pathway. The ORR process normally proceeds through either of two pathways: one is the production of water through a four-electron pathway, and the other is production of hydrogen peroxide through a two-electron pathway (Scheme 10.17). An effective ORR catalyst reduces oxygen molecules to water through the four-electron route. Incomplete reduction of oxygen to hydrogen peroxide not only leads to low energy conversion efficiency, but also produces this reactive intermediate that can further convert to harmful free radical species capable of destroying the catalyst itself and the membrane in the case of a fuel cell.

Certain naturally occurring metal macrocyclic complexes are known to have high electrical conductivity via π-electrons comparable to metals and at the same time are known to catalyze oxygen reduction under physiological conditions. Many researchers have attempted to use these compounds as cathodes in fuel cells. In catalysts which contain metal nanoparticles, only the surface of the metal particles work as the catalytically active sites, and the bulk phase does not participate in the ORR reaction; instead every single metal atom can function as an active site in the case of a metal complex catalyst, leading to a much more effective use of metal-based catalysts. The overpotential at the cathode side in a fuel cell is influenced by the activation overpotential of the catalyst. In the case a metal complex catalyst, the redox properties of the complex will decide the potential where O_2 is reduced.

Characterization of the ORR is commonly carried out by electrochemical techniques such as cyclic voltammetry (CV), measurements using a rotating disk electrode (RDE) or rotating ring disk electrode (RRDE). CV records the current drawn from the electrode as it is cycled between chosen high and low potentials. From the resulting voltammogram the electrochemical surface area and mass- and area-specific activities of oxygen reduction catalysts can be obtained [132].

A promising family of materials, transition metal N_4-macrocycles such as Fe, Co, and Ni porphyrin complexes or Fe, Co, and Ni phthalocyanine complexes, have been one of the most widely studied series of compounds as alternative electrocatalysts for ORR. Other transition metal compounds stabilized by polyaniline or polypyrrole ligands have also been recently included in the search for effective ORR catalysts.

10.6.1 Cathodes Based on Transition Metal Complexes with Phthalocyanine Ligands

Jasinscki [133] reported in 1964 that Cobalt(II) phthalocyanine (CoPc) showed an electrocatalytic activity for ORR comparable to that of Pt or Ag. Since then several transition metal phthalocyanine complexes have been studied. These macrocyclic complexes can bind molecular oxygen to the metal center and transfer electrons via the conjugated aromatic rings surrounding the active center. Randin [134] found a linear correlation between the redox potential of the central metal and the electrochemical activity of several metallophthalocyanines toward ORR (Fig. 10.9).

The central metals should have accessible *d*-orbitals located at the level between the highest occupied molecular orbital (HOMO) and lowest unoccupied molecular orbital (LUMO) of the macrocycles [135]. Metals that meet these two conditions are mostly transition metals with partially filled *d*-electron orbitals (group VIa to VIII). Beck [136] hypothesized a mechanism for ORR on MPc catalysts. It was supposed that the central metal(II) cation was oxidized when O_2 was adsorbed and the molecular O_2 was then partially reduced to the superoxide state.

$$M(II) + O_2 = [M(III) - O_2]^- \qquad (10.1)$$

$$[M(III) - O_2]^- + e^- = M(II) + \text{intermediates} \qquad (10.2)$$

Zagal [135] found that the second step is the rds for the ORR reaction: the M(III)/M(II) potential was an important parameter in determining the electrocatalytic activity of metallophthalocyanine compounds. The more positive the M(III)/M(II) potential, the higher is the activity of the MPc. In acid solutions, the initial step of the ORR on iron phthalocyanine complexes is believed to be the chemisorption of the oxygen molecule by the central transition metal atom [137].

Of the various phthalocyanines evaluated for the reduction of oxygen for fuel cells, the complexes with cobalt and copper appear to be the most stable, while those with iron and cobalt have the best combination of activity and stability. Moreover, Fe, Mn, and Cr phthalocyanine complexes promote a four-electron reduction of oxygen, unlike CoPc which follows a two-electron process. However, Lu and Reddy [138] reported that CoPc not only exhibits a high electrocatalytic

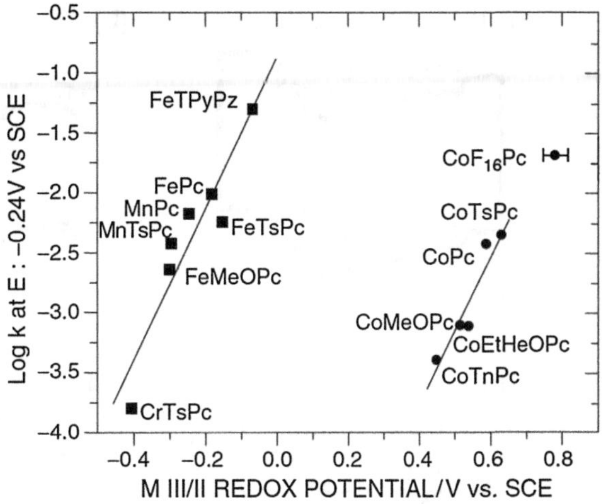

Fig. 10.9 ORR activity of various metallophthalocyanines in 0.1 M NaOH plotted against the first oxidation potential. Adapted from Ref. [134] with permission from Elsevier

activity toward ORR but also demonstrated high tolerance toward methanol cross over in direct methanol fuel cells. Like CoPc, FePc also exhibits a high resistance toward fuel cell crossover in direct alcohol fuel cells. [139]; Ma et al. [140] reported that FePc shows an acceptable ORR activity and a borohydride tolerance as well as a remarkable stability in alkaline media in direct borohydride fuel cells.

Recent investigation of iron phthalocyanines for PEMFCs has been carried by Baker et al. [141] who prepared four types of FePcs: Fe(III) phtalocyanine (FePc), Fe(III) phtalocyanine-4,4′,4″,4‴-tetrasulfonic acid (FePc $(SO_3)4$), Fe(II) 1,2,3, 4,8,9,10,11,15,16,17,18,22,23,24,25-hexadecachloro-29*H*, 31*H*-phtalocyanine (Fe-PcCl16), and 2,11,20,29-tetra-tert-butyl- 2,3-naphtalocyanine (NpPc (tBu)4. The study showed that the nature of the substituent had a strong effect on the O_2 adsorption of the FePc species.

The results showed that in addition to substituent, the temperature also had an influence on the ORR mechanism. Of the four FePcs tested, FePcCl16 was the most stable in an acidic ORR environment but it showed a low kinetic.

10.6.2 Transition Metal Complexes with Porphyrin Ligands

Porphyrins are the second major class of macrocyclics investigated as ORR nonprecious metal catalysts. Indeed, nature shows us that cytochrome *c* oxidase and related heme/copper terminal oxidases catalyze the complete four-proton, four-electron conversion of oxygen to water without releasing partially reduced peroxide intermediates that are toxic to cells. Attempts to mimic such processes

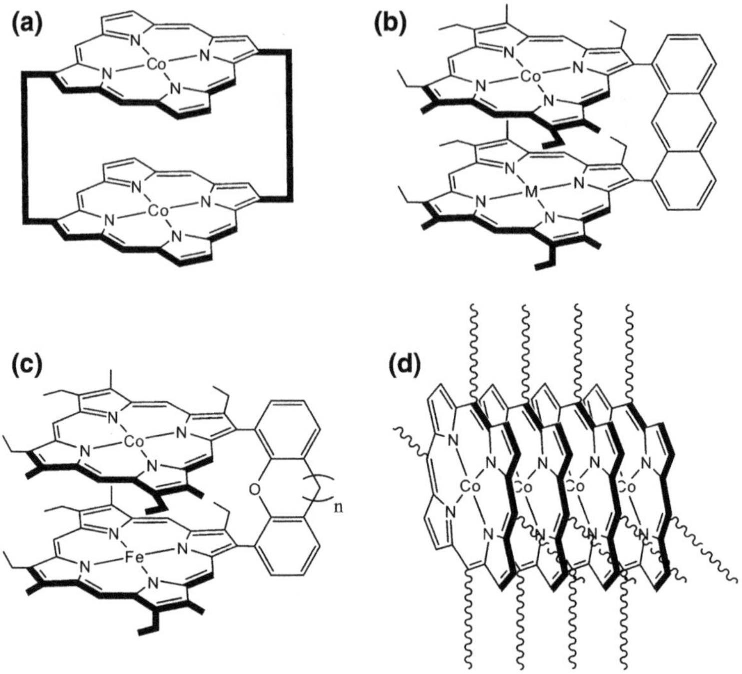

Fig. 10.10 Porphyrin complexes designed as O_2 reduction electrocatalysts

has led to the design of porphyrin platforms with precise control over the hydrogen-bonding functional nature [142]. A mononuclear complex of cobalt porphyrin (CoP) is known to function as a two-electron reduction catalyst of O_2 [143]. On the other hand, the cofacial diporphyrin complexes (Fig. 10.10a), capable of forming μ-peroxo complexes (PCo–O–O–CoP) by sandwiching O_2 with the two metal atoms, show higher selectivity for the four-electron reduction of O_2 due to the decrease in the O=O bond order upon the formation of the μ-peroxo bridge which allows the cleavage of O_2 [144, 145]. When a rigid pillar such as anthracene (Fig. 10.10b) is introduced at one side of the two porphyrin rings to bind them, the steric repulsion upon the coordination of O_2 is facilitated, the turnover speed of ORR increases considerably [146]. Diporphyrin complexes, which have a flexible linkage between the two porphyrins are found to catalyze the four-electron reduction of O_2 characterized by fast kinetics (Fig. 10.10c) [147]. Moreover, porphyrin assemblies by intermolecular interaction have shown a high activity for four-electron reduction of oxygen (Fig. 10.10d) [148].

10.6.3 Carbon-Supported Metal Chelates for ORR Synthesized at High Temperature

Treatment at high temperature of carbon-supported metal chelates under an inert atmosphere has been found to improve both the activity and stability of catalysts for ORR. The pyrolysis treatment changes the nature of the starting material, which are generally macrocyclic compounds, modifying the morphology of the active sites. Yeager et al. [149], reported that nonprecious metal catalysts can indeed be prepared by pyrolyzing a mixture of polyacrylonitrile, cobalt and iron salts instead of the conventional transition metal macrocycles. Since then, several Fe- and Co-based catalysts for ORR in acid media have been prepared using separate metal and nitrogen precursors that are brought together with a carbonaceous support in a heat-treatment step at 600–1,000 °C [150–153]. Recently Dodelet et al. [19, 21], has shown that a careful choice of nitrogen ligands and metal precursors, combined with an appropriate carbon support and optimized heat-treatment conditions, can yield very active Me/N/C-catalysts that perform nearly as well as commercial Pt/C-materials. This group produced catalysts by impregnating iron acetate (FeAc) onto a microporous carbon black that was subsequently pyrolyzed in ammonia (NH_3), thereby creating activated micropores that may host catalytic sites. To obtain high content of microporous carbon blacks a mixture of pore filler and iron precursor was introduced into the micropores. Planetary ball-milling was used to fill in the pore-filler/iron precursor mixture to prevent limitations of solubility and absorbability normally associated with simple impregnation. The same research group found that the N-bearing, 1,10-phenanthroline (phen) pore filler after being pyrolyzed firstly with Ar at 1,050 °C and secondly with NH_3 at 950 °C resulted in a very high catalytic activity. With a nominal 1 wt% Fe content, the current density of a cathode made with the iron-based electrocatalyst reported was the same with that of a platinum-based cathode with a loading of 0.4 mg of platinum per square centimeter at a cell voltage of $\geq$0.9 V. The most surprising thing is the presence of iron not as metallic particles. It has been proposed that most of the Fe/N/C catalytic sites consist of Fe^{2+} cations coordinated by four pyridinic functionalities attached to the edges of two graphitic sheets, each belonging to adjacent crystallites on either side of a slit pore in the carbon support (Fig. 10.11). It was found that the ORR activity depends both of the iron content and on the concentration of nitrogen functionalities on the support surface, as well as on the microporous surface area. However, no consensus has yet been reached regarding the exact identity of the active catalytic species in these types of catalysts [154–156].

In general for NPMCs the simultaneous presence of a transition metal, carbon, and nitrogen plays an important role in formation of active site of the catalyst. Moreover, the catalyst prepared with Co and/or Fe precursors is more active compared to other nonprecious transition metals.

Zelenay et al. [22] reported that metal nanoparticles were formed when polyaniline was used as a precursor to a carbon–nitrogen template, treated at high

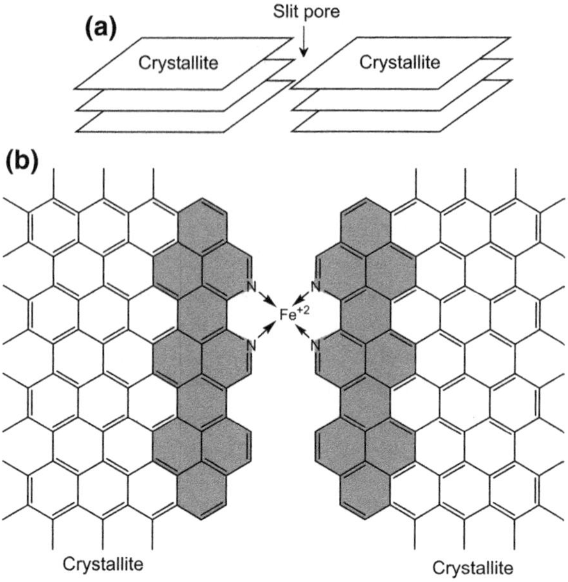

Fig. 10.11 Schematic representation of catalytic site formation in the micropores of the carbon support (*A*); Plan view of the presumed catalytic site (incomplete) and graphitic sheet growth (*shaded aromatic cycles*) between two crystallites after pyrolysis

temperatures in the presence of cobalt and iron salts. The electrocatalysts obtained showed a remarkable performance stability for a nonprecious metal catalyst (700 h at a fuel cell voltage of 0.4 V) as well as excellent four-electron selectivity with a very low hydrogen peroxide formation (< 1.00 %).

The same research group, synthesized a Polypirrole (PPy)-Co/C [17] catalyst without heat treatment. The most important property of the catalyst was that no apparent degradation was observed for 100 h in an H_2-air fuel cell, but its reduction activity remained relatively low. The ORR activity of the nonheat-treated (PPy)-Co/C catalyst was better than that of the one heat treated at 800 °C.

Despite the extensive efforts made in recent years to find catalysts able to replace platinum, poor stability in acid electrolytes has remained a major problem of nonprecious metal catalysts. This is generally attributed to the oxidative corrosion of the carbon support and active sites of the catalysts caused by the hydrogen peroxide (H_2O_2) that is formed during the two-electron reduction of oxygen [157, 158]. It has been shown that the H_2O_2 would be an intermediate in the formation of another more reactive species, the HO radical released by the reaction of free Fe^{2+} with H_2O_2. Recently, a protonation reaction model has also been proposed to explain the poor stability of the NPMCs in acid electrolytes [159].

Most studies confirm that the pyrolysis temperature is a crucial parameter in determining the structure of metallophthalocyanine-derived electrocatalysts and, consequently, their electrochemical performance. It is now apparent that pyrolysis temperatures around 600 °C lead to the prevalent formation of $M–N_4$ units with the metals predominantly in the +2 oxidation state, while metal particles are

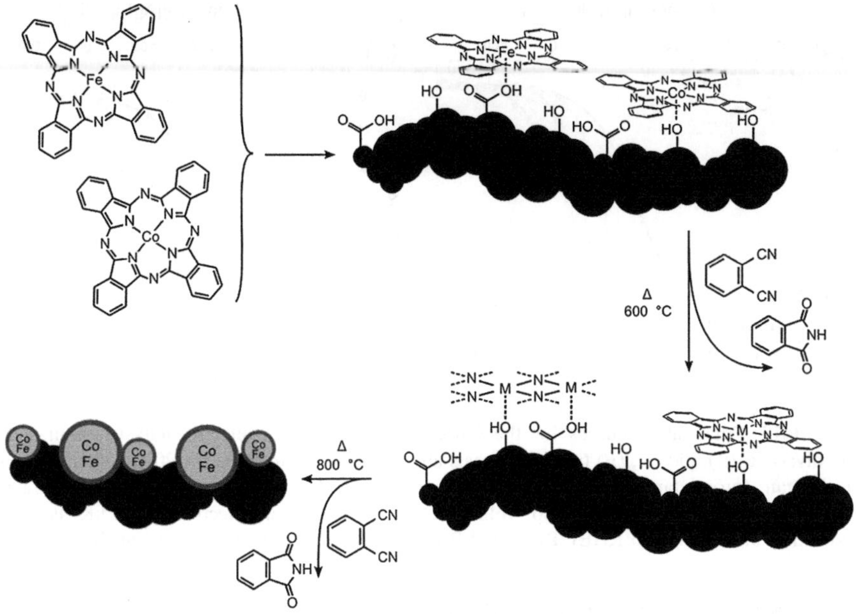

Fig. 10.12 Proposed mechanism for the pyrolysis of Fe and Co phthalocyanines supported on Ketjenblack. Reprinted from Ref. [166] with permission from Elsevier

obtained at higher temperatures, above 800 °C [160, 161]. Both catalysts are active for the ORR in both alkaline and acidic media, yet the electrocatalysts are generally more stable in the alkaline environment [162]. Despite concerted efforts to obtain *platinum free* cathode electrocatalysts for use in acidic media such as in PEMFC devices (H_2/O_2), question marks still remain regarding both long-term stability and the true nature of the catalytic sites. Until such issues are resolved these materials although low cost cannot be taken into serious consideration as alternatives to the state-of-the-art nanostructured platinum electrocatalysts.

However, as an alternative, these nitrogen containing transition metal catalysts have been successfully used for oxygen reduction in alkaline media [163, 164]. The advantages of using alkaline media are high kinetics of electrocatalytic processes and improved stability of the catalyst materials themselves. In particular, recent progress in the development of solid anion-exchange membranes has contributed to a resurgence of interest in alkaline electrocatalytic reactions [165]. The most commonly utilized catalysts for ORR in alkaline media are the Co and Fe materials. Recently, Vizza et al. [166] reported that the impregnation of Ketjenblack (C) with iron and cobalt phthalocyanines (MPc) individually or in a 1:1 stoichiometric mixture, followed by heat treatment at 600 °C under inert atmosphere, gave materials (FePc/C(600), CoPc/C(600), and FePc–CoPc/C(600)) that contain single metal ions coordinated by four nitrogen atoms (M–N4 units). Increasing the pyrolysis temperature to 800 °C resulted in the prevailing formation

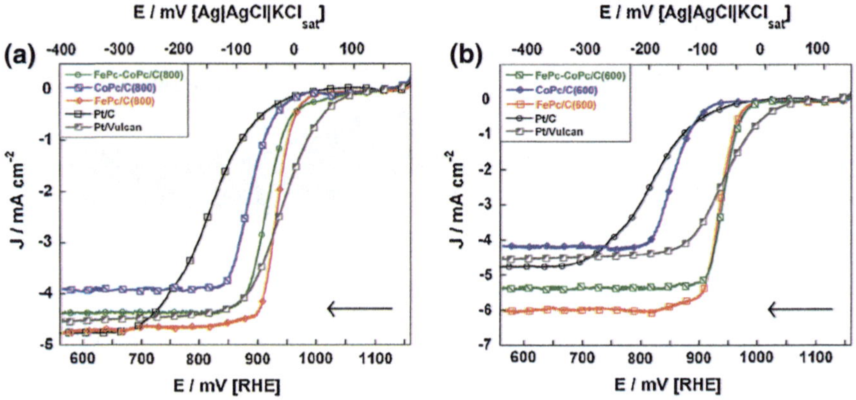

Fig. 10.13 Polarization curves for the ORR on the Fe and Co electrocatalysts obtained by pyrolysis at: (*A*) 800 °C, (*B*) 600 °C. Curves are also reported for Pt/C (Pt 20 wt%/Vulcan) and Pt/Vulcan (proprietary 3.2 wt% Pt). Experimental conditions: KOH 0.1 M, O_2 saturated, RDE″ = 1,600 rpm, linear sweep voltammetry 5 mVs^{-1}, ref. electrode AgCl|KClsat (all potentials are referred to RHE). Reprinted from Ref. [166] with permission from Elsevier

of carbon-supported, nanosized metal particles (CoPc/C(800) and FePc–CoPc/C(800)) or metal oxide (FePc/C(800)). A key role of the carbon support in determining the material structure at either temperature investigated was demonstrated by TPD, EXAFS, XANES, and XRPD studies. These also showed that a Fe–Co alloy is obtained at 800 °C when a mixture of FePc and CoPc was used (Fig. 10.12).

Electrodes coated with the different Fe, Co, and Fe–Co materials, containing ca. 3 wt% metal loadings, were scrutinized for the ORR in alkaline media by linear sweep voltammetry (Fig. 10.13). For comparative purposes, two Pt electrocatalysts containing 3 and 20 wt% metal were investigated. The electrochemical activity of all materials was analyzed by Tafel and Koutecky–Levich plots as well as chronopotentiometry.

The Fe-containing electrocatalysts have been found to be highly active for the ORR in alkaline media with convective limiting currents as high as 600 $AgFe^{-1}$ at room temperature and onset potentials as high as 1.02 V versus RHE. It has been found that (i) the ORR mass activity of the Pc-derived electrocatalysts is superior to that of the Pt catalysts investigated; (ii) the activity of FePc and FePc–CoPc/C, heat treated at either 600 or 800 °C, is superior to that of the corresponding Co materials; (iii) the electrocatalysts obtained at 600 °C are fairly more active than those obtained at 800 °C.

It is worth noticing that the mixed-metal catalyst FePc–CoPc/C(600) is as active as FePc/C(600) which, however, contains about three times more iron. This finding, in conjunction with the number of electron involved in the ORR process suggests the existence of a cooperative effect between the Fe and Co sites in FePc–CoPc/C(600). Such a cooperation seems to hold also for the Fe–Co material pyrolyzed at 800 °C. While for the latter, the Fe–Co cooperation may be attributed

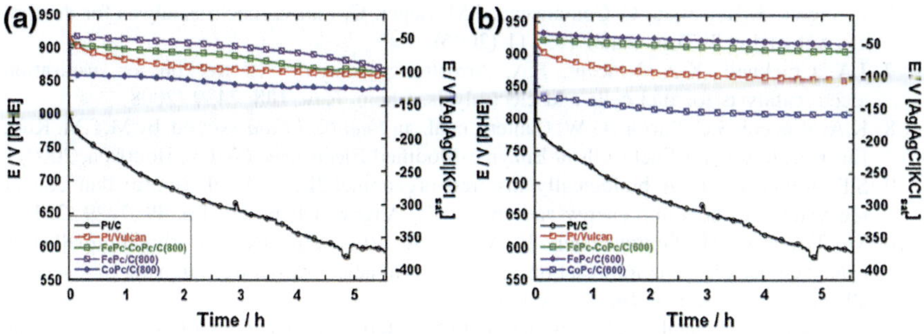

Fig. 10.14 Chronopotentiometric curves for the ORR catalyzed by the Fe and Co phthalocyanines heat treated at: (*A*) 800 °C, (*B*) 600 °C. Curves obtained with Pt/C and Pt/Vulcan are reported for comparative purposes. Experimental conditions: KOH 0.1 M, O_2 saturated, RDE″ = 1,600 rpm, constant current chronopotentiometry at 2.5 mAcm^{-2}, ref. electrode AgCllKClsat (all potentials are referred to RHE). Reprinted from Ref. [166] with permission from Elsevier

to the formation of a Fe–Co alloy, no sound explanation can be forwarded for the materials obtained at 600 °C where the carbon surface is likely covered by a random distribution of M–N4 sites. As pointed out by several authors, nitrogen functionalities in the carbon support (Vulcan) may have a role in the ORR activity in alkaline media [167–170]. Therefore, we cannot rule out that the functionalized Ketjenblack may contribute to the observed ORR activity, though no specific report has ever been reported.

The excellent stability (Fig. 10.14), the selective reduction of oxygen to water (4e$^-$ pathway) and the inability to oxidize alcohols makes the Pc-derived catalysts excellent candidates for manufacturing cathodes for DAFCs operating in alkaline media with anion-exchange membranes. Studies by the same researchers fully confirm the excellent performance of this class of catalysts in DAFCs fed with ethanol, glycerol, and ethylene glycol [171].

References

1. E. Antolini, Catalysts for direct ethanol fuel cells. J. Power Sources **170**, 1 (2007)
2. V. Bambagioni et al., Pd and Pt-Ru anode electrocatalysts supported on multi-walled carbon nanotubes and their use in passive and active direct alcohol fuel cells with an anion-exchange membrane (alcohol = methanol, ethanol, glycerol). J. Power Sources **190**, 241 (2009)
3. C. Bianchini et al., Selective oxidation of ethanol to acetic acid in highly efficient polymer electrolyte membrane-direct ethanol fuel cells. Electrochem. Commun. **11**, 1077 (2009)
4. C. Bianchini, P.K. Shen, Palladium-based electrocatalysts for alcohol oxidation in half cells and in direct alcohol fuel cells. Chem. Rev. **109**, 4183 (2009)
5. K. Matsuoka, Y. Iriyama, T. Abe, M. Matsuoka, Z. Ogumi, Alkaline direct alcohol fuel cells using an anion exchange membrane. J. Power Sources **150**, 27 (2005)

6. F. Vigier, S. Rousseau, C. Coutanceau, J.M. Leger, C. Lamy, Electrocatalysis for the direct alcohol fuel cell. Top. Catal. **40**, 111 (2006)
7. J.A. Cracknell, K.A. Vincent, F.A. Armstrong, Enzymes as working or inspirational electrocatalysts for fuel cells and electrolysis. Chem. Rev. **108**, 2439 (2008)
8. K.A. Vincent, S.C. Barton, G.W. Canters, et al., in *Fuel Cell Catalysis,* ed. by M.T.M. Koper. Electrocatalysis for Fuel Cells at Enzyme-Modified Electrodes, (Wiley, Hoboken, 2009)
9. S.P. Annen et al., A biologically inspired organometallic fuel cell (omfc) that converts renewable alcohols into energy and chemicals. Angew. Chem. Int. Ed. **49**, 7229 (2010)
10. M. Trincado, H. Grutzmacher, F. Vizza, C. Bianchini, Domino rhodium/palladium-catalyzed dehydrogenation reactions of alcohols to acids by hydrogen transfer to inactivated alkenes. Chem. Eur. J. **16**, 2751 (2010)
11. T. Zweifel, J.V. Naubron, T. Buttner, T. Ott, H. Grutzmacher, Ethanol as hydrogen donor: Highly efficient transfer hydrogenations with rhodium(I) amides. Angew. Chem. Int. Ed. **47**, 3245 (2008)
12. T. Zweifel, J.V. Naubron, H. Grutzmacher, Catalyzed dehydrogenative coupling of primary alcohols with water, methanol, or amines. Angew. Chem. Int. Ed. **48**, 559 (2009)
13. V. Bambagioni et al., Energy efficiency enhancement of ethanol electrooxidation on Pd-CeO2/C in passive and active polymer electrolyte-membrane fuel cells. Chemsuschem **5**, 1266 (2012)
14. M. Bevilacqua et al., Improvement in the efficiency of an OrganoMetallic Fuel Cell by tuning the molecular architecture of the anode electrocatalyst and the nature of the carbon support. Energy Environ. Sci. **5**, 8608 (2012)
15. E. Katz, A.F. Buckmann, I. Willner, Self-powered enzyme-based biosensors. J. Am. Chem. Soc. **123**, 10752 (2001)
16. K. Asazawa et al., A platinum-free zero-carbon-emission easy fuelling direct hydrazine fuel cell for vehicles. Angew. Chem. Int. Ed. **46**, 8024 (2007)
17. R. Bashyam, P. Zelenay, A class of non-precious metal composite catalysts for fuel cells. Nature **443**, 63 (2006)
18. F.R. Brushett et al., A carbon-supported copper complex of 3,5-Diamino-1,2,4-triazole as a cathode catalyst for alkaline fuel cell applications. J. Am. Chem. Soc. **132**, 12185 (2010)
19. M. Lefevre, E. Proietti, F. Jaouen, J.P. Dodelet, Iron-based catalysts with improved oxygen reduction activity in polymer electrolyte fuel cells. Science **324**, 71 (2009)
20. S.F. Lu, J. Pan, A.B. Huang, L. Zhuang, J.T. Lu, Alkaline polymer electrolyte fuel cells completely free from noble metal catalysts. Proc. Nat. Acad. Sci. U.S.A. **105**, 20611 (2008)
21. E. Proietti et al., Iron-based cathode catalyst with enhanced power density in polymer electrolyte membrane fuel cells. Nat. Commun. **2**(416), 1–9 (2011)
22. G. Wu, K.L. More, C.M. Johnston, P. Zelenay, High-performance electrocatalysts for oxygen reduction derived from polyaniline, iron, and cobalt. Science **332**, 443 (2011)
23. S. Yamazaki, T. Ioroi, Y. Yamada, K. Yasuda, T. Kobayashi, A direct CO polymer electrolyte membrane fuel cell. Angew. Chem. Int. Ed. **45**, 3120 (2006)
24. Y. Hasegawa, M. Watanabe, I.D. Gridnev, T. Ikariya, Enantioselective direct amination of alpha-cyanoacetates catalyzed by bifunctional chiral Ru and Ir amido complexes. J. Am. Chem. Soc. **130**, 2158 (2008)
25. T. Ikariya, A.J. Blacker, Asymmetric transfer hydrogenation of ketones with bifunctional transition metal-based molecular. Acc. Chem. Res. **40**, 1300 (2007)
26. M. Ito, T. Ikariya, Catalytic hydrogenation of polar organic functionalities based on Ru-mediated heterolytic dihydrogen cleavage. Chem. Commun. 5134–5142 (2007)
27. M. Ito et al., Hydrogenation of N-Acylcarbamates and N-Acylsulfonamides catalyzed by a bifunctional Cp*Ru(PN) complex. Angew. Chem. Int. Ed. **48**, 1324 (2009)
28. M. Ito, A. Osaku, C. Kobayashi, A. Shiibashi, T. Ikariya, A convenient method for the synthesis of Protic 2-(Tertiary phosphino)-1-amines and their CP*RuCl complexes. Organometallics **28**, 390 (2009)

29. S. Shirai, H. Nara, Y. Kayaki, T. Ikariya, Remarkable positive effect of silver salts on asymmetric hydrogenation of acyclic imines with Cp*Ir complexes bearing chiral N-sulfonylated diamine ligands. Organometallics **28**, 802 (2009)
30. V. Artero, M. Fontecave, Some general principles for designing electrocatalysts with hydrogenase activity. Coord. Chem. Rev. **249**, 1518 (2005)
31. S. Canaguier, M. Fontecave, V. Artero, Cp*(-)-Ruthenium-Nickel-Based H-2-evolving electrocatalysts as bio-inspired models of NiFe hydrogenases. Eur. J. Inorg. Chem. **2011**, 1094–1099 (2011)
32. Z.M. Heiden, T.B. Rauchfuss, Homogeneous catalytic reduction of dioxygen using transfer hydrogenation catalysts. J. Am. Chem. Soc.**129**, 14303 (2007)
33. T. Matsumoto, K. Kim, S. Ogo, Molecular catalysis in a fuel cell. Angew. Chem. Int. Ed. **50**, 11202 (2011)
34. T. Matsumoto et al., Organometallic Catalysts for Use in a Fuel Cell. ChemCatChem. **5**, 1368–1373 (2013)
35. S. Ogo, Electrons from hydrogen. Chem. Commun. 3317–3325 (2009)
36. S. Ogo et al., A dinuclear Ni(mu-H)Ru complex derived from H-2. Science **316**, 585 (2007)
37. J.C. Fontecilla-Camps, A. Volbeda, C. Cavazza, Y. Nicolet, Structure/function relationships of NiFe - and FeFe -hydrogenases. Chem. Rev. **107**, 4273 (2007)
38. J.C. Gordon, G.J. Kubas, Perspectives on how nature employs the principles of organometallic chemistry in dihydrogen activation in hydrogenases. Organometallics **29**, 4682 (2010)
39. Y. Nicolet, C. Piras, P. Legrand, C.E. Hatchikian, J.C. Fontecilla-Camps, Desulfovibrio desulfuricans iron hydrogenase: The structure shows unusual coordination to an active site Fe binuclear center. Struct. Fold. Des. **7**, 13 (1999)
40. P.M. Vignais, B. Billoud, Occurrence, classification, and biological function of hydrogenases: An overview. Chem. Rev. **107**, 4206 (2007)
41. B. Askevold, H.W. Roesky, S. Schneider, Learning from the neighbors: Improving homogeneous catalysts with functional ligands motivated by heterogeneous and biocatalysis. Chemcatchem **4**, 307 (2012)
42. X.L. Hu, B.S. Brunschwig, J.C. Peters, Electrocatalytic hydrogen evolution at low overpotentials by cobalt macrocyclic glyoxime and tetraimine complexes. J. Am. Chem. Soc. **129**, 8988 (2007)
43. U.J. Kilgore et al., Ni((P2N2C6H4X)-N-Ph)(2) (2+) complexes as electrocatalysts for H-2 production: Effect of substituents, acids, and water on catalytic rates. J. Am. Chem. Soc. **133**, 5861 (2011)
44. A.D. Wilson et al., Hydrogen oxidation and production using nickel-based molecular catalysts with positioned proton relays. J. Am. Chem. Soc. **128**, 358 (2006)
45. A. Le Goff et al., From hydrogenases to noble metal-free catalytic nanomaterials for H-2 production and uptake. Science **326**, 1384 (2009)
46. M.R. Dubois, D.L. Dubois, Development of molecular electrocatalysts for CO_2 reduction and H-2 production/oxidation. Acc. Chem. Res. **42**, 1974 (2009)
47. M.R. DuBois, D.L. DuBois, The roles of the first and second coordination spheres in the design of molecular catalysts for H-2 production and oxidation. Chem. Soc. Rev. **38**, 62 (2009)
48. A.D. Wilson et al., Nature of hydrogen interactions with Ni(II) complexes containing cyclic phosphine ligands with pendant nitrogen bases. Proc. Nat. Acad. Sci. U.S.A. **104**, 6951 (2007)
49. M.L. Helm, M.P. Stewart, R.M. Bullock, M.R. DuBois, D.L. DuBois, A synthetic nickel electrocatalyst with a turnover frequency above 100,000 s(-1) for H-2 production. Science **333**, 863 (2011)
50. C. Bianchini, P. Fornasiero, A synthetic nickel electrocatalyst with a turnover frequency above 100,000 s-1 for H_2 production. Chemcatchem **4**, 45 (2012)
51. M. Frey, Hydrogenases: Hydrogen-activating enzymes. Chembiochem **3**, 153 (2002)

52. E.E. Benson, C.P. Kubiak, A.J. Sathrum, J.M. Smieja, Electrocatalytic and homogeneous approaches to conversion of CO_2 to liquid fuels. Chem. Soc. Rev. **38**, 89 (2009)
53. M.A. Scibioh, B. Viswanatan, Electrochemical reduction of Carbon dioxide. Proc. Indian Nat. Sci. Acad. A. **70**, 1–56 (2004)
54. T. Abe et al., Electrochemical CO(2) reduction catalysed by cobalt octacyanophthalocyanine and its mechanism. J. Porphyrins Phthalocyanines **1**, 315 (1997, 1997)
55. P.A. Christensen, A. Hamnett, A.V.G. Muir, An insitu ftir study of the electroreduction of CO_2 by copc-coated edge graphite-electrodes. J. Electroanal. Chem. **241**, 361 (1988)
56. M.N. Mahmood, D. Masheder, C.J. Harty, Use of gas-diffusion electrodes for high-rate electrochemical reduction of carbon-dioxide .2. Reduction at metal phthalocyanine-impregnated electrodes. J. Appl. Electrochem. **17**, 1223 (1987)
57. H. Tanabe, K. Ohno, Electrocatalysis of metal phthalocyanine thin-film prepared by the plasma-assisted deposition on a glassy-carbon in the reduction of carbon-dioxide. Electrochim. Acta **32**, 1121 (1987)
58. S. Meshitsuka, M. Ichikawa, K. Tamaru, Electrocatalysis by metal phthalocyanines in the reduction of carbon dioxide. J. Chem. Soc. Chem, Commun. **0**, 158 (1974)
59. C.M. Lieber, N.S. Lewis, Catalytic reduction of carbon dioxide at carbon electrodes modified with cobalt phthalocyanine. J. Am. Chem. Soc. **106**, 5033 (1984)
60. B.R. Eggins, J.T.S. Irvine, J. Grimshaw, The voltammetry of mixed solutions of carbon dioxide and metal phthalocyanines in DMSO. J. Electroanal. Chem. Interfacial Electrochem. **266**, 125 (1989)
61. G. Fachinetti, G. Fochi, T. Funaioli, P.F. Zanazzi, Isolation of a tetranuclear iron cluster promoting CO_2 reduction and intermediate to iron carbonyl disproportionation. J. Chem. Soc. Chem. Commun. 89 (1987)
62. P.A. Vigato, S. Tamburini, D.E. Fenton, The activation of small molecules by dinuclear complexes of copper and other metals. Coord. Chem. Rev. **106**, 25 (1990)
63. J.-M. Savéant, Molecular catalysis of electrochemical reactions. Mechanistic Aspects Chem. Rev. **108**, 2348 (2008)
64. T. Abe et al., Electrocatalytic CO_2 reduction by cobalt octabutoxyphthalocyanine coated on graphite electrode. J. Mol. Catal. Chem. **112**, 55 (1996)
65. B.J. Fisher, R. Eisenberg, Electrocatalytic reduction of carbon dioxide by using macrocycles of nickel and cobalt. J. Am. Chem. Soc. **102**, 7361 (1980)
66. M. Beley, J.P. Collin, R. Ruppert, J.P. Sauvage, Electrocatalytic reduction of carbon dioxide by nickel cyclam2+ in water: Study of the factors affecting the efficiency and the selectivity of the process. J. Am. Chem. Soc. **108**, 7461 (1986)
67. M. Beley, J.-P. Collin, R. Ruppert, J.-P. Sauvage, Nickel(II)-cyclam: An extremely selective electrocatalyst for reduction of CO_2 in water. J. Chem. Soc. Chem. Commun. **0**, 1315 (1984)
68. J.P. Collin, A. Jouaiti, J.P. Sauvage, Electrocatalytic properties of (tetraazacyclotetradecane)nickel(2+) and Ni2(biscyclam)4+ with respect to carbon dioxide and water reduction. Inorg. Chem. **27**, 1986 (1988)
69. G.B. Balazs, F.C. Anson, Effects of co on the electrocatalytic activity of ni (cyclam)(2+) toward the reduction of CO_2. J. Electroanal. Chem. **361**, 149 (1993)
70. J.D. Froehlich, C.P. Kubiak, Homogeneous CO_2 reduction by Ni(cyclam) at a glassy carbon electrode. Inorg. Chem. **51**, 3932 (2012)
71. C.M. Bolinger, N. Story, B.P. Sullivan, T.J. Meyer, Electrocatalytic reduction of carbon dioxide by 2,2′-bipyridine complexes of rhodium and iridium. Inorg. Chem. **27**, 4582 (1988)
72. J. Grodkowski et al., Reduction of cobalt and iron corroles and catalyzed reduction of CO_2. J. Phys. Chem. A **106**, 4772 (2002)
73. A.R. Guadalupe et al., Novel chemical pathways and charge-transport dynamics of electrodes modified with electropolymerized layers of [Co(v-terpy)2]2+. J. Am. Chem. Soc. **110**, 3462 (1988)

74. I. Bhugun, D. Lexa, J.M. Saveant, Catalysis of the electrochemical reduction of carbon dioxide by iron(O) porphyrins: Synergystic effect of weak Bronsted acids. J. Am. Chem. Soc. **118**, 1769 (1996)
75. S. Daniele, P. Ugo, G. Bontempelli, M. Fiorani, An electroanalytical investigation on the nickel-promoted electrochemical conversion of CO_2 to co. J. Electroanal. Chem. **219**, 259 (1987)
76. F.R. Keene, C. Creutz, N. Sutin, Reduction of carbon-dioxide by tris(2,2′-bipyridine) cobalt(i). Coord. Chem. Rev. **64**, 247 (1985, 1985)
77. J. Hawecker, J.-M. Lehn, R. Ziessel, Efficient photochemical reduction of CO_2 to CO by visible light irradiation of systems containing Re(bipy)(CO)3X or Ru(bipy)32+-CO_2+ combinations as homogeneous catalysts. J. Chem. Soc. Chem. Commun. **0**, 536 (1983)
78. J. Hawecker, J.-M. Lehn, R. Ziessel, Electrocatalytic reduction of carbon dioxide mediated by Re(bipy)(CO)3Cl (bipy = 2,2′-bipyridine). J. Chem. Soc. Chem. Commun. **0**, 328 (1984)
79. B.P. Sullivan, C.M. Bolinger, D. Conrad, W.J. Vining, T.J. Meyer, One-electron and 2-electron pathways in the electrocatalytic reduction of CO_2 by fac-re(2,2′-bipyridine)(co)3cl. J. Chem. Soc. Chem. Commun. 1414 (1985, 1985)
80. B.P. Sullivan, T.J. Meyer, Photoinduced irreversible insertion of CO_2 into a metal-hydride bond. J. Chem. Soc. Chem. Commun. **0**, 1244 (1984)
81. B.P. Sullivan, T.J. Meyer, Kinetics and mechanism of CO_2 insertion into a metal-hydride bond - a large solvent effect and an inverse kinetic isotope effect. Organometallics **5**, 1500 (1986)
82. H. Ishida, H. Tanaka, K. Tanaka, T. Tanaka, Selective formation of hcoo- in the electrochemical CO_2 reduction catalyzed by ru(bpy)2(co)2 2+ (bpy = 2,2′-bipyridine). J. Chem. Soc. Chem. Commun. 131 (1987)
83. H. Ishida, K. Tanaka, T. Tanaka, Electrochemical CO_2 reduction catalyzed by ru(bpy)2(co)2 2+ and ru(bpy)2(Co)cl+- the effect of ph on the formation of co and hcoo. Organometallics **6**, 181 (1987)
84. C.M. Bolinger et al., Electrocatalytic reduction of CO_2 based on polypyridyl complexes of rhodium and ruthenium. J. Chem. Soc. Chem. Commun. 796 (1985, 1985)
85. F.P.A. Johnson, M.W. George, F. Hartl, J.J. Turner, Electrocatalytic reduction of CO_2 using the complexes Re(bpy)(CO)(3)L (n) (n=+1, L=P(OEt)(3), CH_3CN; n=0, L=Cl-, Otf(-); bpy=2,2′-bipyridine; Otf(-)=CF_3SO_3) as catalyst precursors: Infrared spectroelectrochemical investigation. Organometallics **15**, 3374 (1996)
86. Y. Hayashi, S. Kita, B.S. Brunschwig, E. Fujita, Involvement of a binuclear species with the Re-C(O)O-Re moiety in CO_2 reduction catalyzed by tricarbonyl rhenium(I) complexes with diimine ligands: Strikingly slow formation of the Re-Re and Re-C(O)O-Re species from Re(dmb)(CO)(3)S (dmb=4,4′-dimethyl-2,2′-bipyridine, S=solvent). J. Am. Chem. Soc. **125**, 11976 (2003)
87. H. Takeda, K. Koike, H. Inoue, O. Ishitani, Development of an efficient photocatalytic system for CO_2 reduction using rhenium(l) complexes based on mechanistic studies. J. Am. Chem. Soc. **130**, 2023 (2008)
88. E.E. Benson, C.P. Kubiak, Structural investigations into the deactivation pathway of the CO_2 reduction electrocatalyst Re(bpy)(CO)(3)Cl. Chem. Commun. **48**, 7374 (2012, 2012)
89. J.M. Smieja, C.P. Kubiak, Re(bipy-tBu)(CO)(3)Cl-improved catalytic activity for reduction of carbon dioxide: IR-Spectroelectrochemical and mechanistic studies. Inorg. Chem. **49**, 9283 (2010)
90. F.A. Armstrong, J. Hirst, Reversibility and efficiency in electrocatalytic energy conversion and lessons from enzymes. Proc. Nat. Acad. Sci. U.S.A. **108**, 14049 (2011)
91. V.S. Thoi, C.J. Chang, Nickel N-heterocyclic carbene-pyridine complexes that exhibit selectivity for electrocatalytic reduction of carbon dioxide over water. Chem. Commun. **47**, 6578 (2011, 2011)

92. R. Alvarez et al., Synthesis and X-ray crystal structure of [Mo(CO2)2(PMe3)3(CNPri)]: The first structurally characterized bis(carbon dioxide) adduct of a transition metal. J. Chem. Soc. Chem. Commun. 1326 (1984)
93. R. Alvarez et al., Carbon-dioxide chemistry - synthesis, properties, and structural characterization of stable bis(carbon dioxide) adducts of molybdenum. J. Am. Chem. Soc. **108**, 2286 (1986)
94. M. Aresta, C.F. Nobile, V.G. Albano, E. Forni, M. Manassero, New nickel-carbon dioxide complex: Synthesis, properties, and crystallographic characterization of (carbon dioxide)-bis(tricyclohexylphosphine)nickel. J. Chem. Soc. Chem. Commun. **0**, 636 (1975)
95. G.S. Bristow, P.B. Hitchcock, M.F. Lappert, A novel carbon dioxide complex: synthesis and crystal structure of [Nb([eta]-$C_5H_4Me)_2(CH_2SiMe_3)$([eta]$_2$-CO_2)]. J. Chem. Soc. Chem. Commun. 1145 (1981)
96. S. Slater, J.H. Wagenknecht, Electrochemical reduction of carbon dioxide catalyzed by Rh(diphos)2Cl. J. Am. Chem. Soc. **106**, 5367 (1984)
97. D.L. Dubois, Development of transition metal phosphine complexes as electrocatalysts for CO_2 and CO reduction. Comments Inorg. Chem. **19**, 307 (1997, 1997)
98. P.R. Bernatis, A. Miedaner, R.C. Haltiwanger, D.L. Dubois, Exclusion of 6-coordinate intermediates in the electrochemical reduction of CO_2 catalyzed by pd(triphosphine)(CH_3CN) (BF_4)(2) complexes. Organometallics **13**, 4835 (1994)
99. D.L. Dubois, A. Miedaner, R.C. Haltiwanger, Electrochemical reduction of CO_2 catalyzed by pd(triphosphine)(solvent) (bf4)2 complexes - synthetic and mechanistic studies. J. Am. Chem. Soc. **113**, 8753 (1991)
100. A. Miedaner, B.C. Noll, D.L. DuBois, Synthesis and characterization of palladium and nickel complexes with positively charged triphosphine ligands and their use as electrochemical CO_2-reduction catalysts. Organometallics **16**, 5779 (1997)
101. D.L. DuBois, in *Encyclopedia of Electrochemistry* ed. by A.J. Bard, M. Stratmann, F. Scholz, et al. Electrochemical reactions of carbon dioxide, (Wiley-VCH, Weinheim, 2006)
102. J.W. Raebiger et al., Electrochemical reduction of CO_2 to CO catalyzed by a bimetallic palladium complex. Organometallics **25**, 3345 (2006)
103. J.H. Jeoung, H. Dobbek, Carbon dioxide activation at the Ni,Fe-cluster of anaerobic carbon monoxide dehydrogenase. Science **318**, 1461 (2007)
104. B.D. Steffey, C.J. Curtis, D.L. Dubois, Electrochemical reduction of CO_2 catalyzed by a dinuclear palladium complex containing a bridging hexaphosphine ligand - evidence for cooperativity. Organometallics **14**, 4937 (1995)
105. C. Caix, S. ChardonNoblat, A. Deronzier, Electrocatalytic reduction of CO_2 into formate with (eta(5)-Me_5C_5)M(L)Cl (+) complexes (L=2,2$'$-bipyridine ligands; M=Rh(III) and Ir(III)). J. Electroanal. Chem. **434**, 163 (1997)
106. T. Reda, C.M. Plugge, N.J. Abram, J. Hirst, Reversible interconversion of carbon dioxide and formate by an electroactive enzyme. Proc. Nat. Acad. Sci. U.S.A. **105**, 10654 (2008)
107. T.C. Johnson, D.J. Morris, M. Wills, Hydrogen generation from formic acid and alcohols using homogeneous catalysts. Chem. Soc. Rev. **39**, 81 (2010)
108. J.H. Jiang, A. Wieckowski, Prospective direct formate fuel cell. Electrochem. Commun. **18**, 41 (2012)
109. P. Kang et al., Selective electrocatalytic reduction of CO_2 to formate by water-stable iridium dihydride pincer complexes. J. Am. Chem. Soc. **134**, 5500 (2012)
110. J.O.M. Bockris, J.C. Wass, On the photoelectrocatalytic reduction of carbon dioxide. Mater. Chem. Phys. **22**, 249 (1989)
111. M. Halmann, B. Aurianblajeni, Electrochemical reduction of carbon-dioxide at elevated pressure on semiconductor electrodes in aqueous-solution. J. Electroanal. Chem. **375**, 379 (1994)
112. K. Hirota et al., Photoelectrochemical reduction of CO_2 in a high-pressure CO_2 plus methanol medium at p-type semiconductor electrodes. J. Phys. Chem. B **102**, 9834 (1998)

113. S. Kaneco, H. Katsumata, T. Suzuki, K. Ohta, Photoelectrochemical reduction of carbon dioxide at p-type gallium arsenide and p-type indium phosphide electrodes in methanol. Chem. Eng. J. **116**, 227 (2006)
114. M. Le et al., Electrochemical reduction of CO_2 to CH_3OH at copper oxide surfaces. J. Electrochem. Soc. **158**, E45 (2011)
115. T. Cottineau, M. Morin, D. Bélanger, Modification of p-type silicon for the photoelectrochemical reduction of CO_2. ECS Trans. **19**, 1 (2009)
116. R. Hinogami, Y. Nakamura, S. Yae, Y. Nakato, An approach to ideal semiconductor electrodes for efficient photoelectrochemical reduction of carbon dioxide by modification with small metal particles. J. Phys. Chem. B **102**, 974 (1998)
117. E.E. Barton, D.M. Rampulla, A.B. Bocarsly, Selective solar-driven reduction of CO_2 to methanol using a catalyzed p-GaP based photoelectrochemical cell. J. Am. Chem. Soc. **130**, 6342 (2008)
118. M.G. Bradley, T. Tysak, D.J. Graves, N.A. Viachiopoulos, Electrocatalytic reduction of carbon dioxide at illuminated p-type silicon semiconduccting electrodes. J. Chem. Soc. Chem. Commun. 349 (1983)
119. T. Arai et al., Photoelectrochemical reduction of CO_2 in water under visible-light irradiation by a p-type InP photocathode modified with an electropolymerized ruthenium complex. Chem. Commun. **46**, 6944 (2010)
120. B. Kumar, J.M. Smieja, C.P. Kubiak, Photoreduction of CO_2 on p-type silicon using Re(bipy-But)$(CO)_3$Cl: Photovoltages exceeding 600 mV for the selective reduction of CO_2 to CO. J. Phys. Chem. C **114**, 14220 (2010)
121. J.M. Smieja et al., Kinetic and structural studies, origins of selectivity, and interfacial charge transfer in the artificial photosynthesis of CO. Proc. Nat. Acad. Sci. U.S.A. **109**, 15646 (2012)
122. O. Khaselev, A. Bansal, J.A. Turner, High-efficiency integrated multijunction photovoltaic/electrolysis systems for hydrogen production. Int. J. Hydrogen Energy **26**, 127 (2001)
123. S. Licht et al., Efficient solar water splitting, exemplified by RuO2-catalyzed AlGaAs/Si photoelectrolysis. J. Phys. Chem. B **104**, 8920 (2000)
124. Y. Yamada et al., One chip photovoltaic water electrolysis device. Int. J. Hydrogen Energy **28**, 1167 (2003)
125. S. Yamane et al., Efficient solar water splitting with a composite "n-Si/p-CuI/n-i-p a-Si/n-p GaP/RuO2" semiconductor electrode. J. Phys. Chem. C **113**, 14575 (2009)
126. J.J. Concepcion, R.L. House, J.M. Papanikolas, T.J. Meyer, Chemical approaches to artificial photosynthesis. Proc. Nat. Acad. Sci. U.S.A. **109**, 15560 (2012)
127. J.A. Treadway, J.A. Moss, T.J. Meyer, Visible region photooxidation on TiO2 with a chromophore-catalyst molecular assembly. Inorg. Chem. **38**, 4386 (1999)
128. W.J. Youngblood et al., Photoassisted overall water splitting in a visible light-absorbing dye-sensitized photoelectrochemical cell. J. Am. Chem. Soc. **131**, 926 (2009)
129. W.J. Youngblood, S.H. A. Lee, K. Maeda, T.E. Mallouk, Visible light water splitting using dye-sensitized oxide semiconductors. Acc. Chem. Res. **42**, 1966 (2009)
130. J. Zhang, *PEM Fuel Cell Electrocatalysts and Catalyst Layers. Fundamentals and Applications* (Springer London, London, 2008)
131. B.E. Logan, *Microbial Fuel Cells* (Wiley, Hoboken, New Jersey, 2008)
132. Y. Garsany, O.A. Baturina, K.E. Swider-Lyons, S.S. Kocha, Experimental methods for quantifying the activity of platinum electrocatalysts for the oxygen reduction reaction. Anal. Chem. **82**, 6321 (2010)
133. R. Jasinski, A new fuel cell cathode catalyst. Nature **201**, 1212 (1964)
134. J.-P. Randin, Interpretation of the relative electrochemical activity of various metal phthalocyanines for the oxygen reduction reaction. Electrochim. Acta **19**, 83 (1974)
135. J.H. Zagal, Metallophthalocyanines as catalysts in electrochemical reactions. Coord. Chem. Rev. **119**, 89 (1992)
136. F. Beck, The redox mechanism of the chelate-catalysed oxygen cathode. J. Appl. Electrochem. **7**, 239 (1977)

137. B. Wang, Recent development of non-platinum catalysts for oxygen reduction reaction. J. Power Sources **152**, 1 (2005)
138. Y.H. Lu, R.G. Reddy, The electrochemical behavior of cobalt phthalocyanine/platinum as methanol-resistant oxygen-reduction electrocatalysts for DMFC. Electrochim. Acta **52**, 2562 (2007)
139. S. Baranton, C. Coutanceau, C. Roux, F. Hahn, J.M. Leger, Oxygen reduction reaction in acid medium at iron phthalocyanine dispersed on high surface area carbon substrate: tolerance to methanol, stability and kinetics. J. Electroanal. Chem. **577**, 223 (2005)
140. J.F. Ma, Y.N. Liu, P. Zhang, J. Wang, A simple direct borohydride fuel cell with a cobalt phthalocyanine catalyzed cathode. Electrochem. Commun. **10**, 100 (2008)
141. R. Baker, D.P. Wilkinson, J.J. Zhang, Electrocatalytic activity and stability of substituted iron phthalocyanines towards oxygen reduction evaluated at different temperatures. Electrochim. Acta **53**, 6906 (2008)
142. C.J. Chang, L.L. Chng, D.G. Nocera, Proton-coupled O-O activation on a redox platform bearing a hydrogen-bonding scaffold. J. Am. Chem. Soc. **125**, 1866 (2003)
143. R.R. Durand Jr, F.C. Anson, Catalysis of dioxygen reduction at graphite electrodes by an adsorbed cobalt(ii) porphyrin. J. Electroanal. Chem. Interfacial Electrochem. **134**, 273 (1982)
144. J.P. Collman et al., Electrode catalysis of the four-electron reduction of oxygen to water by dicobalt face-to-face porphyrins. J. Am. Chem. Soc. **102**, 6027 (1980)
145. R.R. Durand, C.S. Bencosme, J.P. Collman, F.C. Anson, Mechanistic aspects of the catalytic reduction of dioxygen by cofacial metalloporphyrins. J. Am. Chem. Soc. **105**, 2710 (1983)
146. D. Thompsett, in *Handbook of Fuel Cells*, ed. by W. Vielstich, A. Lamm, H.A. Gasteiger Fundamentals, Technology and Applications (Wiley, New York, 2006)
147. C.J. Chang et al., Electrocatalytic four-electron reduction of oxygen to water by a highly flexible cofacial cobalt bisporphyrin. Chem. Commun. 1355 (2000)
148. K. Oyaizu, H. Murata, M. Yuasa, in *Molecular Catalysts for Energy Conversion,* ed. by T. Okada, M. Kaneko (Springer Berlin Heidelberg, Berlin, 2009)
149. S. Gupta, D. Tryk, I. Bae, W. Aldred, E. Yeager, Heat-treated polyacrylonitrile-based catalysts for oxygen electroreduction. J. Appl. Electrochem. **19**, 19 (1989)
150. J.-P. Dodelet, in *N4-Macrocyclic Metal Complexes*, (Springer, New York, 2006), pp. 83–147
151. A. Garsuch et al., in *Handbook of Fuel Cells* ed. by W. Vielstich, A. Lamm, H.A. Gasteiger, Fundamentals, Technology and Applications (Wiley, New York, 2006)
152. C.M. Johnston, P. Piela, P. Zelenay, in *Handbook of Fuel Cells* ed. by W. Vielstich, A. Lamm, H. A. Gasteiger, Fundamentals, Technology and Applications, (Wiley, New York, 2006)
153. J.A.R. van Veen, J.F. van Baar, K.J. Kroese, Effect of heat treatment on the performance of carbon-supported transition-metal chelates in the electrochemical reduction of oxygen. J. Chem. Soc. Faraday Trans. 1 Phys. Chem. Condens. Phases **77**, 2827 (1981)
154. F. Charreteur, F. Jaouen, S. Ruggeri, J.P. Dodelet, Fe/N/C non-precious catalysts for PEM fuel cells: Influence of the structural parameters of pristine commercial carbon blacks on their activity for oxygen reduction. Electrochim. Acta **53**, 2925 (2008)
155. F. Jaouen, J.P. Dodelet, Non-noble electrocatalysts for O-2 reduction: How does heat treatment affect their activity and structure? Part I. Model for carbon black gasification by NH3: Parametric calibration and electrochemical validation. J. Phys. Chem. C **111**, 5963 (2007)
156. F. Jaouen, A.M. Serventi, M. Lefevre, J.P. Dodelet, P. Bertrand, Non-noble electrocatalysts for O-2 reduction: How does heat treatment affect their activity and structure? Part II. Structural changes observed by electron microscopy, Raman, and mass spectroscopy. J. Phys. Chem. C **111**, 5971 (2007)

157. X.G. Li, C.P. Liu, W. Xing, T.H. Lu, Development of durable carbon black/titanium dioxide supported macrocycle catalysts for oxygen reduction reaction. J. Power Sources **193**, 470 (2009)
158. G. Wu et al., Performance durability of polyaniline-derived non-precious cathode catalysts. ECS Trans. **25**, 1299 (2009)
159. G. Liu, X.G. Li, P. Ganesan, B.N. Popov, Development of non-precious metal oxygen-reduction catalysts for PEM fuel cells based on N-doped ordered porous carbon. Appl. Catal. B Environ. **93**, 156 (2009)
160. Y.H. Lu, R.G. Reddy, Electrocatalytic properties of carbon supported cobalt phthalocyanine-platinum for methanol electro-oxidation. Int. J. Hydrogen Energy **33**, 3930 (2008)
161. H.J. Zhang et al., Electrochemical performance of a novel CoTETA/C catalyst for the oxygen reduction reaction. Electrochem. Commun. **11**, 206 (2009)
162. X.G. Li, G. Liu, B.N. Popov, Activity and stability of non-precious metal catalysts for oxygen reduction in acid and alkaline electrolytes. J. Power Sources **195**, 6373 (2010)
163. R.R. Chen, H.X. Li, D. Chu, G.F. Wang, Unraveling oxygen reduction reaction mechanisms on carbon-supported Fe-Phthalocyanine and Co-Phthalocyanine catalysts in alkaline solutions. J. Phys. Chem. C **113**, 20689 (2009)
164. I. Kruusenberg, L. Matisen, Q. Shah, A.M. Kannan, K. Tammeveski, Non-platinum cathode catalysts for alkaline membrane fuel cells. Int. J. Hydrogen Energy **37**, 4406 (2012)
165. G. Merle, M. Wessling, K. Nijmeijer, Anion exchange membranes for alkaline fuel cells: A review. J. Membr. Sci. **377**, 1 (2011)
166. V. Bambagioni et al., Single-site and nanosized Fe-Co electrocatalysts for oxygen reduction: Synthesis, characterization and catalytic performance. J. Power Sources **196**, 2519 (2011)
167. Z. Chen, D. Higgins, Z.W. Chen, Electrocatalytic activity of nitrogen doped carbon nanotubes with different morphologies for oxygen reduction reaction. Electrochim. Acta **55**, 4799 (2010)
168. Z. Chen, D. Higgins, Z.W. Chen, Nitrogen doped carbon nanotubes and their impact on the oxygen reduction reaction in fuel cells. Carbon **48**, 3057 (2010)
169. K.P. Gong, F. Du, Z.H. Xia, M. Durstock, L.M. Dai, Nitrogen-doped carbon nanotube arrays with high electrocatalytic activity for oxygen reduction. Science **323**, 760 (2009)
170. T.C. Nagaiah, S. Kundu, M. Bron, M. Muhler, W. Schuhmann, Nitrogen-doped carbon nanotubes as a cathode catalyst for the oxygen reduction reaction in alkaline medium. Electrochem. Commun. **12**, 338 (2010)
171. A. Marchionni et al., Electrooxidation of ethylene glycol and glycerol on Pd-(Ni-Zn)/C anodes in direct alcohol fuel cells. Chemsuschem **6**, 518 (2013)

Chapter 11
Concluding Remarks

11.1 Summary

The goal of this book was to explore the range of nanostructured materials being utilized in low temperature energy-related electrocatalysis. Our intention was to first present the paradigm of renewable energy, describing what drives research in this area and follow this with a rundown of the relevant electrochemical devices in which nanotechnology plays a key role, i.e., fuel cells, electrolyzers, and in CO_2 electroreduction. This would then be followed by a section-by-section look at the key materials where nanotechnology comes into play; i.e., electrocatalysts and their support materials (carbon blacks, carbon nanotubes, etc.).

Chapter 1 started with a short review of the current world energy and resources situation, focusing also on environmental issues, stressing the need for a transition to a new energy system which is completely renewable but at the same time does not poison our planet. The concept of *renewables* was defined and some of the most relevant renewable energy resources were reviewed, with a special emphasis on the connection between their application in energy harvesting and electrochemical energy conversion technologies. The concepts of Energy Returned On Energy Invested (EROEI) and net energy were introduced, together with a discussion of Life Cycle Analysis approach. Next, we introduced the concept of the hydrogen economy and the use of energy vectors as part of a sustainable energy paradigm. A short introduction to fuel cells and electrolyzers followed, thus defining the devices where the materials focused on in this book will be potentially exploited. At the end of this chapter we introduced the exciting challenge of CO_2 electroreduction.

Chapter 2 is entitled "*A bird's eye view of energy related electrochemistry*," and indeed this is what was offered. The fundamentals of electrochemistry related to electrocatalytic materials were introduced. The chapter then went on to discuss the most important electrochemical reactions that are encountered in electrochemical devices like fuel cells and electrolyzers. That is, Hydrogen Oxidation, Hydrogen Evolution, Oxygen Reduction, Oxygen Evolution, and Alcohol Oxidation. The last section of Chap. 2 introduced in detail the altogether new topic of CO_2 electroreduction. A review of the recent scientific literature in this field was provided to give the reader a feel for the state of the art in this quickly expanding research area.

A. Lavacchi et al., *Nanotechnology in Electrocatalysis for Energy*,
Nanostructure Science and Technology 170, DOI: 10.1007/978-1-4899-8059-5_11,
© Springer Science+Business Media New York

After having reviewed the electrochemical concepts behind electrochemical energy conversion in Chaps. 2 and 3 described in some detail the architectures for the most commonly used devices. The chapter covered the two main fields, Fuel Cells and Electrolyzers. Regarding fuel cells, first a short review of the technological background was given introducing the main constitutive elements and the key parameters for performance assessment. Then a somewhat detailed description of fuel cell designs was given, reporting relevant data such as stability and maximum power outputs. As fuels we focused on hydrogen, methanol, and other alcohols and small organic molecules.

In the second part we provided a general background for electrolysis. Electrolysis is of prime importance to the hydrogen economy as it can directly convert electrical energy into chemical energy by using renewable power sources such as photovoltaics or wind power. The chapter ended by describing a new class of electrolyzers that have recently appeared on the electrolysis panorama, namely devices using sacrificial agents.

The first section of this book then closed with a detailed discussion of the factors that affect the design of electrocatalysts (Chap. 4). This chapter also discussed the role of nanotechnology in addressing the targets for effective electrocatalyst development.

The rest of the book was devoted to describing the different classes of materials used. We divided these chapters between the support materials and the active phase materials. Chapters 5 and 6 discussed in detail the materials used as supports for the electrocatalysts. Chapter 5 dealt with the most common type of support material that has been used till date, the carbon blacks, but also touches on the new and up and coming carbon-based materials like mesoporous and gel compounds as well as graphene and carbon nanotubes. Chapter 6 had a close-up look at alternative support materials like titania nanotubes and other conducting oxides, which are especially important for fuel cells fed with liquid fuels. Their use as both innocent support and kinetic promoter of the kinetics of a variety of other nanomaterials completed the discussion.

We then moved on to the active materials. Chapter 7 described the main approaches to metal nanoparticle synthesis and the main commercial electrocatalysts. The large variety of nanostructured metals with shape and structure control was discussed in Chap. 8. A special emphasis was given to surface structure control, with a discussion of the recent discovery of new synthetic routes to high index faceting for activity enhancement. Chapter 9 reported the new class of nanoengineered nanoparticles, the development of which has the primary objective of reducing the noble metal loading. The focus was on "hollow" and "core–shell nanoparticles." Saving the best for last, in Chap. 10 we discussed the new "molecular" approach to electrocatalysis. The use of macrocycles and heat treated macrocycles, was extensively reviewed with a special emphasis on the most recent findings. In this chapter, we also introduced the breakthrough discovery of organometallic complexes employed in hydrogen or energy production, as well as in electroreduction of CO_2 to fuels or raw chemicals. This exciting area promises to indeed extend nanotechnology to the single-site scale where a single metal atom is involved in the reaction in contrast to what occurs on nanosized electrocatalysts.

11.2 Considerations

Polymer electrolyte fuel cells have been primarily developed as a power source for automotive applications. Nevertheless, as pointed out by the US Department of Energy they still fail to hit the targets defined for the application. Platinum content and durability of the devices under intermittent operation are still not completely resolved issues and a lot of work is still needed. On the other hand the use of fuel cells requires a transition to a hydrogen economy. The use of hydrogen as a fuel requires a completely new distribution grid, as well as new production processes based on the coupling of renewable energy resources, such as PV and wind energy, to electrolytic processes. Hydrogen storage is also still an open issue as hydrogen is well known to be difficult to store and transport. The target compression pressure for hydrogen cylinders for automotive use is 700 bar. This is what is considered the minimum to allow a car powered with it to have sufficient autonomy to compete with current internal combustion engine devices. Electric cars powered by batteries still cannot compete with state-of-the-art technology in terms of autonomy. Furthermore, battery charging is a time-consuming process, not allowing fast refueling required for long trips. Fuel cells resolve this problem as they can be quickly refueled, this of course once the right hydrogen distribution network is set up. We also need improvements in electrolysis efficiency to support the sustainability of the automotive sector. Research in electrolysis is actually focused on the search for better electrocatalytic materials and to new supports that stabilize current electrolyzer architectures (i.e., “zero gap” and PEM electrolyzers). The optimal situation would be to have alkaline membranes with stability comparable to that of acidic membranes currently employed in fuel cells. This is because the kinetics of the OER reaction is much faster in alkaline media as compared to acidic resulting in a lower activation overpotential and in the end an increase in the energy efficiency of the process. This is also important for electrocatalysis as basic electrolytes allow the use of non-noble metal compounds, reducing the capital cost (one of the major issue in electrolysis) of the device.

Alternatives to hydrogen could be liquid fuels that are derived from biomasses. Among the liquids used in fuel cells, methanol has undoubtedly been the most investigated. However, cell efficiency and performance stability cannot compete with those of hydrogen, which, at present remains the most promising vector for the automotive sector. The use of ethanol has also been proposed as an alternative fuel. For ethanol to be useful in fuel cells, we need to have complete oxidation to CO_2 a process that remains an extraordinary demanding task for electrocatalysis.

The use of liquid fuels, due to their ease of storage, is particularly appealing for portable applications. Indeed portable devices powered with fuel cells do not need an electric grid at all, just a cartridge for fuel recharging.

As we have seen electrocatalysis is a central theme in the development of a sustainable energy system entirely based on renewable resources. The result is a scientific field that, even after decades of development, is still quickly evolving. The number of researchers devoting their activity to the application of electrocatalysis to sustainability is still growing. Certainly, this is due to the ever-increasing interest in developing sustainable alternatives to processes based on polluting and resources hungry technologies; but is also due to the fact that institutions provide significant funding for projects in the field. While such a healthy funding situation is certainly positive there are also some negative aspects. The production of scientific information is incredibly large and it's not easy to keep up with all the progress even for specialists. The situation is even more difficult for a nonspecialist or, just as an example, an entrepreneur who wants to understand the true potential of a material. The situation is further complicated by the fact that at least at the material research stage, there are no recognized benchmarks for a material performance evaluation. In this sense a very important role is provided by the DOE targets. By the way this is still not enough. DOE activities are expressed in terms of performance in complete fuel cell devices. Very often with completely new materials, characterization can only be performed at the level of half-cell measurements and not in complete devices. Furthermore, not all research groups, especially those oriented in the synthesis of materials, have the chance to assemble a complete cell. Hence, a lot of literature relies on half-cell measurements alone. There are no recognized standards for this characterization and it is a common practice to perform rather arbitrary and opportune selection of the reference material to which novel materials are compared. This aspect should not be overlooked by experts in the field, and cannot be justified by the need to present results in the best possible light. We have tried, in the various chapters, to mediate between differing reports regarding the performance of the same material, reporting the more evident contradictions.

In the end, and on the basis of the literature we have examined in preparing this book, we auspicate the definition of protocols for the half cell assessment of material performance. This should be done for each of the reactions for which the materials are intended for. We feel this is a task of extraordinary importance. With such a growing field and astonishingly large information available there is significant need for a common base for material selection and evaluation. A common protocol for the publication would certainly support the development of good and easily accessible information in such an incredibly active research field.

11.3 Thinking Outside of the Box

The potential of nanotechnology in electrocatalysis does not end with fuel cells, electrolytic hydrogen production, or CO_2 electroreduction. Such applications have become over the last few decades some of the most relevant for electrocatalysis, but the future will likely lead us to new opportunities for its exploitation. This will give us the chance to use the lessons we have learnt from the development of nanostructured electrocatalytic materials in many other fields.

Among possible other challenges it is likely that agriculture will experience enormous problems in the future due to the gradual reduction of fossil fuel availability and their increasing costs. A future relevant global challenge will consist in how to address a transition involving the gradual phase out of fossil fuels. This has to occur planning a gradual increase of the energy obtained from modern renewable sources such as wind and photovoltaic for agriculture. The concept of switching agriculture to renewable energy sources mainly producing electricity has been recently considered and referred to as "turning electricity into food" [1].

It is important to mention that energy inputs in agriculture are not only limited to what is linked to mechanical labor. Many other relevant components required to guarantee high productivity and ultimately cheap food, also include energy intensive chemicals, mainly fertilizers and pesticides. If we consider this it is obvious that renewable energy cannot provide the same services and products as fossil fuels at the same costs. A problem here is that, at least at present there is no way to directly transform electricity into chemicals such as fertilizers and pesticides. Nevertheless, there is a tremendous potential in the concept of turning electricity into food. We can foresee the possibility, at least in principle, of developing electrochemical processes that transform the electricity provided by renewable energy resources into chemical compounds essential for agriculture.

One of the few electrochemical processes used in agriculture is the one that leads to the production of hydrogen by electrolysis for the synthesis of ammonia and then urea and fertilizers. But this alone is not enough. Indeed a few electrochemical and low temperature approaches to the fixation of atmospheric nitrogen have been reported [2]. This is the first step in the synthesis of fertilizers. While at present at least as far as we know no approaches to pesticide synthesis other than preparation from fossil fuel derived "raw chemicals" have been reported. There is room for research here. Electrocatalysis and our ability to manipulate matter at the nanoscale offer the chance to explore a variety of electrochemical reactions which, in principle, could be used to support agriculture.

We believe that the future could see the rise of a variety of new approaches to the chemical industry entirely based on the use of biomass-derived products and electrochemical processes [3, 4]. This may be done in energetically self-sustainable processes with no need of external energy inputs also resulting in hydrogen production [5]. Hydrogen and important raw chemical production could happen at the expenses of agricultural biomass and electrical energy produced by renewable resources directly using electrochemical processes. Electrolytic processes using biomass-derived alcohols (e.g., ethanol, ethylene glycol, or glycerol) where the alcohols are selectively transformed into higher added-value carboxylic compounds with no need of chemical reagents have been recently developed, at least at the lab stage. These processes also result in the simultaneous production of ultrapure hydrogen, with an energy cost that is one-third of the energy required by a traditional water electrolyzer.

Many other applications can be identified for the principles and the materials outlined in this book. The hard-won lessons and expertise gained in each of the topics covered should give fruit in a wide range of energy-related issues. Agriculture and raw chemical production are just two examples. A challenge for scientists and professionals is not only to continue to develop newer and better materials, but also to think outside the box looking for exploitation in emerging issues for sustainability.

References

1. U. Bardi, T. El Asmar, A. Lavacchi, Turning electricity into food: the role of renewable energy in the future of agriculture. J. Clean. Prod. **53**, 224 (2013)
2. L. Rong, T.S.I John, T. Shanwen, Synthesis of ammonia directly from air and water at ambient temperature and pressure. Sci. Rep. **3**, 427–452 (2013)
3. M. Simões, S. Baranton, C. Coutanceau, Electrochemical valorisation of glycerol. Chemsuschem **5**, 2106 (2012)
4. A. Marchionni et al., Electrooxidation of ethylene glycol and glycerol on Pd-(Ni-Zn)/C anodes in direct alcohol fuel cells. ChemSuschem **6**, 518 (2013)
5. V. Bambagioni et al., Self-sustainable production of hydrogen, chemicals, and energy from renewable alcohols by electrocatalysis. ChemSuschem **3**, 851 (2010)

List of Symbols

$E°$	Standard Reduction potential [V]
ΔG	Gibbs free energy change [J mol^{-1}]
$\Delta G°$	Standard Gibbs free energy change
E	Reduction potential [V]
F	Faraday constant [96845 C mol^{-1}]
T	Temperature [K]
$T°$	Standard temperature [298.15 K]
n	Number of exchanged electrons [dimensionless]
μ_i	Stoichiometric coefficient for the i-th specie
a_n	Activity of the component n [mol l^{-1}]
p_n	Partial pressure of the component n [N m^{-2}]
P	Pressure [N m^{-2}]
R	Ideal gas constant [8.3144621 J mol^{-1} K^{-1}]
ΔS	Entropy change [J K^{-1} mol^{-1}]
c_n	Molar concentration of component n [mol l^{-1}]
C	Molar concentration [mol l^{-1}]
η	Overpotential [V]
α_a	Anodic charge transfer coefficient [dimensionless]
α_c	Cathodic charge transfer coefficient [dimensionless]
J	Current density [A cm^{-2}]
j_c	Cathodic current density [A cm^{-2}]
j_a	Anodic current density [A cm^{-2}]

A. Lavacchi et al., *Nanotechnology in Electrocatalysis for Energy*,
Nanostructure Science and Technology 170, DOI: 10.1007/978-1-4899-8059-5,
© Springer Science+Business Media New York

$\mathbf{j^0}$	Exchange current density [A cm^{-2}]
$\mathbf{j_k}$	Kinetic current density [A cm^{-2}]
$\mathbf{j_l}$	Levich current density [A cm^{-2}]
D	Diffusion coefficient [cm^2 s^{-1}]
ω	Rotation speed [rad s^{-1}]
ν	Kinematic viscosity [cm^2 s^{-1}]
$\mathbf{C_l}$	Reactant concentration in Levich equation [mol cm^{-2}][1]

[1] *Note* The symbol C_l for concentration was defined to avoid confusion about units of measure used in Levich Equation where the cm appears as the base unit for length.

Index

A. Lavacchi et al., *Nanotechnology in Electrocatalysis for Energy*,
Nanostructure Science and Technology 170, DOI: 10.1007/978-1-4899-8059-5,

© Springer Science+Business Media New York

D

E

9781489980588